Springer Proceedings in Geography and Sustainability

Springer Proceedings in Geography and Sustainability features state-of-the-art contributions from conferences, workshops, and similar meetings. The series presents a snapshot of timely developments in fields including but not limited to physical, human, and integrated geography; cross-disciplinary approaches to environmental challenges; and sustainability writ large. It aims to foreground trends emerging from meetings around the world, especially regions that are traditionally underrepresented.

Proposals must include the following:

- A description of the meeting's aims and theme
- The name, place, date, and website of the meeting
- The committee(s) and/or organization(s) responsible for planning the meeting
- A description of the organizer's review process for submissions

Please reach out to the publishing editor, Andrea Brody-Barre (andrea.brody-barre@springer.com), with enquiries.

Srikanta Patnaik · Francesco Vigliarolo ·
Jesús García-Madariaga ·
Chandrakanta Sahoo
Editors

People, Planet and Profit: Advancing Sustainable Business Practices

International Conference on Sustainable Business (ICSB 2025)

Springer

Editors
Srikanta Patnaik
Interscience Institute of Management and Technology
Bhubaneswar, Odisha, India

Jesús García-Madariaga
Faculty of Economics and Business
Universidad Complutense de Madrid
Madrid, Spain

Francesco Vigliarolo
National University of La Plata
Buenos Aires, Argentina

Chandrakanta Sahoo
Institute of Management and Information Technology
Cuttack, Odisha, India

ISSN 3059-3344 ISSN 3059-3352 (electronic)
Springer Proceedings in Geography and Sustainability
ISBN 978-3-032-11295-8 ISBN 978-3-032-11296-5 (eBook)
https://doi.org/10.1007/978-3-032-11296-5

This Springer imprint is published by the registered company Springer Nature Switzerland AG
The registered company address is: Gewerbestrasse 11, 6330 Cham, Switzerland

ICSB-2025 Conference Committee

Patron

Prof. Amiya Kumar Rath

General Chair

Prof. Satya Swaroop Debashish

Organizing Chairs

Prof. Raghunath Sahu
Prof. Manas Ranjan Kabat

Program and Publication Chair

Prof. Srikanta Patnaik

Convener

Dr. Chandrakanta Sahoo

Co-convener

Dr. Swati Mishra

Advisory Committee

Prof. Shiv K. Tripathy, Vice Chancellor, Atmiya University, Rajkot
Dr. Ganesh Abhimanyu Dash, Associate Professor, College of Administrative and Financial Sciences, Riyadh, Saudi Arabia
Prof. P. Narayana Reddy, Director General (South), Association of Indian Management School, India
Dr. Basanna Patagundi, Chief Learning Officer, Swiss Learning Exchange, Bengaluru, Karnatak
Prof. Tankanatha Dhamala, Professor, Central Department of Mathematics, Institute of Science and Technology, Tribhuvan University, Kathmandu, Nepal
Mr. Bobby Varanasi, CEO, Regenerative Futures, Malaysia
Prof. S. Venkataramanaiah, Professor, Chairperson—Admission, IIML (Noida Campus)

Technical Committee

Dr. Chandan Thakur, Delhi School of Business, Delhi
Dr. Monika Srivastav, Department of Humanities and Liberal Art, DIT University, Dehradun
Prof. D. Pradeep Kumar, MITS School of Business, Andhra Pradesh
Prof. Subhasree Kar, Reva University, Bengaluru
Dr. S. Md. Karimulla Basha, United Skill Development Corporation, Bengaluru
Dr. Prakash Chandra Dash, Institute of Management and Information Science, Bhubaneswar
Prof. Surjeet Kar, Biju Patnaik Institute of Information Technology and Management, (BIITM), Bhubaneswar
Prof. Dhananjay Sahoo, School of Commerce, Banaras Hindu University, UP
Dr. Omar Faruk, Department of Business Studies, Daffodils International University, Bangladesh
Prof. Chandan Kumar Sahoo, School of Management, NIT, Rourkela
Dr. Bijaya Kumar Sundaray, Department of PMIR, Utkal University, Bhubaneswar
Dr. Alok Chhajer, United Skill Development Corporation, Bengaluru
Dr. Chandra Sekhar, NIT, Kurukshetra
Dr. Subhendu Kumar Pradhan, SKCG Autonomous College, Paralakhimundi, Gajapati
Dr. K. V. Geetha Devi, MITS School of Business, Andhra Pradesh

Organizing Committee

Dr. Suvendra Kumar Jayasingh, IMIT, Cuttack
Dr. Alok Ranjan Tripathy, IMIT, Cuttack
Dr. Chinmaya Kumar Swain, IMIT, Cuttack

Dr. Kunal Pattnaik, IMIT, Cuttack
Dr. Bamadev Mahapatra, IMIT, Cuttack
Dr. Jayanta Kumar Dash, IMIT, Cuttack
Dr. Dhaneshwar Mardi, IMIT, Cuttack
Mr. Satyaprakash Swain, IMIT, Cuttack
Dr. Srutipragyan Swain, IMIT, Cuttack
Mrs. Sujata Ray, IMIT, Cuttack
Dr. Debasisha Mohanty, CIME, Bhubaneswar
Dr. M. C. Tripathy, CIME, Bhubaneswar
Dr. Amitabh Nanda, CIME, Bhubaneswar
Dr. Bandana Sarangi, CIME, Bhubaneswar
Dr. Bidya Dash, CIME, Bhubaneswar
Dr. Kedareswar Panda, CIME, Bhubaneswar
Dr. P. D. Das, CIME, Bhubaneswar
Debidutta Acharya, CIME, Bhubaneswar

Publicity Committee

Dr. Suvendra Kumar Jayasingh, HOD, CSE
Dr. Swati Mishra, Assistant Professor, MBA
Dr. Kunal Pattnaik, HOD, Humanities
Dr. Bamadev Mahapatra, Assistant Professor, MBA

Finance Committee

Dr. Chandrakanta Sahoo, HOD, MBA
Dr. Suvendra Kumar Jayasingh, HOD, CSE
Dr. Swati Mishra, Assistant Professor, MBA
Mr. Arun Kumar Swain, Technical Assistant
Mr. Ajit Kumar Sitha, Programmer

Hospitality and Logistic Committee

Mr. Arun Kumar Swain, IMIT, Cuttack
Mr. Ajit Kumar Sitha, Programmer
Mr. Gobinda Ballav Mohanty, IMIT, Cuttack
Mrs. Sasmita Mohapatra, IMIT, Cuttack
Mr. Satyaprakash Swain, IMIT, Cuttack
Mrs. Srutipragyan Swain, IMIT, Cuttack
Mrs. Sujata Ray, IMIT, Cuttack
Dr. Kedareswar Panda, CIME, Bhubaneswar

Dr. P. D. Das, CIME, Bhubaneswar
Dr. Bandana Sarangi, CIME, Bhubaneswar
Dr. Bidya Dash, CIME, Bhubaneswar

Preface

The global pursuit of sustainable development has evolved from aspiration to urgency. In this context, the *International Conference on Sustainable Business (ICSB-2025)*, held on **May 5–6, 2025**, at the **World Skill Center, Bhubaneswar**, focused on the theme "*People, Planet and Profit.*" It emphasized the critical intersection of social inclusion, environmental responsibility, and economic viability.

The conference was inaugurated by **Prof. Sabita Acharya** (Vice Chancellor, Utkal University—Chief Guest) and **Prof. Amiya Kumar Rath** (Vice Chancellor, BPUT—Patron), with academic guidance from **Dr. Satya Swaroop Debasish** (General Chair, Utkal University), and organizational leadership from **Prof. M. R. Kabat** (IMIT, Cuttack) and **Prof. R. N. Sahu** (CIME, Bhubaneswar). Keynotes by **Prof. B. Nishantha** (University of Colombo, Sri Lanka), **Dr. Ritesh Kumar Dubey** (XIMB, India), and **Prof. Monika Srivastav** (Chitkara University, India) set the tone for thoughtful discussions on sustainability.

ICSB-2025 brought together a vibrant community of scholars, policymakers, and practitioners from diverse regions and disciplines. The presented papers bridged theory and practice across marketing, finance, environmental science, and HR, showcasing interdisciplinary approaches to sustainability.

Major contributions included insights on ESG disclosures, circular economy adoption, sustainable infrastructure through PPPs, and inclusive, community-driven business models. These discussions highlighted that sustainability is not just a compliance concern—it is a strategic and ethical imperative.

ICSB aspires to become a recurring global platform fostering transdisciplinary research and innovation in sustainable business. We believe the ideas exchanged here will inspire continued scholarly engagement and practical solutions.

We sincerely thank all contributors, reviewers, session chairs, and participants for making ICSB-2025 a meaningful and successful endeavor.

Bhubaneswar, India

Srikanta Patnaik
Publication and Program Chair

Editorial

The intersection of ethical responsibility, economic viability, and environmental stewardship has never been more important than it is today. The International Conference on Sustainable Business: People, Planet, and Profit (ICSB-2025), held from 5 to 6 May 2025, stood as a vibrant platform for intellectual engagement, academic discourse, and cross-sector collaboration. This landmark event was jointly organized by the **Institute of Management and Information Technology (IMIT), Cuttack**, and the **College of IT and Management Education (CIME), Bhubaneswar**, in collaboration with the **Interscience Institute of Management and Technology (IIMT), Bhubaneswar**. Bringing together a distinguished gathering of researchers, academicians, policymakers, industry experts, and sustainability advocates, ICSB-2025 showcased the urgency and relevance of embedding sustainability into the heart of modern business practices.

As global economies grapple with the complexities of climate change, resource scarcity, social inequality, and rapid technological disruption, sustainable business has emerged as a powerful paradigm—one that transcends traditional profit-centric models and embraces the Triple Bottom Line of People, Planet, and Profit. The conference explored how organizations can integrate human rights, environmental ethics, and economic strategy to drive long-term resilience and inclusive growth. ICSB-2025 offered a comprehensive view of sustainability across multiple disciplines. From responsible finance and ethical governance to green human resource practices, circular economy models, and AI-enabled decision-making, the conference reflected the multifaceted nature of sustainability. It emphasized not only the moral imperative of environmental and social responsibility but also the strategic advantage sustainability can offer in a competitive business landscape.

The proceedings of ICSB-2025 reflect the intellectual depth and practical relevance of the discussions held during the conference. This year, the organizing committee received over 90 scholarly contributions from institutions and professionals across India and abroad. After a stringent peer-review process, 29 high-quality papers were accepted for presentation and publication. Of these, 23 authors completed the registration process and are featured in this volume. These papers represent a rich tapestry of perspectives and methodologies, unified by a commitment to reimagining business as a force for good.

Through plenary talks, research presentations, and interactive sessions, ICSB-2025 provided a platform to explore innovative models of corporate governance, sustainable finance, inclusive technology, environmental responsibility, and employee well-being. The conference also provided a space to align research efforts with international frameworks such as the United Nations Sustainable Development Goals (SDGs), the UN Guiding Principles on Business and Human Rights, and UNESCO's focus on economic and social justice. As the world advances toward a more interconnected, data-driven, and ethically conscious era, the relevance of sustainable business will continue to grow. ICSB-2025 served not just as an academic event, but as a movement to inspire leadership, ignite interdisciplinary research, and create pathways for responsible innovation. The papers published herein are a testament to that vision—capturing critical insights, practical strategies, and visionary thinking for shaping the future of sustainable business.

The conference brought together a series of insightful and empirically grounded studies that span across various domains, including human resource management, corporate governance, financial inclusion, AI applications, circular economy, and sustainable finance. The following papers, selected through a rigorous peer-review process, exemplify the conference's core themes of innovation, responsibility, and sustainability. Each article contributes to a deeper understanding of how businesses and institutions can balance profit with purpose, and growth with equity, in an era of global complexity and transformation.

The first chapter entitled "**Exploring the Impact of Green Human Resource Management Practices on Organizational Sustainability: A Case Study of EcoLuxe Hotels**" investigates the effectiveness of Green HRM in the hospitality sector. Focusing on EcoLuxe Hotels, the study reveals that green recruiting, eco-conscious training, and sustainable performance management have enhanced employee behavior, fostered an environmental culture, and improved the hotel's overall triple bottom line in a post-pandemic environment.

The second chapter entitled "**Transforming HR: The Evolution from Operational Tasks to Data-Driven Strategic Insights**" presents a case study from the Dutch Police Force to demonstrate how HR analytics can reduce attrition and improve alignment between personal identity and professional roles. The research underscores the strategic use of predictive analytics for sustainable workforce planning in public sector institutions.

The third chapter entitled "**Corporate Governance and Corporate Sustainability with Reference to the New Companies Act, 2013**" examines how regulatory mandates in India have influenced sustainability strategies in Tata Steel. The findings reveal that compliance with CSR mandates, gender diversity provisions, and board transparency requirements significantly bolstered the company's environmental and community initiatives.

The fourth chapter entitled "**Financial Inclusion of the Slum Dwellers of Kolkata with Fintech Services: One More Step Towards SDG Goals**" analyzes how PMJDY and fintech tools have empowered financially marginalized women in

Kolkata. The study confirms that formal banking access has led to improved saving behavior, access to welfare schemes, and alignment with broader SDG targets.

The fifth chapter entitled "**Customer Retention: PCA-Enhanced AI Model**" employs Principal Component Analysis and Support Vector Machines to predict customer churn. The study demonstrates that AI-enhanced models, when supported by high-quality data, can optimize campaign outcomes and increase customer lifetime value through data-driven decision-making.

The sixth chapter entitled "**Mapping the Landscape of Sustainable Tourism in India a Bibliometric Review and Systematic Synthesis**" uses bibliometric tools to explore the evolution of sustainable tourism research. The study identifies global collaborations, influential institutions, and critical research gaps—particularly the need for technological integration and more robust policy evaluation in the Indian context.

The seventh chapter entitled "**Chemical Fertilizer Usage in Indian Agriculture: Trade-Off Between Carbon Emissions and Food Security**" explores the complex relationship between fertilizer use, agricultural productivity, and environmental sustainability. Using econometric modeling, the study reveals short- and long-term linkages among food production, emissions, and climate change.

The eighth chapter entitled "**The Impact of CSR Perception on Employee Performance: Exploring the Mediating Role of Person-Organization Fit in Public Sector Organizations**" delves into the psychological mechanisms that connect CSR perception and employee performance. Through structural equation modeling, it establishes person-organization fit as a critical mediating factor in aligning individual and institutional values.

The ninth chapter entitled "**Strategic Integration of Corporate Social Responsibility: Linking Theory and Practice for Sustainable Development**" evaluates CSR practices across six major corporations, including Unilever and ITC. The study demonstrates that strategic CSR not only drives environmental and social outcomes but also enhances brand credibility and operational resilience.

The tenth chapter entitled "**A Study of Volatility Spillover of Sustainable Financial Markets on Indian Stock Market**" uses GARCH and VECM models to assess how global ESG-linked financial instruments affect volatility in India's stock market. The findings emphasize the importance of understanding cross-border financial risks in the era of sustainable finance.

The eleventh chapter entitled "**Circular Economy and Sustainability: Strategies, Challenges, and Implications**" analyzes circular economy practices using case studies from the EU and China. The research identifies regulatory, cultural, and technological barriers to CE implementation but confirms its transformative potential for reducing resource consumption and promoting green employment.

The twelfth chapter entitled "**Advancing ESG Practices: Trends, Challenges, and Future Directions in Sustainable Finance**" combines stakeholder theory and the Triple Bottom Line to evaluate ESG adoption. With case studies of Microsoft and Unilever, the paper reveals inconsistencies in ESG reporting and advocates for ethical AI usage and sector-specific metrics to increase ESG accountability.

The thirteenth chapter entitled "**Predicting Cryptocurrency Trends: Approaches to Diversification and Optimization**" uses machine learning to optimize cryptocurrency portfolios. It explores the intersection of financial inclusion, volatility management, and ESG-aligned investment strategies, offering practical insights for sustainable digital finance.

The fourteenth chapter entitled "**The Effectiveness of Social Media Networking as a Marketing Tool for Promoting Consumer Durable Goods in Tourism Industry**" investigates how content quality, trust, and interactivity affect consumer purchasing behavior. It recommends that marketers adopt nuanced, platform-specific strategies to boost engagement and sustainability in the tourism-retail interface.

The fifteenth chapter entitled "**Stock Market Investments: An Exploration of Investor Perceptions**" evaluates how demographic factors influence investor sentiment in the Indian equity market. The study proposes a model that can guide targeted financial education and risk communication strategies.

The sixteenth chapter entitled "**Intrinsic Motivation in the Workplace: The Roles of Age, Gender, and Tenure**" highlights how intrinsic motivators like autonomy and meaningful work drive engagement, particularly among Gen Y employees in India's IT sector. The research encourages HR departments to shift from extrinsic to purpose-driven incentive models.

The seventeenth chapter entitled "**Impact of Knowledge Management on Employee Engagement in the Information Technology Industry**" confirms a strong relationship between structured knowledge-sharing systems and employee engagement. It proposes knowledge audits and mentoring as key strategies to retain talent in high-skill environments.

The eighteenth chapter entitled "**Impact of Hybrid Work Model on Job Satisfaction of Tech Professionals in Information Technology Industry**" explores the dual effects of flexibility and digital fatigue in hybrid work environments. It calls for personalized hybrid work strategies and gender-sensitive HR policies.

The nineteenth chapter entitled "**HR Analytics in Predicting Employee Turnover in IT Industry: A Structural Equation Modeling Approach**" identifies leadership support and data infrastructure as critical predictors of turnover. The study validates the use of analytics to design pre-emptive retention strategies.

The twentieth chapter entitled "**Toward Sustainable Work-Life Balance in the Indian IT Industry: A Predictive Analysis of Challenges, Organizational Initiatives, and Technology-Enabled Interventions**" assesses how Infosys, TCS, and Wipro approach employee well-being. Despite progressive policies, issues like digital exhaustion and blurred work-life boundaries persist—especially for women—highlighting the need for integrated wellness frameworks.

The twenty-first chapter entitled "**Impact of Compensation Management Practices on Job Satisfaction in Indian Information Technology Industry**" uses regression analysis to show that fringe benefits and incentive structures significantly influence employee satisfaction. The findings advocate for transparent, inclusive compensation policies.

The twenty-second chapter entitled "**The Role of HR Analytics in Enhancing Talent Management: Insights from Bangalore's IT Industry**" examines the drivers behind effective HR analytics adoption. Leadership support, organizational readiness, and workforce upskilling emerge as essential for optimizing recruitment, retention, and training outcomes.

The final or the twenty-third chapter entitled "**Unleashing Digital Innovation for Circular Economy Transformation: A Thematic Analysis of Industry Case Studies**" examines how digital transformation can be leveraged to drive circular economy outcomes. Analyzing five global industry leaders—Philips, Unilever, BMW, Interface, and Nestlé—the study applies thematic analysis to explore how digital technologies such as IoT, blockchain, AI, and 3D printing are utilized to optimize resources, reduce waste, and extend product life cycles. The findings reveal common themes of technological advancement, operational efficiency, collaboration, and shared sustainability challenges. While digital tools are found to enhance transparency and resource tracking, barriers like high implementation costs and complex supply chains remain significant. This study contributes valuable insights for both academia and industry, highlighting how businesses can strategically adopt digital innovation to foster sustainable transformation within a circular economy framework.

Each research paper presented at ICSB-2025 adds a vital layer of understanding to the evolving landscape of sustainable business. From green HR practices to ESG integration, fintech-driven inclusion, and AI-powered decision-making, the conference showcases how innovation, regulation, and ethics converge to shape a resilient and inclusive global economy. These multidisciplinary perspectives demonstrate that sustainability is no longer an auxiliary concern—it is the foundation of future business leadership.

Acknowledgements

We are deeply thankful to the review committee members who dedicated their time and specialized expertise to ensure the quality of submissions. Their constructive feedback not only enhanced the quality of the papers but also mentored many budding scholars.

Special acknowledgment is extended to the session chairs—Dr. Chinmaya Kumar Dash, Dr. Prakash Chandra Dash, Dr. Swapnamayee Palit, Dr. Swati Mishra, Dr. Bamdev Mohapatra, and Prof. Surya Dev—whose intellectual stewardship and moderation brought rationality and clarity to the technical sessions.

We also wish to appreciate the support staff, volunteers, and student coordinators from IMIT, CIME, and IIMT, who worked behind the scenes to ensure the smooth organization of the event.

We express our sincere gratitude to the esteemed dignitaries who graced the occasion and elevated the stature of ICSB-2025. We especially acknowledge Prof. Sabita Acharya, Vice Chancellor, Utkal University (Chief Guest); Prof. Amiya Kumar Rath, Vice Chancellor, BPUT (Patron); Dr. Satya Swaroop Debasish, Professor, Utkal University (General Chair); Prof. M. R. Kabat, Principal, IMIT, Cuttack; and Prof. R. N. Sahu, Principal, CIME, Bhubaneswar (Organizing Chairs) for their academic leadership and visionary support.

We are especially indebted to our keynote speakers, Prof. B. Nishantha (University of Colombo, Sri Lanka), Dr. Ritesh Kumar Dubey (XIMB, Bhubaneswar, India), and Prof. Monika Srivastava (Chitkara University, Punjab, India), for sharing global insights on the ICSB-2025 platform and engaging the audience with forward-looking ideas that transcend disciplines and borders.

We thank our publication team and acknowledge the valuable contribution of the Interscience Institute of Management and Technology (IIMT), Bhubaneswar. Under the guidance of Prof. Srikanta Patnaik, the institute provided critical editorial and logistical support, which together with Springer Nature facilitated the timely compilation of this Book of Proceedings. We hope it will serve as both an academic resource and a catalyst for future research in the domain of sustainability.

Finally, we extend our heartfelt appreciation to the academic leadership of BPUT and Utkal University. Their unwavering support, guidance, and encouragement were instrumental in shaping ICSB-2025 into a high-impact, interdisciplinary platform committed to advancing the global sustainability agenda.

Srikanta Patnaik
Francesco Vigliarolo
Jesús García-Madariaga
Chandrakanta Sahoo

Contents

About the Editors

Prof. (Dr.) Srikanta Patnaik is a Director of Interscience Institute of Management and Technology, Bhubaneswar, Odisha, India. He has received his Ph.D. (Engineering) on Computational Intelligence from Jadavpur University, India in 1999 and supervised 12 Ph.D. theses and more than 30 M.Tech. theses in the area of Machine Intelligence, Soft Computing Applications and Re-Engineering. Dr. Patnaik has published more than 60 research papers in international journals and conference proceedings. He is author of two text books and edited 12 books and few invited book chapters, published by leading international publisher like Springer-Verlag, Kluwer Academic, etc. Dr. Patnaik was the Principal Investigator of TAPTEC project "Building Cognition for Intelligent Robot" sponsored by All India Council for Technical Education, New Delhi and Major Research Project "Machine Learning and Perception Using Cognition Methods" sponsored by University Grant Commission. He is the Editors-in-Chief of *International Journal of Information and Communication Technology* and *International Journal of Computational Vision and Robotics* published from Inderscience Publishing House, England and also Editors-in-Chief of Book Series on *Modeling and Optimization in Science and Technology* published from Springer, Germany.

Prof. (Dr.) Francesco Vigliarolo did his Ph.D. in Social Sciences, specializing in economic sociology, ethical finance, and social economics. He is a full-time Professor of Regional Economy at the Catholic University of La Plata. Additionally, he also serves as the Director of the UNESCO Chair in "Economic Systems and Human Rights" at the National University of La Plata. He is also an Associate Researcher at the CESOT (Center for the Study of Sociology of Work) at the University of Buenos Aires. His primary focus lies in local development processes, particularly the construction of territorial identities that involve asserting rights (ontological reason) through associative forms of production and ethical finance tools. Throughout his career, he had published numerous books and scientific articles delving into concepts such as "Ethical social capital," "Economic socialization and democratization," "The principles of economic phenomenology," "Ontological reason of economy and demand for rights of economy," and "Meso-economic approach."

Prof. (Dr.) Jesús García-Madariaga is a Full Professor in Marketing at the Faculty of Economics and Business Sciences of Complutense University in Madrid (Spain). He is chair of the Markco2 research group, former Editor-in-Chief of *International Journal of Internet Marketing and Advertising* (Inderscience) and visiting scholar at different international universities in extended stays (California Polytechnic State University) and temporary. His research interests encompass Digital Marketing, Neuromarketing, Customer Value Management, Tourism, and CSR. His research has been published in refereed international journals, including Journal of Consumer Behavior, Decision Support Systems, Technological and Economic Development of Economy, Journal of Hospitality and Tourism Management, Journal of Retailing and Consumer services, Physiology and Behavior and Journal of Business Ethics among others.

Dr. Chandrakanta Sahoo is a distinguished academician with 17 years of experience in teaching and research across various reputed institutions in Andhra Pradesh and Odisha. Currently serving as an Associate Professor at the Institute of Management and Information Technology, Cuttack, Dr. Sahoo specializes in Organizational Behavior, Human Resource Management, and Business Research. He is proficient in data visualization using Microsoft Power BI and Google Data Studio. Dr. Sahoo holds a Ph.D. in Business Administration from Utkal University, with his thesis focusing on Corporate Social Responsibility. He is also an ISO-certified internal auditor and an AICTE-certified Universal Human Values teacher. His professional background includes significant roles in compliance committees and administrative positions, notably as Head of the Department at Sanskrithi School of Business and at present in Department of MBA at Institute of Management and Information Technology, Cuttack. He has authored numerous journal articles and book chapters on HRM, CSR, and organizational behavior, with publications in leading journals and edited volumes. Dr. Sahoo has organized international conferences, conducted workshops, and had an active online presence with educational resources on YouTube. His broad expertise and extensive contributions to academia make Dr. Sahoo an excellent candidate for the Editorial Board of the International Book Series.

Exploring the Impact of Green Human Resource Management Practices on Organizational Sustainability: A Case Study of EcoLuxe Hotels

Debjani Palai, Kocherla Srinivasulu, and Yimer Ayalew Ahmed

Abstract

This research analyzes the impact of Green Human Resource Management (GHRM) practices at EcoLuxe Hotels, a global hotel chain committed to achieving excellence in environmental sustainability in the hospitality sector. The purpose of this study is to analyze the impact of GHRM practices of EcoLuxe Hotels on employee behavior, organizational sustainability, and Triple Bottom Line (TBL) performance, particularly in the post-pandemic hospitality environment. This research utilizes a qualitative case study approach to analyze the implementation and effectiveness of GHRM practices at EcoLuxe Hotels. The research showed that the adoption of GHRM practices such as green recruiting, environmental awareness training, performance evaluation assessment, and reward systems substantially improved employee pro-environmental behavior and inculcated a robust green corporate culture at EcoLuxe Hotels. Furthermore, post-pandemic changes such as digital training and resource-efficient operations improved the hotel's sustainability performance across environmental, social, and economic aspects. Further study should apply mixed-method both quantitative and qualitative approaches to evaluate the enduring impacts of GHRM practices on environmental performance metrics, staff retention, and operational

D. Palai (✉)
Institute of Professional Studies and Research, Cuttack, India
e-mail: debajani.palai@gmail.com

K. Srinivasulu
Department of Business Administration (UG), P B Siddhartha College of Arts and Science, Vijayawada, Andhra Pradesh, India

Y. A. Ahmed
College of Business and Economics, Wollo University, Dessie, Ethiopia
e-mail: yimer.ayalew@wu.edu.et

S. Patnaik et al. (eds.), *People, Planet and Profit: Advancing Sustainable Business Practices*, Springer Proceedings in Geography and Sustainability,
https://doi.org/10.1007/978-3-032-11296-5_1

efficacy. The originality of this research lies in its analysis of the hotel sector and its integration of post-pandemic factors into the analysis of GHRM practices.

Keywords

Green HRM • Organizational sustainability • Pro-environmental employee behavior • Hospitality industry • Post-pandemic adaptation

1 Introduction

The need of environmental sustainability has compelled corporate houses across all sectors to reevaluate their operational strategies, with HRM becoming a significant catalyst for green change. Organizational Sustainability refers to maintain a consistent growth and steady operation even in volatile environment. It focuses on organizational stability and adaptability by predicting and controlling environmental challenges. It helps to bring stability with job security of employees and believing in strong leadership [3]. The GHRM practices help to sustain an organization by well-defined planning on go-green concept and allocating resources accordingly. At the same time, it demands pro-environmental behavior of employees. Pro-Environmental Behavior of Employees refers to focus on conserving resources of environment and reducing wastage. It is a conscious effort to protect our environment and sustainability of organizations [4]. To enhance pro-environmental behavior organizations continuously provide training and developmental programs to employees. As a result, employees learn green practices and use it in their daily operation. Practice of green behavior is the part of organization's focus on prevailing of green culture. Green Culture helps the organization to be environmentally friendly and enhances sustainability. It has a deep-rooted impact on employee morale and commitment. Employees give value to the practices of green behavior and aim to reduce ecological footprint [1]. They promote environmental stewardship. Environmental awareness, organizational sustainability, green culture and pro-environmental behavior are the basic management tools of GHRM which enhance performance and profit of organizations.

The hotel sector, known for its resource-based operations, faces pressing demands from environmentally concerned customers, regulatory authorities, and global sustainability initiatives to embrace eco-friendly practices. GHRM has emerged as a business strategy that connects environmental goals into HR operations, including recruiting, training, performance assessment, and reward systems. By balancing employee behavior with firm sustainability goals, GHRM offers a means to inculcate a culture of environmental accountability.

EcoLuxe Hotels, a global hotel chain committed to sustainability illustrates innovation in the hospitality sector with its complete GHRM approach. These activities, particularly in the post-pandemic context like digital green training, eco-friendly recruiting, and performance management systems linked green key performance indicators (KPIs). Despite growing interest in GHRM, empirical research on its implementation and effectiveness in the hotel industry are very few, particularly in post-COVID organizational adjustments.

This research examines the impact of GHRM policies at EcoLuxe Hotels on pro-environmental employee behavior and their contribution to the firms' TBL performance, offering insights into sustainable workforce within a competitive and dynamic environment.

2 Review of Literature

The literature on GHRM exemplifies the progression from theoretical underpinnings to real implementations across several industries and regions. Studies, like Renwick et al. (2013) offer a conceptual framework by categorizing GHRM into green recruiting, training, performance evaluation and reward mechanism. This research integrates GHRM to sustainability by promoting pro-environmental behavior (PEB), offering a solid foundation for future theoretical development. Pham et al. (2020) improves the theoretical foundations of GHRM and highlight the importance of employee attitudes and behaviors in achieving environmental objectives [10]. These studies are foundational in establishing GHRM as a strategic function; however, they are limited empirical rigor and applicability across many organizational environments.

The research conducted by Zaid Pham et al. (2018) and Awaad et al. (2022) contribute to the body of literature by linking GHRM practices with green supply chain management (GSCM) and green innovation, respectively. Zaid Pham et al. utilizes data from 150 manufacturing firms to show that green recruiting, training, and environmentally sustainable incentives significantly impact supply chain sustainability [13]. Awaad et al. organized a quantitative study 335 Small Medium Enterprises (SME) in Saudi Arabia. The findings show that GHRM supports green innovation, hence contributing to the TB [3] L. These studies integrate HRM with cross-functional sustainability initiatives. However, the most of the studies are limited to some geographical locations or sectors. Kou et al. (2022) stated that green innovation may serve as a prelude to GHRM transformation, signifying a reciprocal link. Nevertheless, the limited geographic scope and lack of industry diversity minimizes the generalizability of the results [8].

Numerous studies redirect attention to employee conduct and intrinsic motivation. Ardiza et al. (2021) investigate the influence of Green Human Resource Management (GHRM) on Organizational Citizenship Behavior for the Environment (OCBE) [2], whereas Abiwu and Nunoo (2021) examine its effect on psychological commitment [1]. Their results indicate that ecologically focused performance evaluations and incentives promote sustainable behavior. Iftikar et al. (2022) examine five-star hotels and find substantial connections between GHRM procedures and pro-environmental actions among staff. Although these studies highlight the behavioral impacts of green HRM, they mostly rely on small or single-sector samples, hence limiting conclusions to large extent [6]. Mirahsani et al. (2023) offer a macro-level approach, arguing for the incorporation of green culture in HRM systems to promote long-term sustainability; hitherto, the lack of real-world frameworks minimizes the study's relevance.

Sectoral and contextual factors are required to contribute the literature [9]. Darvishmotevali and Altinay (2022) examine the hotel industry and focus on how servant leadership in hotels promotes employee environmental involvement [4] whereas Khaskhely et al. (2022) examine the industrial sector to supports green recruiting and pay in large-scale manufacturing to minimize carbon emissions [7]. These studies offer region specific insights into the implementation of GHRM amid sector-specific environmental challenges. Abiwu and Nunoo (2021) and Kou et al. (2022) offer a post-pandemic perspective, illustrating how changes in HRM result from COVID-19 have reinforced sustainability and resilience measures. This new subject shows how challenges may stimulate green reforms in HR [1, 8].

GHRM strategies on green training, performance management and reward mechanism are consistently lined with improved environmental and organizational performance. Employee engagement and behavioral commitment are necessary mediators of the effectiveness of GHRM. The integration of GHRM with innovation and supply chain operations improves strategic complexity. However, the field continues to face challenges like limited sample sizes, longitudinal research limitation, the lack of defined measures, and the underrepresentation of certain geographical regions and sectors like technology, agriculture, and services (Table 1).

Table 1 An overview of review of literature

Criteria	Strongest contributors	Originality	Weaknesses
Theoretical clarity	Renwick et al. (2013) [11], Pham et al. (2020) [10]	Defined GHRM constructs and linked with EGB	Limited empirical grounding
Empirical rigor	Zaid Pham et al. (2018) [13], Awaad et al. (2022) [3]	Large-scale industry data; integration with innovation	Regionally confined
Behavioral focus	Ardiza et al. (2021) [2], Abiwu and Nunoo (2021) [1]	Explores OCBE and psychological commitment	Small samples, no cross-sectoral insights
Innovation linkage	Kou et al. (2022) [8], Zaid Pham et al. (2018) [13]	Shows bidirectional impact of innovation on GHRM	Ignores cost or conflict implications
Context-specific application	Darvishmotevali and Altinay (2022) [4], Khaskhely et al. (2022) [7]	Studies sector-specific environmental pressures	Results not generalizable beyond region

Source Authors' own compilation

3 Research Gap

Despite the extensive research on GHRM in the last few two decades, notable gaps exist, especially on empirical validation, sectoral diversity and HR reforms in post-pandemic period. There is a paucity of quantitative assessment methods to evaluate the actual results of GHRM operations. Most of the studies, including the case of the EcoLuxe Hotels depend on qualitative approaches and anecdotal evidence to analyze the efficacy without offering measurable metrics such as decrease in energy usage, cost savings, or employee retention. The absence of empirical rigor obscures the establishment of causal relationships between GHRM practices and sustainable performance.

There is a lack of comparative sectoral studies analyzing implementation of GHRM across diverse organizational set up, mostly in service sectors such as hospitality, which are often underrepresented in comparison to manufacturing and industrial sectors. Despite the substantial changes in organizational processes due to the effect of COVID-19 pandemic, there is a paucity of scholarly research into how post-pandemic environment have redefined GHRM strategies like digital training, work from home and hybrid engagement models. Besides, studies rarely analyze how cultural and institutional environments mediate the effectiveness of GHRM practices. Addressing these gaps will be pertinent for evolving GHRM from a conceptual model into a measurable, transferable framework across industries and regions.

4 Objectives of the Research

This research has following primary objectives:

1. To explore the effectiveness and strategic relevance of GHRM practices in the hospitality sector, using EcoLuxe Hotels as a focal case.
2. To examine the effectiveness of GHRM practices in promoting pro-environmental behavior among employees. This involves assessing how green recruitment, training, and performance management systems impact individual and collective green behaviors reflected in energy-saving practices, waste reduction, and participation in sustainability programs.
3. To examine the relationship between GHRM activities and sustainability performance at EcoLuxe Hotels, mostly in the post-pandemic environment. The study explores how HR strategies adapted in post-COVID-19 period. These strategies could be digital onboarding, virtual training, and reduced physical resource usage which have improved or impeded the organization's green objectives and resilience.
4. To evaluate how GHRM practices contribute to the firms' Triple Bottom Line (TBL): people, planet and profit. This includes understanding how GHRM links with cost-efficiency, employee engagement, stakeholders' expectations, and corporate environmental responsibility.

With these objectives, this research would offer insights not only into the operational outcomes of GHRM but also into its strategic potential to improve organizational culture and sustainability performance in a highly competitive service industry.

5 Methodology

This study adopts a qualitative case study approach to investigate the planning, implementation, and impact of GHRM practices at EcoLuxe Hotels, a pioneering hotel chain with sustainability embedded into its corporate mission. A single-case is selected to enable an in-depth exploration of how GHRM policies are operationalized in a real-world hospitality environment. This case was selected purposively due to its commitment to sustainability and its incorporation of HR systems with environmental goals. Data has been obtained from company relevant documents concerning internal HR policy manuals, sustainability reports, training modules, and performance appraisal frameworks available in public domain and it helps to trace the formal structure of GHRM practices. Data analysis was made using thematic coding with the support of NVivo software to emanate recurring themes.

6 Case Analysis

Case Study: Implementation of Green HRM Practices in a Hotel Chain: a Pathway to Sustainable Hospitality

The hospitality industry, particularly hotels, has faced increasing pressure to adopt sustainable practices in response to growing environmental concerns, consumer demand for eco-friendly services, and regulatory requirements. The competitive nature of the industry necessitates that hotels not only offer quality services but also demonstrate environmental responsibility, both to attract environmentally conscious customers and to comply with sustainability guidelines. Green HRM (GHRM) practices can play a critical role in integrating sustainability into the core operations of a hotel, especially through employee engagement, training, and performance management.

One prominent hotel chain, EcoLuxe Hotels, with operations across multiple countries, has recently embarked on a comprehensive GHRM strategy to enhance its sustainability performance. The hotel chain has adopted various green initiatives, ranging from energy conservation measures to waste reduction, and has introduced specific GHRM practices to embed environmental responsibility within its organizational culture. These practices include green recruitment, eco-friendly training programs for employees, performance management systems aligned with sustainability goals, and reward mechanisms for employees who demonstrate pro-environmental behaviors.

Problem Statement in the Context of the Case

EcoLuxe Hotels, like many organizations in the hospitality sector, faces the challenge of effectively embedding sustainability into their organizational culture through HRM practices. Although the company has implemented various GHRM strategies, such as green recruitment and training, the true impact of these practices on employee behavior and organizational performance remains uncertain. Specifically, the hotel chain seeks to understand the following:

1. How effective are the implemented GHRM practices in promoting environmental sustainability within the hotel?
2. What is the relationship between practices of GHRM and pro-environmental behavior of employees?
3. To what extent have post-pandemic changes influenced the hotel's approach to GHRM practices and its impact on sustainability?
4. How do these GHRM practices contribute to EcoLuxe Hotels' overall environmental, social, and economic performance (the Triple Bottom Line)?

GHRM Implementation in EcoLuxe Hotels

EcoLuxe Hotels introduced the following key GHRM practices as part of its green strategy:

1. *Green Recruitment*: The hotel has actively sought to hire employees who are environmentally conscious and passionate about sustainability. During the recruitment and selection process, candidates are examined on the basis of both their qualifications and environmental awareness which develops commitment to sustainable green practices.
2. *Green Training and Development*: New hires and existing employees undergo extensive training programs focused on sustainable practices, including energy conservation, waste reduction, eco-friendly cleaning practices, and the use of environmentally sustainable products. The hotel also provides continuous learning opportunities to encourage employees to suggest innovative green solutions.
3. *Green Performance Management*: The hotel's performance appraisal system incorporates environmental goals, rewarding employees who demonstrate eco-friendly behaviors, such as reducing resource consumption or contributing to sustainability initiatives. Green performance indicators are integrated into the annual reviews of employees across departments.
4. *Green Compensation and Rewards*: Employees who actively contribute to the hotel's environmental goals receive bonuses, promotions, or recognition through the company's reward system. This includes providing rewards for employees who introduce energy-saving innovations or engage in community-based sustainability projects.
5. *Post-Pandemic Adaptations*: In response to the COVID-19 pandemic, EcoLuxe Hotels adapted its GHRM practices by incorporating remote working options and emphasizing digital training methods. The hotel also adjusted its environmental policies, focusing on sustainability in a more resource-efficient manner,

such as reducing waste from single-use items and encouraging virtual meetings to cut down on travel emissions. By focusing on environmental conscious behavior after pandemic, EcoLuxe Hotels are focusing on employee training and developmental programs to create awareness and practice green behavior. To maintain constant environmentally friendly behavior EcoLuxe Hotels, add rewards and incentives with green performance management system. The management has equipped energy conservation practices with its employees. As a result, the green goal of the organization is achieved and employee performance is enhanced. The hotels prioritize green recruitment, green performance evaluation, green employee involvement, green reward and recognition policies in post-pandemic era. The green teams and green committees are formed to provide ideas for further improvement of green environment, check green practices and solicit employee feedback on green practices. By following these norms EcoLuxe e Hotels protect environment, save energy and create a sustainable workforce which plays very vital role in volatile environment.

7 Integrating Literature and Case

The EcoLuxe Hotels case study illustrates how GHRM function as a strategic instrument to link sustainability into company culture, particularly in the hospitality industry. This is aligned with the theoretical foundations established by Renwick et al. (2013), who defined GHRM as a collection of HR practices such as green recruiting, training, performance management and reward mechanism [11]. All these functions are intended to support environmental sustainability via employee behavior. EcoLuxe consistently applies these functions, illustrating a strong connection with this theoretical framework. Furthermore, Pham et al. (2020) emphasize the need for strategic alignment between environmental goals and HR responsibilities- a relationship distinctly illustrated in EcoLuxe's integration of sustainability metrics into recruiting standards, training initiatives, and performance evaluations [10]. This highlights the strategic change in HRM from administrative support to sustainable leadership.

A crucial element focused in both reviews and the case analysis is the role of employee behavior as a mediator of environmental outcomes. The behavioral approach in GHRM, as illustrated by Ardiza et al. (2021) and Iftikar et al. (2022), shows that green training and reward mechanism can remarkably impact Organizational Citizenship Behavior for the Environment (OCBE) and Employee Green Behavior (EGB). EcoLuxe links environmental outcomes and employee onboarding and offers recognition and rewards for eco-friendly actions [2, 6]. The formation of green teams and committees supports the implementation of a participatory and behavior-based approach to sustainability, illustrating the employee-focused frameworks proposed in recent studies. These approaches illustrate how individual behavioral changes, supported by HR interventions, may result in collective cultural transformation- a concept most commonly supported by literature but less often studied with such precision in reality.

Besides, the instance advances the current discussion on the role of GHRM in achieving TB outcomes. Awaad et al. (2022) state that GHRM approaches supports in achieving ecological objectives while simultaneously strengthening employee engagement and improving operational efficiency. EcoLuxe's initiatives like energy saving, minimization of operational waste, and increasing community-based sustainability initiatives illustrate adherence to these TBL principles. However, this research highlights the need for performance assessment metrices and quantitative validation and mostly showcases qualitative results. The paucity of quantifiable metrics like decrease in energy use or cost savings limits the potential to generalize or benchmark performance, a recognized limitation in both the case study and most of the current GHRM studies [3]

The EcoLuxe case is unique for its explicit adjustment with post-pandemic adaptation. Abiwu and Nunoo (2021) state that the COVID-19 pandemic compelled the corporate houses to reevaluate their HR strategy and promote resilience and sustainability [1]. EcoLuxe experienced this transition by digitizing training, implementing work from home practices, minimizing travel-related emissions, and increasing initiatives to minimize single-use waste. These adaptive practices offer a novel aspect to GHRM, which remains nascent in scholarly discourses. Although the literature acknowledges the pandemic's impact on HR goals, yet, few research has studied how service-oriented companies operationalize implemented green recovery strategies. EcoLuxe offers innovative insights into the evolution of GHRM in reaction to global disruptions, thereby, presenting a forwarding looking perspective on sustainability resilience.

The sectoral setting amplifies the value of the case. The hospitality sector, characterized by significant consumption and emissions, has unique sustainability issues. Darvishmotevali and Altinay (2022) examine green behavior in hotels from the perspective of leadership impact whereas Iftikar et al. (2022) linked GHRM with employee environmental initiatives [4]. EcoLuxe advances this body of work by offering a more comprehensive and holistic approach. The formation of green committees, implementation of performance-based incentives, and cross-departmental collaboration offers a robust framework that other service-sector companies may replicate. Besides, the case connects theoretical frameworks with context-specific applications, thereby, advancing the discourse on how industry-specific contexts impact the design and effectiveness of GHRM interventions (Table 2).

The EcoLuxe Hotels case substantiates critical theoretical grounding from the literature, namely on the strategic role of HR in sustainability, the behavioral effects of GHRM practices, and the significance of GHRM in relation to TBL performance. Despite its lack use of quantitative metrics, the case stands out as an original contribution, offering an actionable blueprint for integrating GHRM into an ever-changing and highly customer-centric industry. As such, it strengthens the position that GHRM is not only an operational adjustment but also a transformative organizational philosophy that could link people, processes, and environmental responsibility in pursuit of sustainability in the run.

Table 2 EcoLuxe hotels and green HRM: case-literature comparison

Aspect	EcoLuxe hotels case	Related studies
Sector	Hospitality	Darvishmotevali and Altinay (2022), Iftikar et al. (2022)
GHRM scope	Comprehensive (Recruitment → Reward)	Pham et al. (2020), Zaid Pham et al. (2018) [4, 6, 10, 13
Post-pandemic adaptation	Yes (digital training, eco-conscious behavior)	Abiwu and Nunoo (2021) [1]
Employee-centric innovation	Yes (green committees, suggestion systems)	Kou et al. (2022) [8]
TBL focus	Implicit	Awaad et al. (2022) [3]

8 Implications for Policy

The adoption of GHRM approaches at EcoLuxe Hotels offers several practical insights for policy implications at both organizational and institutional level. The incorporation of sustainability criteria into HR policies like recruitment, training, and performance management can serve as a replicable framework for the hotel and other service industries looking to accomplish environmental goal. The business needs to develop green recruiting criteria and competence frameworks that evaluate applicants' environmental awareness, therefore, matching workforce values with sustainability goals from the beginning.

Secondly, performance evaluations and reward structures should officially comprise environmental key performance indicators (KPIs) to support responsibility and motivation. Regulatory bodies and industry association may enable this by formulating sector-specific green HRM guidelines and sustainability reporting standards that promote consistent implementation. Additionally, governments and tourism boards might contemplate providing tax incentives benefits or recognition schemes to firms that effectively implement GHRM practices and show quantifiable sustainability outcomes, thereby reinforcing public–private collaboration in promoting green workforce development.

9 Directions for Further Research

Despite the comprehensive applications GHRM at EcoLuxe Hotels, several areas need more scholarly investigation. Future research should focus on quantifying the direct impacts of GHRM policies and practices on firm performance metrics like energy cost saving, guest satisfaction, staff retention, and brand equity. Longitudinal study is required to evaluate whether environmentally responsible behavior sustained through GHRM practices continues over time or declines in the without continuous motivation.

Furthermore, comparative analyses across industries, mostly of high-emission sectors (e.g., hospitality, manufacturing) and low-emission sectors (e.g., IT, finance) may be organized to shed light on how context impact the design and effectiveness of GHRM. Researchers need to investigate the mediating impacts of business culture, leadership style, and national institutional frameworks on employee green behavior. Considering the growing reliance on technology, future research may focus on how digital HR instruments such as including e-learning platforms and HR analytics, either support or obstruct the adoption of green HRM in a post-pandemic period.

10 Conclusion

The EcoLuxe Hotels case illustrates an ever-changing, integrated strategy for GHRM that links corporate HR actions with environmental sustainability goals. The hotel industry has integrated sustainability into its core business operations and HR culture through green recruiting, training, performance management, and reward mechanism. This indicates necessary theoretical proposition from GHRM literature and strengthens the strategic significance of HR in accomplish TBL performance.

Significantly, EcoLuxe adapts its strategy to link with post-pandemic environment, showing the dynamic nature of GHRM and its significance in boosting organizational resilience. The case effectively substantiates several theoretical assertions, although it highlights the need for empirical investigations and sector-wise comparisons to strengthen the academic and practical contributions of GHRM. The case ultimately emphasizes that sustainability in service sectors cannot be seen as a standalone initiative; rather, it must be systematically integrated into HR processes, with HR serving as a pivotal driver of change.

References

1. Abiwu, L., & Nunoo, G. N. (2021). Green selection practices. In *Human resource management practices for promoting sustainability* (Chapter 6). https://doi.org/10.4018/978-1-7998-4522-5.ch006
2. Ardiza, F., Nawangsari, L. C., & Sutawidjaya, A. H. (2021). The influence of green performance appraisal & green compensation to improve employee performance through OCBE. *International Review of Management & Marketing, 11*(4), 13–22. https://doi.org/10.32479/irmm.11632
3. Awwad Al-Shammari, A. S., Alshammrei, S., Nawaz, N., & Tayyab, M. (2022). Green human resource management and sustainable performance with the mediating role of green innovation: A perspective of new technological era. *Frontiers in Environmental Science, 10*, Article 901235. https://doi.org/10.3389/fenvs.2022.901235
4. Darvishmotevali, M., & Altinay, L. (2022). Green HRM, environmental awareness and green behaviors: The moderating role of servant leadership. *Tourism Management, 88*, Article 104401. https://doi.org/10.1016/j.tourman.2021.104401
5. Elkington, J. (1997). *Cannibals with forks: The triple bottom line of 21st century business.* Capstone Publishing.
6. Iftikar, T., Hussain, S., Malik, M. I., Hyder, S., Kaleem, M., & Saqib, A. (2022). Green human resource management and pro-environmental behaviour nexus with the lens of AMO theory.

Cogent Business & Management, 9(1), 2124603. https://doi.org/10.1080/23311975.2022.2124603
7. Khaskhely, M. K., et al. (2022). Understanding the impact of green human resource management practices and dynamic sustainable capabilities on corporate sustainable performance: Evidence from the manufacturing sector. *Frontiers in Psychology, 13*, 84448. https://doi.org/10.3389/fpsyg.2022.844488
8. Kuo, Y. K., Khan, T. I., Islam, S. U., Abdullah, F. Z., Pradana, M., & Kaewsaeng-On, R. (2022). Impact of green HRM practices on environmental performance: The mediating role of green innovation. *Frontiers in Psychology, 13*, Article 916723. https://doi.org/10.3389/fpsyg.2022.916723
9. Mirahsani, N., Azizan, O., Shahriari, M., Rexhepi, G., & Najmi, A. (2024). Green culture toward employee green behaviour; the mediation roles of perceived support and green identity. *Environment, Development and Sustainability, 26*(6), 16149–16172.
10. Pham, N. T., Hoang, H. T., & Phan, Q. P. T. (2020). Green human resource management: a comprehensive review and future research agenda. *International Journal of Manpower, 41*(7), 845–878. https://doi.org/10.1108/IJM-07-2019-0350
11. Renwick, D. W., Redman, T., & Maguire, S. (2013). Green human resource management: A review and research agenda. *International journal of management reviews, 15*(1), 1–14. https://doi.org/10.1111/j.1468-2370.2011.00328.x
12. Lakshmi, P. V., & Battu, N. (2018). A study on green HRM–an emerging trend in HR practices. *International Journal of Management, 9*(3), 74–82.
13. Zaid, A. A., Jaaron, A. A. M., & Bon, A. T. (2018). The impact of green human resource management and green supply chain management practices on sustainable performance: An empirical study. *Journal of Cleaner Production, 204*, 965–979. https://doi.org/10.1016/j.jclepro.2018.09.062

Transforming HR: The Evolution from Operational Tasks to Data-Driven Strategic Insights

Chandrakanta Sahoo, Deepak Kumar Sahoo, and R. Durga Prasad

Abstract

This research work examines the Dutch Police Force's use of HR analytics to significantly minimize attrition rates among pilots in its aviation unit by linking recruitment practices and retention strategies with firms' goals. The purpose of this study is to validate the strategic implication of data-driven HR practices in increasing employees' stability and operational efficiency in the public sector. The research utilizes a qualitative case study approach to analyze secondary data such as HR analytics reports, exit interviews, and organizational records. The finding reveals that attrition rates of pilots recruited directly from aviation school was 40% in comparison to the pilots who were recruited into police force as police, received pilot training and promoted as pilot to fly Dutch police helicopters. The results show that internal recruits illustrated increased organizational commitment owing to intrinsic motivation and alignment with professional identity, resulting in significant reduction of cost in training and onboarding. The originality of present research lies in its practical application of HR analytics in a public institution, linking both academic theory with practice. It demonstrates how predictive analytics can provide informed evidence-based

C. Sahoo (✉)
Department of Business Administration, Institute of Management and Information Technology, Cuttack, Odisha, India
e-mail: csahoo@imit.ac.in

D. K. Sahoo
College of IT & Management Education, Bhubaneswar, Odisha, India
e-mail: principal@cime.ac.in

R. Durga Prasad
School of Business Management, IBS University, Port Moresby, Papua New Guinea
e-mail: durga.prasad@ibs.ac.pg

Southern Cross University, Lismore, Australia

S. Patnaik et al. (eds.), *People, Planet and Profit: Advancing Sustainable Business Practices*, Springer Proceedings in Geography and Sustainability,
https://doi.org/10.1007/978-3-032-11296-5_2

decision-making and foster sustainable HR strategy. Longitudinal studies across industries may be conducted in future to evaluate long-term outcomes, examine ethical implications of HR analytics, and explore employee sentiment analysis in predictive HR models.

Keywords

HR analytics • Employee retention • Dutch police force • Predictive analytics • Workforce management

1 Introduction

Human Resource Management (HRM) has transitioned from fundamental administrative roles to a strategic, data-driven partner role in a company's decision-making. An important factor in this shift has been the rise of HR analytics, which allowed HR to shift its focus from operational efficiency to drawing strategic insights from data. HR analytics is the systematic identification and quantifications of the human drivers of business outcomes [11]. HR analytics involves the systematic collection, analysis, and interpretation of HR data. It identifies the HR trends, patterns to measure the impact of workforce factors on business success. This supports firms to make data-driven decisions that improve productivity, employee engagement, and overall organizational effectiveness. The transition from operational strategies to that of strategic, data-driven HR strategy indicates a significant transformation with broader implications for organizational agility, talent competitiveness, and long-term sustainability. Nevertheless, the execution of data-driven strategies is not free from challenges; HR department often encounters resistance to opposition to change, inadequate analytical skills, data integration challenges with law, and issues concerning data quality and privacy. In spite of these challenges, HR analytics can play a commendable role in overcoming them by supporting evidence-based decision-making, giving predictive insights into workforce trends and pattern, and aligning HR practices with corporate objectives. This present study examines the progression of HR analytics across three stages: operational HR, strategic HR, and data-driven HR. It focuses on the important changes, measurements, and technologies that have affected HR analytics, showing how it can improve workforce management, make the most of talent, and deliver business results through a case study of the Dutch Police Force.

2 Review of Literature

The evolution of HRM from a mere administrative support role to a strategic partnership role in a firm's performance has garnered substantial scholarly attention. One of the seminal definitions offered by Wright and McMahan (1992) on SHRM describes it as "the pattern of planned human resource deployments and activities

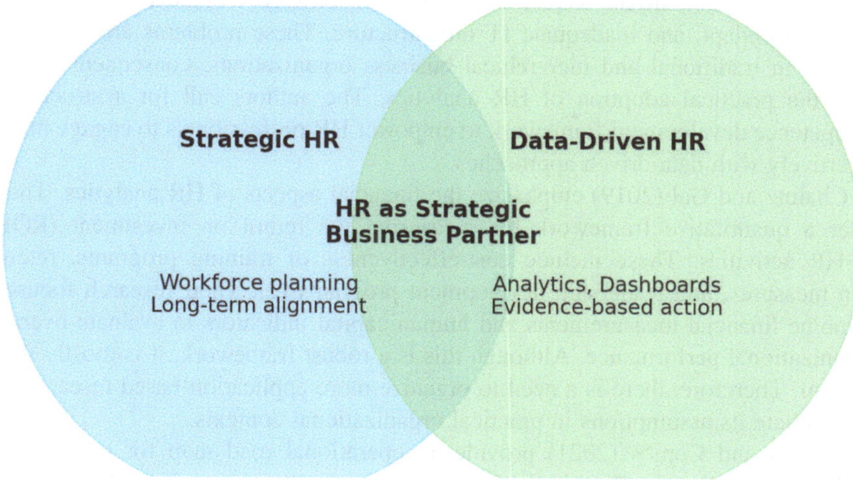

Fig. 1 Integration of strategic HR and data-driven HR framework. *Source* Authors' own creation

intended to enable an organization to achieve its goals [12]." This definition positions HR at the core of business strategy implementation, implying that human capital is not merely a cost center, rather a crucial enabler of a firm's competitive advantage. Therefore, HR practices are now accepted as a cohesive framework for achieving long-term firm goals rather than as separate administrative functions.

Complementing SHRM, Marler and Boudreau (2017) explain Data-Driven HR as the use of workforce analytics and systematic data analysis to guide HR decisions. It relies on evidence-based practices instead of traditionally used decisions based on intuition, increasingly focusing on technologies and tools like predictive analytics, HR dashboards, and artificial intelligence (AI) to explain employee behavior, improve talent management, and measure organizational efficiency. SHRM and Data-Driven HR together offer a dual paradigm for businesses aiming to leverage human capital for strategic benefit [6] (Fig. 1).

Several studies have focused on strategic implications of HR analytics. Zebua et al. (2024) contend that HR analytics act as a conduit between strategic planning and traditional HR operations by using evidence-based assessment of HR policies. Their outcomes validate that the businesses adopting the data-driven approach for recruitment, training, and employee engagement often report improved productivity and employee satisfaction [13]. This research is mostly descriptive and suffers from paucity of comprehensive sectoral analysis, mostly in non-corporate areas such as public administration or law enforcement.

McCartney and Fu (2022) emphasize that there is an increasing need to bridge academic theory and HR practice [7]. They argue that while the usage of HR analytics is widely accepted, nevertheless, it suffers from limitations in implementation capabilities within HR department. Their analysis highlights structural difficulties in HR analytics implementation like insufficient technical skills, unwillingness to adapt, and inadequate IT infrastructure. These problems are basically evident in traditional and hierarchical business organizations, consequently limiting the practical adoption of HR analytics. The authors call for a structured competence development framework to empower HR professionals to engage more assertively with data-driven approaches.

Chalutz and Gal (2019) emphasize the financial aspects of HR analytics. They offer a quantitative framework for measuring the return on investment (ROI) of HR activities. These include cost-effectiveness of training programs, retention measures, and leadership development programs [2]. Their research focuses combine financial measurements and human capital indicators to evaluate overall organizational performance. Although this is a robust framework, it is mostly theoretical. Therefore, there is a need to organize more application-based research to substantiate its assumptions in practical organizational contexts.

Falletta and Combs (2021) provide an operational road map for use of HR analytics. Their comprehensive guide encompasses all aspects from data collection to selection of metrics and visualization techniques. They endorse the use of HR dashboards and real-time key performance indicators (KPIs) to guide strategic workforce planning. A key finding reveals that there is a need for linking analytics into the decision-making culture of HR departments. This vision accepts a degree of digital maturity and leadership commitment that some firms, particularly in the public sector, have not yet attained [3].

The study by Ravesangar and Narayanan (2024) focuses on the behavioral effects of analytics-driven human resources [10]. Their research analyzes how predictive algorithms might detect workers' risk of disengagement or turnover. The researchers contend that firms might proactively devise interventions by examining indicators such as absenteeism, productivity metrics, and sentiment analysis. This proactive approach to retention is particularly important in knowledge-intensive industries; yet, the work fails to include ethical issues concerning employee monitoring and consent.

Mishra et al. (2016) augment this discourse by investigating the function of AI and machine learning in predictive HR analytics [8]. Their research exhibits how algorithmic models can estimate talent requirement, identify leadership potential, and strategize succession pathways. However, the use of AI in HR engenders controversies concerning algorithmic bias, transparency, and the potential loss of human judgement in serious personnel decisions. These concerns call for ethical governance systems that balance analytical accuracy with fairness and accountability.

Adoption of data-driven HR necessitates a change of organizational culture to accept it. Okatta et al. (2024) emphasize that the implementation of HR analytics is not just a technological transition but also a cultural transformation. They argue that the firms should inculcate a data-centric mindset among HR professionals by investing resources for training and forming cross-functional teams that link domain expertise with data science capabilities. However, the authors identify that

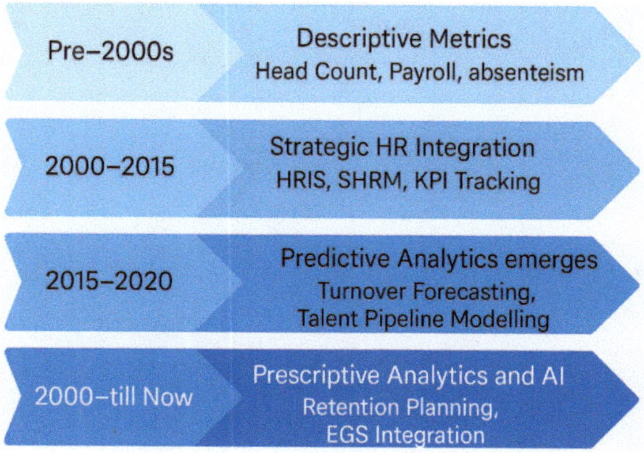

Fig. 2 Evolution of HR analytics. *Source* Authors' own creation based on reviews

this shift is resource-intensive and may pose challenges for small- and medium-sized firms (SMEs) and public sector enterprises with limited resources [9] (Fig. 2).

Harshini and Balu (2023) explain the historical progression of HR analytics, highlighting its beginning from rudimentary reporting capabilities to sophisticated, AI-enhanced analytics [4]. Their categorization of development comprises three phases: descriptive, predictive, and prescriptive analytics. Descriptive analytics explains the historical occurrences; predictive analytics forecasts future trends; and prescriptive analytics recommends certain courses of action. Although this evolutionary approach is beneficial, the researchers inadequately investigate the contextual variables such as regulatory compliance, organizational maturity, and industry dynamics which can affect organizational advancement through these phases (Fig. 3).

A significant limitation in the literature is the inadequate representation of empirical research from the public sector. Most of the existing research originates from corporate environments, which often possess more resources and flexibility to explore new technologies. Subsequently, there is a limited understanding on how efficiently HR analytics can be implemented in public institutions marked by bureaucratic rigidity and regulatory scrutiny. This research studies the case of the Dutch Police Force, a government department that effectively used HR analytics to minimize attrition rates among helicopter pilots. This example offers essential empirical evidence to a subject primarily characterized by theoretical proposition by showcasing concrete outcomes, including increased pilot retention and saving costs.

A significant limitation in existing corpus of literature is the absence of longitudinal studies. Most contemporary researches evaluate the impacts of HR analytics using cross-sectional data, which inadequately reflect the long-term sustainability of analytics-driven solutions. Longitudinal research is necessary for evaluating the sustainability of workforce improvements, monitoring changing employee sentiments, and measuring cumulative financial advantages over time. In the absence of such analysis, the strategic importance of HR analytics remains only partly validated.

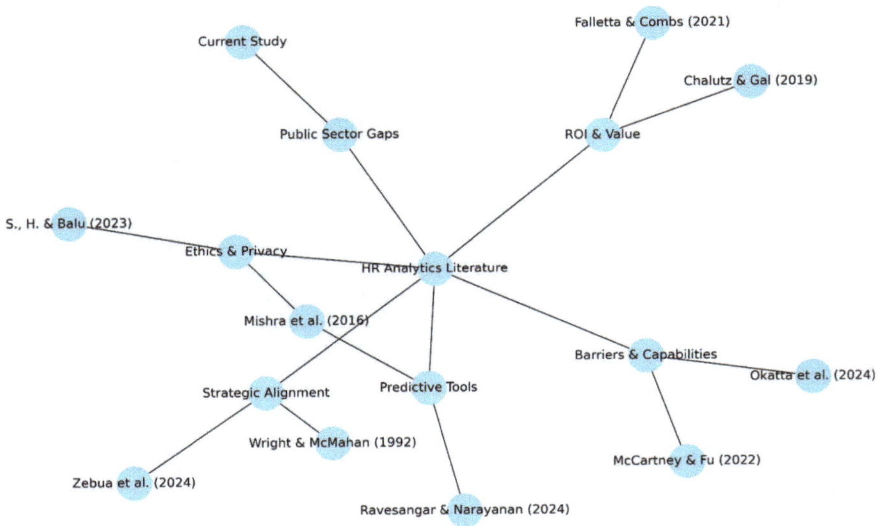

Fig. 3 Thematic concept map: key dimensions of HR analytics literature

Ethical issues in HR analytics also necessitate further scrutiny. Although several research studies highlight the advantages of AI and big data in augmenting HR performance, they often neglect challenges related to privacy, data security, and informed consent. As predictive and prescriptive models gain importance, the risk of misuse and overreach can escalate challenges, particularly in contexts characterized by power asymmetries between employers and workers. A subsequent study should focus on providing ethical norms and compliance frameworks to ensure the proper use of employee data.

3 Objectives

- To study the strategic use of HR analytics in workforce planning and retention in the public sector, precisely focusing on the Dutch Police Force.
- To assess measurable outcomes such as attrition rates, tenure, and cost-per-hire related to internal versus external recruiting techniques.
- To demonstrate how data triangulation and evidence-based insights from HR dashboards, departure interviews, and organizational records can guide HR decision-making and better employee retention.

4 Methodology

This research uses a qualitative case-based approach utilizing secondary data, with deliberate efforts to improve its empirical basis. The investigation focuses on measurable outcome results by comparing pre- and post-intervention HR metrics such as attrition rates, cost-per-hire, and average tenure within the aviation unit of the Dutch Police Force. These measurable metrices were derived from organizational records and HR analytics dashboards to evaluate the strategic impact of internal recruiting strategies. The present study used triangulation by including many sources such as HR data reports, departure interview summaries, internal policy documents, and strategic workforce analytics. This enabled thematic coherence and strengthened the credibility of the results. The present analysis uses extensive secondary data but accepts the limitation of excluding original data. Consequently, subsequent research may take a mixed-methods approach to include semi-structured interviews with HR professionals, survey of respondents directly impacted by policy changes, and focus group discussions comparing internally and externally recruited police personnel. A pilot project integrating predictive modeling into HR dashboards may also be conducted to evaluate the effectiveness of real-time decision-making in public sector organizations. Future research may add to the existing results and offer longitudinal insights, hence improving generalizability and practical applicability for both academics and practitioners.

5 Findings and Discussion

Case Study: Using HR Analytics to Lower Attrition in the Dutch Police Force

As the nation's law enforcement organization, the Dutch Police Force is in charge of upholding public order, enforcing the law, and guaranteeing the protection and security of its residents. With a decentralized system that permits local and regional police forces to oversee certain regions while being coordinated by a Central National Agency, it functions under the Ministry of Justice and Security. Among the many tasks assigned to the force are traffic management, counterterrorism, criminal investigations, and public safety during major events.

Community policing, which encourages close collaboration between police and local communities, is a hallmark of the Dutch Police Force. Additionally, the organization has led the way in incorporating technology, such as via the use of sophisticated police tools and data analytics. For instance, they apply HR analytics to improve workforce management, minimize attrition, and promote employee engagement, especially in specialized groups such as the helicopter pilot division. This strategy not only benefits operational efficiency but also helps maintain a trained staff within the police force.

The aviation unit, responsible for aerial surveillance and emergency support, confronted a growing challenge in retaining helicopter pilots. Two main sources were used in the recruitment process: (a) hiring trained pilots directly from aviation schools and (b) hiring police officers from inside the force, sending them to

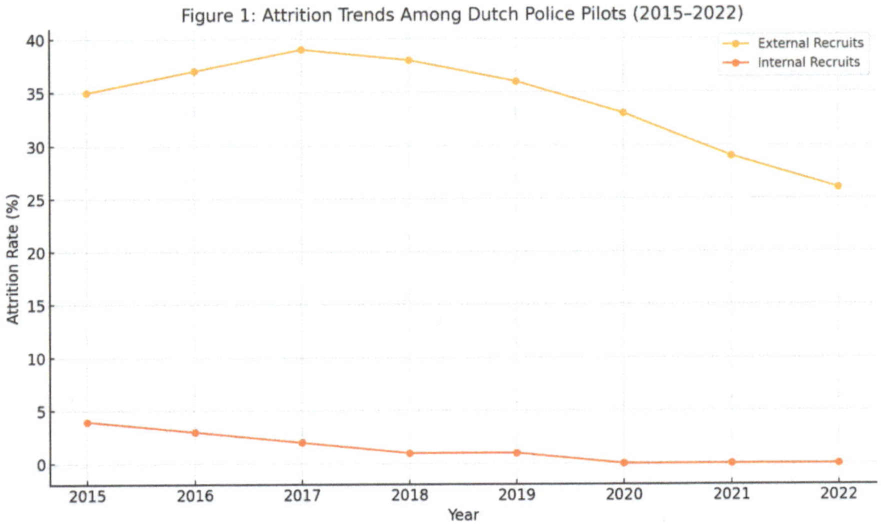

Fig. 4 HR analytics report during 2015–2022

aviation school for training, and then inducting them as police pilots once they graduated.

As time went on, the Dutch Police identified a pattern of high pilot turnover rates, especially for those directly hired from the aviation schools. The police leadership made the decision to examine the pilot attrition statistics in order to look into this issue.

After further examination, HR analytics showed a distinct trend. When compared to pilots hired from inside the police department, attrition rates were much higher for pilots directly recruited from aviation schools. It's fascinating to note that pilots promoted from the ranks of current police officers saw almost no attrition (Fig. 4).

Exit interviews with departing pilots yielded crucial information. We observed that the pleasure of working for the police force motivated pilots hired from aviation schools. These people did not see the job as a long-term profession and were often drawn to the thrill and adventure it brought. They departed the department in search of better career opportunities elsewhere after gaining significant exposure and experience.

However, pilots hired from within the police department demonstrated a stronger sense of dedication. Due to their advancement in the police force, these pilots already had a close shared bond with the organization. They were deeply committed to the police force and its purpose. Often characterized as "once a cop, always a cop," this group of pilots had no plans to leave the force after completing their aviation training. They were more likely to remain for a long time because they had a feeling of belonging to the force and were dedicated to serving the country.

HR Analytics Insights: The Dutch Police Force was able to make the following conclusions by using HR analytics:

(a) Employee Motivation and Loyalty: After their expectations were satisfied, the direct recruits from aviation schools were more likely to leave since they were driven by outside considerations like adventure and professional advancement. Police officers who chose to become pilots, on the other hand, were driven by internal factors such as duty, service, and a desire to continue working in law enforcement.
(b) The Cost-Effectiveness of Internal Recruitment: HR analytics revealed that training current police officers not only reduced attrition but also saved the department a substantial amount of money on hiring, training, and turnover.
(c) Retention Strategies: The Dutch Police Force determined that it needed to update its recruitment strategy after examining the data. In order to lower turnover in the aviation unit, they discovered that hiring from within the police force produced a more stable workforce.

Solution Put into Practice: To address the attrition problem, the Dutch Police Force redefined its recruitment strategy. They opted to concentrate more on promoting police officers into aviation professions rather than hiring directly from aviation schools. For internal applicants, the department additionally highlighted the dedication to service and long-term career possibilities, which more closely aligned with their beliefs. Over and above, the force also implemented retention programs to recognize and reward officers loyal to the department, further strengthening their commitment to it.

Result: After implementing these changes, the Dutch Police Force saw a notable decrease in pilot attrition. Internal recruit attrition rates fell to zero, and the Police Aviation Department as a whole became more stable and economical. The department's HR analytics systems monitored and predicted attrition trends to ensure long-term staff sustainability.

6 Linking Literature and HR Analytics Practice—The Case of the Dutch Police Force

The literature on HR analytics reveals its rising importance as a strategic tool for alignment of HR operations with business objectives. Research organized by Zebua et al. (2024), McCartney and Fu (2022), and Chalutz and Gal (2019) found that HR analytics supports data-driven decision-making, linking HR processes such as recruiting, retention, performance management, and workforce planning to measurable organizational outcomes. This is evident in the case of the Dutch Police Force, where workforce analytics was used to address the specific issue of increased turnover rate among pilots in the aviation sector [2, 7, 13].

The review of literature reiteratively focuses on evidence-based decision-making and its progression of HR from a support function to a strategic business partner. The story of the Dutch Police illustrates this transition. Through the analysis of attrition data and the exit interviews, the organization explored the motivational differences between the pilots recruited directly from aviation school and those trained and promoted from within the police force. This supports McCartney and Fu's (2022) claim that organized frameworks can help HR practitioners in aligning employee motivation with firm's goals more effectively [7].

Chalutz and Gal (2019)'s emphasis on measuring ROI of HR financial involvement aligns directly with the Dutch Police Force's cost–benefit analysis [2]. The finding reveals that internal promotions significantly reduce turnover and, therefore, the expenses related to recruiting and training reveal the capacity of HR analytics to explain financial implication of human capital strategy. This relationship explains that data, when effectively analyzed, transcends descriptive statistics to provide actionable insights.

Additionally, research by Ravesangar and Narayanan (2024) and Mishra et al. (2016) focus on the power of predictive analytics in identifying turnover patterns and employee behaviors [8, 10]. The Dutch Police Force uses HR analytics to track and predict attrition patterns, using a proactive approach to Human Capital Management (HCM) instead of a reactive approach. By forecasting which recruiting channels are most likely to create long-term, committed professionals, the department successfully changes its approach to maintain stability of a sustained workforce.

This research highlights the significance of employee engagement, retention strategies, and organizational alignment [3, 10]. These factors were vital in the Dutch Police revised recruiting strategy, which prioritized organizational sense of belongingness and intrinsic motivation—the attributes more often noticed in internal recruits. The case substantiates the theory that employee alignment with company value may profoundly impact employee retention, as illustrated by research advocating for cultural fit and intrinsic motivation as predictors of longevity.

The literature has strong theoretical foundations but often lacks validation though a real-world application—a weakness vividly presented in the review summary. This case study illustrates the use of HR analytics, substantiating theoretical claims with quantifiable outcomes such as reduced attrition and increased departmental stability.

7 Policy Implications

The findings from the Dutch Police Force case study offer significant policy implications for public and private sector companies looking for improving employee stability and strategic human resource alignment. The policymakers are required to internalize the use of HR analytics in recruiting, retention, and manpower planning strategies. Policies should specifically promote internal mobility and career progression opportunities, recognizing that talent from within the firm often pays greater commitment resulting in lesser manpower attrition. Furthermore, standardized protocols for data collection, exit interviews, and predictive modeling must be used to enable evidence-based HR decision-making. Investing on development of analytical competencies of HR professionals should be a key priority under the policy, so that the HR department can offer key insights from HR data. The firms should ensure that their analytics practices comply with legal and ethical standards as they obtain more granular data on employee behavior, therefore safeguarding employee trust while facilitating strategic insights. Collectively, this policy direction can transform HR from a transactional role into a strategic role, increasing organizational performance and workforce sustainability, as evident in the Dutch Police Force's effective use of HR analytics.

8 Conclusion

The present study demonstrates how strategically implemented HR analytics can provide measurable and sustainable improvement in public sector human capital management. The case study of the Dutch Police Force offers specific and evidence-based findings. Three pragmatic recommendations arise for HR professionals, especially within governmental or institutional contexts:

- Invest on internal talent by recognizing high-potential employees and provide them with structured career advancement opportunities, mostly for critical or specialized positions.
- Use predictive analytics to evaluate attrition risks, anticipate future manpower requirements, and guide proactive succession planning.
- Triangulate HR data sources like HR dashboards, employees exit interviews, and HR records to understand HR data patterns before taking strategic decisions.

The Dutch Police case highlights the need for aligning recruiting strategy with company culture and mission. By prioritizing long-term commitment over short-term skill development, employers may cultivate resilient, value-driven teams. Future research may be conducted to expand upon these findings by organizing longitudinal and cross-sectoral research that evaluate predictive HR models across diverse organizational environment. The message is unequivocal for HR professionals: HR analytics must not be seen technical add on, but as an essential tool for strategic workforce transformation.

References

1. Cascio, W. F., & Boudreau, J. W. (2016). The search for global competence: From international HR to talent management. *Journal of World Business, 51*(1), 103–114.

2. Chalutz, H., & Gal, B. (2019). An ROI-based review of HR analytics: Practical implementation tools. *Personnel Review, 48*(6), 1429–1448. https://doi.org/10.1108/PR-11-2017-0362
3. Falletta, S. V., & Combs, W. L. (2021). The HR analytics cycle: A seven-step process for building evidence-based and ethical HR analytics capabilities. *Journal of Work-Applied Management, 13*(1), 1–68. https://doi.org/10.1108/JWAM-03-2020-0020
4. Harshini, S., & Balu, L. (2023). The role of people analytics in driving evidence-based decision-making. In *Proceedings of the 11th International Conference on HR for Organizational Sustainability* (pp. 1–13). Mysuru, India. ISBN 978-93-83302-65-9.
5. Levenson, A. (2017). Using workforce analytics to improve strategy execution. *Human Resource Management, 57*(3), 685–700. https://doi.org/10.1002/hrm.21850
6. Marler, J. H., & Boudreau, J. W. (2017). An evidence-based review of HR analytics. *The International Journal of Human Resource Management, 28*(1), 3–26.
7. McCartney, S., & Fu, N. (2022). Bridging the gap: Why, how, and when HR analytics can impact organizational performance. *Management Decision, 60*(13), 25–47. https://doi.org/10.1108/MD-12-2020-1581
8. Mishra, S. N., Lama, D. R., & Pal, Y. (2016). Human resource predictive analytics (HRPA) for HR management in organizations. *International Journal of Scientific & Technology Research, 5*(5), 33–35.
9. Okatta, C. G., Ajayi, F. A., & Olawale, O. (2024). Leveraging HR analytics for strategic decision-making: Opportunities and challenges. *International Journal of Management & Entrepreneurship Research, 6*(4), 1304–1325. https://doi.org/10.51594/ijmer.v6i4.1060
10. Ravesangar, K., & Narayanan, S. (2024). Adoption of HR analytics to enhance employee retention in the workplace: A review. *Human Resources Management and Services, 6*(3), 1–13. https://doi.org/10.18282/hrms.v6i3.3481
11. Van den Heuvel, S., & Bondarouk, T. (2017). The rise (and fall?) of HR analytics. *Journal of Organizational Effectiveness: People and Performance, 4*(2), 157–178. https://doi.org/10.1108/JOEPP-03-2017-0022
12. Wright, P. M., & McMahan, G. C. (1992). Theoretical perspectives for strategic human resource management. *Journal of Management, 18*(2), 295–320. https://doi.org/10.1177/014920639201800205
13. Zebua, D. K., Santosa, T. A., & Putra, F. D. (2024). The role of HR analytics in enhancing organizational performance: A review of literature. *Indonesia Journal of Engineering and Education Technology (IJEET), 2*(2), 362–368. https://doi.org/10.61991/ijeet.v2i2.69

Corporate Governance and Corporate Sustainability with Reference to the New Companies Act, 2013

Alle Yashoda, Srinivas Karri, and Ahmed Fantaw Amanu

Abstract

This study examines the relationship between corporate governance (CG) and corporate sustainability (CS) in the framework of the New Companies Act, 2013, in India, using a Tata Steel case study. The objectives of this research work is to examine statutory governance mandates-impacted sustainability strategies, evaluate the integration of CSR and environmental activities in business operations, and assess their impact on organizational performance. This research uses a qualitative case study approach, focusing on a comprehensive review of corporate reports, regulatory frameworks, and secondary literature. The results indicate that Tata Steel's compliance with important sections of the Companies Act such as board independence, gender diversity, CSR provision, and transparency initiatives has remarkably improved its performance on financial, environmental, and social dimensions. Prominent projects such as the 'Thousand Schools Project' and water neutrality project have augmented the company's community involvement and sustainability image. The research has some important policy implications, signifying that a legislative framework aligned with ESG principles can guide Indian companies towards enduring sustainability while maintaining stakeholder accountability. The originality of this research stems from its case study analysis of a leading Indian firm's strategy

A. Yashoda (✉)
Anwarul-Uloom College of Business Management, Hyderabad, India
e-mail: sriyashoda26@gmail.com

S. Karri
Loyola Academy (Degree & PG College), Hyderabad, India
e-mail: talktoksv@gmail.com

A. F. Amanu
Premier College, Addis Ababa, Ethiopia
e-mail: ahmedfantaw@premiercollege.edu.et

S. Patnaik et al. (eds.), *People, Planet and Profit: Advancing Sustainable Business Practices*, Springer Proceedings in Geography and Sustainability,
https://doi.org/10.1007/978-3-032-11296-5_3

alignment with statutory governance for sustainable development—an aspect inadequately addressed in the present literature. Future research can adopt a comparative approach across sectors to provide 360-degree insights into sector-specific issues, impact assessment frameworks, and the development of ESG practices in developing countries.

Keywords

Corporate governance • Corporate sustainability • Tata steel • Companies act 2013 • CSR

1 Introduction

Corporae governance ensures that corporate houses operate responsibly, aligning the benefits of all the stakeholders directly or indirectly affected by business operations. Sustainability, on the other hand, focuses on the creation of long-term economic, social, and environmental values. In India, the Companies Act, 2013, institutionalized corporate governance and corporate sustainability principles, mandating greater transparency, accountability, and stakeholder engagement. The objective of this paper is to examine the role of the New Company Act, 2013, in aligning governance and sustainability practices. It assesses the efficacy of Sections 135, 149, and other related provisions and analyses real-world corporate compliance to demonstrate business impacts.

2 Review of Literature

The convergence of corporate governance (CG) and corporate sustainability (CS) has emerged as an important subject of scholarly enquiry since corporate houses face critical challenges pertinent to transparency, accountability, and sustainable value creation. The shift from a compliance-based governance framework to a value-centric, sustainability-driven model highlights the growing significance of Environmental, Social, and Governance (ESG) components in strategic decision-making of business. The Companies Act, 2013, in India provides a strong statutory framework to institutionalize these principles; yet, empirical studies on its effectiveness, especially in emerging economies, are scanty.

2.1 Corporate Governance as a Driver of Sustainability

CG framework provides formal legislative mechanism to direct and control the corporate houses on the matters concerning board composition, roles and responsibilities of the board, risk management systems, and stakeholder engagement. Many scholars have studied the effect of these elements on the sustainability agenda inside the corporate houses.

Adams and Frost (2008) emphasize that sustainability reporting need to be integrated into the company's governance frameworks in order to make it impactful. This research work reveals that board-level supervision and the presence of sustainability committees are significant for incorporating ESG elements into core corporate strategies. The board responsibility should move beyond financial stewardship but also include social and environmental effect.

Michelon and Kumar (2013) find a positive correlation between board diversity, specifically relating to gender and independence and the quality of sustainability disclosures. Their empirical study reveals that a diverse board offers a variety of perspectives to improve the company's understanding of social and environmental challenges. The board reputation and ethical commitment are equally important in impacting the performance of the firms in the long run.

2.2 ESG Integration and Financial Performance

The integration of governance with sustainability is not only a moral and reputational imperative, but also a catalyst for financial value creation. Khan and Yoon (2016) find that the businesses focusing on tangible ESG issues relevant to their industry outperform their competitors. The study reveals that strategically governed ESG issues do not minimize shareholder value but rather improve it.

Hristov and Searcy (2024) offer a governance-integrated balanced scorecard that synchronizes sustainability key performance indicators with financial performance measurements. They contend that integrating ESG performance into CEO compensation frameworks and board evaluations fosters an institutional commitment to sustainability. This approach aligns with the notion of 'integrated governance,' whereby ESG components are just fundamental, but the essential level of corporate strategy.

2.3 Stakeholder-Centric Governance

The transition from shareholder priority to stakeholder inclusivity in the governance framework has reshaped board governance. Freeman and Wicks (2007) advance the stakeholder theory, asserting that the sustaining firm performance depends largely on balancing the interests of all stakeholders such as workers, communities, consumers, suppliers, and the natural environment. Governance, in this context, comprises of not just board obligations to shareholders but also a wider social responsibility.

This perspective has been extensively endorsed in corporate sustainability discourse, where stakeholder engagement is viewed as the hallmark of ethical corporate governance. Firms that follow structured processes for community engagement, employee involvement, and environmental audit tend to illustrate more resilience and legitimacy among stakeholders.

2.4 Trends in Global Governance and ESG Best Practices

Recent reports by global consultants on corporate governance explain how ESG is operationalized by corporate houses. The EY Global Board Report (2023) prioritizes that the corporate boards in the present days must possess 'ESG competence,' and appoint directors with specific expertise in climate risk, social equity, and sustainability metrics. The report highlights a global trend towards mandatory sustainability disclosure which also shows a shift from voluntary corporate social responsibility to legally binding ESG norms.

Russell Reynolds Associates (2024) also reports that the global boards are also investing on technical governance tools such as sustainability dashboards and AI-driven risk monitors, to better accountability and real-time ESG reporting. The KPMG Sustainability Report (2023) also moves a step further to integrate governance frameworks with the United Nations Sustainable Development Goals (SDGs). It observes that governance is advancing to include not just formal compliance mechanism but also ethical leadership, stakeholder engagement, and inclusive policy formulation (Fig. 1).

2.5 Governance Within the Indian Context: From the Securities and Exchange Board of India to the Companies Act of 2013

India's progress towards institutionalizing good governance started with the Kumar Mangalam Birla Committee Report (2000) which added Clause 49 in the SEBI Listing Agreement. These policies prioritized on board independence, audit committees, and financial disclosures. However, the compliance remained erratic till the New Companies Act, 2013, came into being. This Act brings about a significant change by embedding sustainability into legal mandates through the following sections:

- Section 135: Mandated CSR for designated companies.
- Section 149: Mandated inclusion of independent and female directors on boards.
- Section 134: Mandated sustainability and CSR disclosures in board reports.
- Sections 177–178: Formation of audit and nomination committees.

Despite these advancements, there is a lack of scholarly research available which systematically evaluates the impact of these provisions on sustainability outcomes

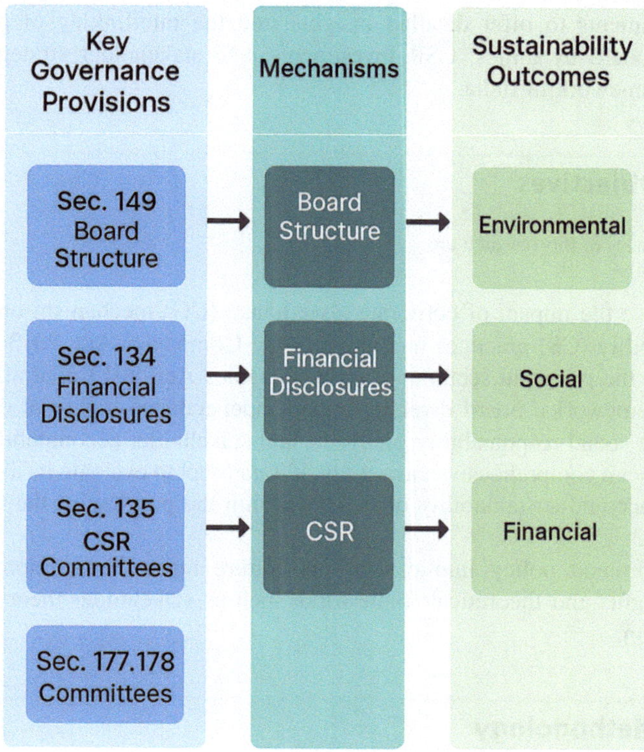

Fig. 1 Governance-sustainability integration framework under the company act, 2013

at the corporate level in India. Most studies focus on voluntary CSR or general ESG performance without explicitly integrating them to statutory governance requirements.

2.6 Research Gap and Study Contribution

The literature highlights a robust theoretical and empirical linkage between CG and CS. However, a substantial gap remains in understanding how the legislation like the Companies Act, 2013, institutionalizes ESG practices within Indian corporate houses. Besides, a significant portion of the existing literature focuses on developed economies or cross-sectional datasets, with very few case-based evidence illustrating how particular firms internalize and operationalize governance mandates for sustainability.

This research addresses these limitations by an in-depth case analysis of Tata Steel, a corporate house widely renowned for its leadership in sustainability. This study uses qualitative content analysis of business reports and aligns with statutory requirements to offer detailed insights into the interlinking of governance structures, board dynamics, CSR investments, and stakeholder strategies within the legal framework in India.

3 Objectives

The objectives of this research are

- To analyse the impact of corporate governance (CG) mechanism on corporate sustainability (CS) practices in context of the Companies Act, 2013, in India.
- To study the principal sections of the Companies Act, 2013, that affect governance frameworks, board diversity, board independence, financial disclosures, corporate social responsibility practices, and stakeholder accountability.
- To conduct a comprehensive case study of Tata Steel to evaluate its alignment of governance and sustainability objectives within the purview of the Companies Act, 2013.
- To recommend policy implications and future research direction based on case insights and theoretical frameworks such as stakeholder theory and ESG integration.

4 Methodology

This study uses a qualitative, exploratory research design based on secondary data sources such as the Companies Act, 2013, Tata Steel's annual and sustainability reports, and publications from EY, KPMG, and academic journals. A case study approach is employed to evaluate Tata Steel's governance and sustainability practices, reinforced by stakeholder theory and ESG integration frameworks. Content analysis is used to systematically review and interpret governance structures, CSR disclosures, and environmental outcomes, while a comparative literature review provides broad insights into global best practices. This research assesses the degree of alignment between statutory requirement and Tata Steel practices, especially on key provisions such as Sections 134, 135, 149, 177, and 178. This methodology facilitates for a comprehensive understanding of how governance mechanisms translate into sustainability outcomes within the context of legal requirement and offer valuable contextual insights pertinent to other Indian companies and emerging markets.

5 Findings and Analysis

This section discusses the findings of the research emanated from a comprehensive case analysis of Tata Steel, a leading Indian multinational company known for its commitment to responsible governance and environmental practices. The study analyses how Tata Steel implemented the legislative provisions of the New Companies Act, 2013, and the resultant real impact. Tata Steel's approaches are used here to exhibit the actual implementation of CG and CS mandate in a real-world context.

5.1 Key Sections Impacting Corporate Governance Under the Company Act, 2023 [9]

Sections 149, 134, 135, 177, and 178 of the Companies Act mandate key governance practices for corporate boards. Section 149 necessitates the inclusion of independent directors and mandates at least one woman director. Section 134 applies financial disclosures and demands that boards explain governance mechanisms related to sustainability. Section 135 presents mandatory Corporate Social Responsibility (CSR) for certain category companies, mandating 2% of average net profits spent on CSR. Sections 177 and 178 establish Audit and Nomination, and Remuneration Committees, respectively, ensuring strong oversight, risk management, and fair compensation practices.

5.2 Corporate Governance Provisions Under the Companies Act, 2013: Transparency and Accountability [9]

Mandatory financial disclosures under Section 134 and whistle-blower mechanisms in Section 177 improve transparency, building stakeholder trust. Section 149 advocate board independence by requiring directors free from material ties to the company, ensuring unbiased decision-making. The inclusion of at least one woman director promotes gender diversity, strengthening inclusive governance. Additionally, Section 177 mandates audit committees to assess and manage risks, while whistle-blower provisions support ethical practices and guard against malpractice, adding to effective risk and ethics management within the organization.

5.3 Corporate Sustainability Under the Companies Act, 2013 [9]

Section 135 of the Companies Act, 2013, mandates CSR for firms meeting specific financial criteria, mandating them to allocate a minimum of 2% of their average net profits towards social initiatives. This has most significantly transformed corporate sustainability practices in India, with firms focusing on priority areas such as educare, health care, and rural development. Furthermore, Section 134 entails boards to include a sustainability overview in their financial disclosures, improving transparency in how companies address ESG concerns. These provisions not only advocate corporate accountability but also encourage businesses to contribute actively to social welfare and sustainable development.

5.4 Case Study: Tata Steel Integrating Governance and Sustainability for Business Success [13]

Tata Steel is a global leader in the steel manufacturing industry. Started in 1907, it has consistently balanced its operations with governance and sustainability principles. With the enactment of the Companies Act, 2013, Tata Steel linked its sustainability practices within the governance framework, enhancing stakeholder confidence and achieving business success (Table 1).

5.4.1 Corporate Governance and Adherence to the Companies Act, 2013

The Companies Act, 2013, included several transformational provisions that Tata Steel embraced to improve its corporate governance and sustainability standards. The company's compliance with the Act's requirement was pivotal in fine-tuning its governance and sustainability standard.

5.4.2 Board Autonomy and Gender Diversity

Section 149 of the Companies Act, 2013, introduced a provision entailing that every listed public company must have at least one-third of its total number of directors as independent directors. Tata Steel went beyond the compliance requirement and appointed 50% of its board members as independent directors. The FY 2022–2023 Annual Report revealed that the board consisted of 6 independent directors out of 12 directors, which reflects compliance and commitment beyond the statutory obligation. This action greatly improved the quality of decision-making and ensured that the board did not have conflicts of interest and thereby reinforced accountability. Furthermore, Tata Steel also complied with the provision that entails appointment of a minimum of one female director. Ms. Deepa Gopalan Wadhwa,

Table 1 Tata steel's compliance and sustainability performance (2022–2023)

Provision	Compliance indicator	Performance outcome
Sec. 149	50% independent directors	Improved decision-making, reduced conflict of interest
Sec. 135	₹275 Cr CSR spend	2 million beneficiaries
Sec. 134	ESG reporting	24% CO_2 reduction, water neutrality
Sec. 177	Active audit committee	Strong risk management
Sec. 178	Remuneration policy	Transparent, merit-based appointments

a former Indian diplomat, was appointed as a female Independent Director, contributing to a more diverse and inclusive leadership framework. This fostered a more diverse and inclusive leadership framework [2, 13].

5.4.3 Audit and Nomination, and Remuneration Committees

Tata Steel improved its governance structure by setting up Audit and Nomination, and Remuneration Committees, as per provisions under Sections 177 and 178. These committees made stringent examination of corporate operations, risk management, and ethical governance practices. This comprehensive oversight enabled Tata Steel to confront with legal provisions while imbibing stakeholders' confidence [13].

5.4.4 Financial Disclosures and CSR

Section 134 mandates that the companies provide a comprehensive overview of their CSR and sustainability activities in their financial disclosures. Tata Steel addressed this by enhancing transparency of their environmental initiatives. It promised to allocate 2% of its average net profits to CSR initiatives as mandated under Section 135, therefore aligning its business strategy with social and environmental goals [11].

5.4.5 Corporate Sustainability Initiatives and the Impact of CSR

The Companies Act, 2013, impacted Tata Steel strengthening its corporate sustainability initiatives. It focused more on environmental conservation, education, and community welfare. The corporation leveraged its CSR corpus to generate substantial social and environmental value.

CSR Investment and Key Projects

Tata Steel's CSR spending for the fiscal year 2022–2023 amounted to ₹275 crore. One of the notable projects implemented by Tata Steel was the 'Thousand Schools Project,' which benefited over 100,000 students, and its water conservation endeavours in areas like Jamshedpur and West Singhbhum, mitigating water shortage problems. These initiatives are clearly linked with the Act's focus on community development and environmental sustainability [6].

Environmental Sustainability

Tata Steel experienced a 24% decline in carbon emissions over the last decade—a notable achievement in its environmental sustainability journey. The company achieved water neutrality in its operations, reducing its reliance on external water sources. These changes were due to the governance and CSR provisions, laid down under the Companies Act, 2013, which encouraged the enterprises to address environmental factors in their operations [6].

5.4.6 Beneficial Results of Governance and Sustainability

The integration of governance and sustainability through the Companies Act, 2013, has produced both tangible and intangible benefits to Tata Steel (Table 2).

Table 2 Tata steel's ESG and sustainability dashboard

24% reduction in carbon emissions (2013–2023)
Water Neutrality in Jamshedpur and West Singhbhum
₹275 Cr CSR spend
2 M + lives impacted
50% independent directors

Revenue Growth

Tata Steel had a 15% yearly growth in revenue from FY 2018 to FY 2023. This rise was primarily attributed to increased stakeholder trust and improved brand value recognition due to its robust commitment to sustainability and corporate governance [2].

Social Impact

Tata Steel's CSR programs positively touched nearly 2 million lives through projects on education, health care, and rural livelihoods. This not only supported the companies in developing social capital but also fostered strong relationship with local communities and governmental bodies, facilitating greater acceptance and collaboration [11].

Enhanced Risk Management

Tata Steel formed Audit and Nomination Committees to proficiently alleviate governance and financial risks. The modifications enhanced Tata Steel's capacity to adhere to the Act and more effectively manage Environmental, Social, Governance (ESG) risks, hence enabling the business to handle regulatory and market problems with more efficiency [11].

5.4.7 Challenges and Greenwashing Concerns

Notwithstanding significant milestones achieved in CSR and sustainability, Tata Steel faces challenges in continuously identifying high-impact CSR projects that will produce long-term value, often necessitating more robust evaluation frameworks [6]. Moreover, while the organization diligently tries to prevent greenwashing through transparent public disclosures of its environmental initiatives, continued vigilance is pivotal to preserve credibility [2].

Tata Steel's experience illustrates how the Companies Act, 2013, has supported the integration of governance and sustainability. Tata Steel complied with governance standards, improved its CSR focus, promoted environmental stewardship, and emerged as a global leader in sustainable industrial practices. The company's achievement in governance and sustainability offers significant insights for other firms aiming to enhance their long-term financial performance while positively adding value to society and the environment (Fig. 2).

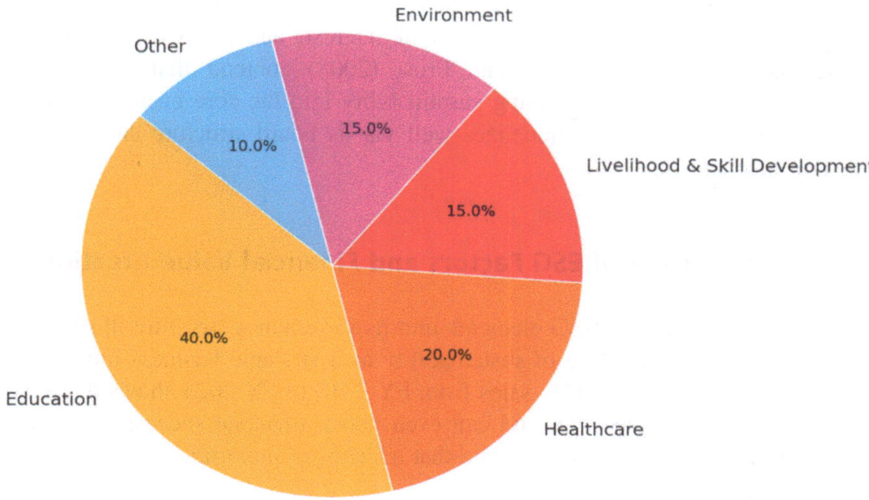

Fig. 2 CSR allocation by Tata steel, FY 2022–2023

6 Discussion

This case of Tata Steel offers some practical implications for incorporating corporate governance (CG) and corporate sustainability (CS) within the framework embedded in the New Companies Act, 2013. The research confirms that legislative compliance integrated with strategic intent and firm's value can provide measurable sustainability outcome. This section evaluates the alignment of Tata Steel's governance practices with stakeholder theory, ESG integration frameworks, and international sustainability best practices, based on the theoretical foundation outlined in the literature.

6.1 Governance Beyond Compliance: Board Composition and Strategic Oversight

Tata Steel's compliance with Section 149 of the New Companies Act, 2013, which requires board independence and gender diversity, shows how legislative directives can catalyse profound structural reforms. The Act mandates one-third of board members to be independent members; however, Tata Steel inducted 50% independent directors into the board, exemplifying a governance culture rooted on ethical leadership and strategic autonomy. The appointment of a female director, Ms. Deepa Gopalan Wadhwa, exemplifies Tata Steel's respect for gender diversity, henceforth refining inclusive decision-making, corroborating the research findings of Michelon et al. (2013) and Mallin and Michelon (2011) that highlight the significance of board diversity in driving sustainability.

These governance improvements are both symbolic and strategic. They facilitate multiple perspectives, minimize conflicts of interest, and improve stakeholder-centric decision-making. Adams and Frost (2008) contend that these board structures are vital for incorporating sustainability into the core business DNA—an idea that Tata Steel operationalizes well via its board structure and strategic oversight mechanism.

6.2 Integration of ESG Factors and Financial Value Creation

Tata Steel's integration of ESG elements into its governance structure illustrates a comprehensive understanding of sustainability as a strategic business imperative. The company's increase of 15% sales from FY 2018 to FY 2023 shows that ESG integration is compatible with, and can even boost, financial success. This confirms Khan et al. (2016), who illustrate that addressing important ESG issues can enhance long-term shareholder value.

Tata Steel's endeavours to align financial disclosures with sustainability metrics in accordance with Section 134 of the New Company Act, 2013, improve transparency. The CSR expenditure of ₹275 crore in FY 2022–2023 and projects like the 'Thousand Schools Project' illustrate both compliances to Section 135 of the Act and a strategic investment in social infrastructure. These activities offer mutual benefits and strengthen community relationship, therefore augmenting the business social license to operate.

6.3 Stakeholder Engagement: Operationalizing Stakeholder Theory

Freeman's stakeholder theory emphasizes that business can prosper by aligning its objectives with the interests of all the stakeholders. Tata Steel illustrates this through its impactful CSR initiatives and commitment to environmental stewardship. Its initiatives such as water neutrality and carbon emissions reduction unswervingly tackle stakeholder worries over environmental degradation, while educational and health programs provided benefits to local communities which fostered goodwill and resilience.

These activities moved beyond corporate philanthropy to exemplify a stakeholder-driven governance framework in which the company's long-term sustainability is intricately linked as a relationship with workers, communities, and the environment. The company's CSR strategy, as outlined under Section 135 of the Act, operationalizes the ethical and strategic dimensions of stakeholder engagement, therefore strengthening its brand value and reducing operational risks.

6.4 Insights from Global ESG Practices: Regional Adaptation with Global Alignment

The journey of Tata Steel's sustainability is interconnected with global trends as highlighted in the research from Russell Reynolds Associates (2024), EY (2023), and KPMG (2023). These studies emphasize board-level ESG expertise, technical tools for impact evaluation, and alignment with SDGs as important elements of Today's governance.

Tata Steel's compliance with global standards, while operating within the regulatory framework of India, illustrates the adaptability and scalability of ESG strategies. The formation of strong audit and nomination committees, the practice of clear sustainability disclosures, and proactive community investments illustrates how Indian firms can successfully localize global governance ideals more effectively.

Hristov and Searcy (2024) emphasize that genuine ESG integration necessitates performance tracking, alignment of incentives, and continuous governance innovation—the domains in which Tata Steel showcases potentialities but also reveals the need for enhanced maturity in impact assessment.

6.5 Synergy Between Governance and Sustainability: Tangible and Intangible Outcomes

The strategic alignment of governance and sustainability at Tata Steel has generated both tangible and intangible results. Tangible results include 24% reduction in carbon emissions in the last 10 years, water neutrality, and increase in revenue. Intangible benefits include improved brand equity, increased investor confidence, and strengthened stakeholder confidence. These outcomes foster sustained competitive advantage for the company.

These findings also confirm that strong board governance is not only a tool for mitigating risk but a catalyst for value creation when aligned with sustainable principles. The audit and nomination committees required under Sections 177 and 178 have strengthened Tata Steel's risk management capabilities, ensuring systematic identification and mitigation of ESG risks.

6.6 Ongoing Challenges and Critical Reflections

Despite these accomplishments, challenges persist. The absence of standardized frameworks for evaluating the impact of CSR is one of the critical challenges. Tata Steel's substantial investment in CSR is impeded by the absence of the sector-specific standards and longitudinal impact evaluations, which restricts the ability to measure original social and environmental outcomes. This corresponds with the findings articulated by KPMG (2023), which advocates for robust ESG assurance mechanism.

The risk of greenwashing, particularly when the firms face increasing tension to publish their environmental achievement remains a critical challenge. Despite Tata Steel's efforts to improve transparency, continued scrutiny and verification by a third party and independent agency are indispensable for sustaining credibility and stakeholder confidence.

Ultimately, while the Companies Act, 2013, offers a comprehensive governance structure, it fails to take into consideration the company size, sectoral variations, or capacity limitations, mostly in small and mid-sized firms (SMEs). Future policy frameworks must provide necessary strategies to foster inclusive sustainable governance across the spectrum of Indian enterprises.

The case of Tata Steel illustrates that the Companies Act, 2013, acts as both a legislative driver and an enabler of sustainable development. The long-term achievement of CG-CS incorporation depends not just on legislative compliance but also on corporate values, board competence, stakeholder engagement, and continuous innovation. The firm's proactive strategy on board governance, ESG integration, and community engagement offers a replicable framework for Indian businesses seeking to link their operations with SDGs.

7 Direction for Future Research

Integrating corporate governance (CG) with sustainability, mostly in accordance with the Companies Act, 2013, has produced positive results. Yet, some gaps persist that necessitate further academic investigation. Longitudinal studies are required to evaluate the enduring impact of governance structures on corporate sustainability (CS) over a long period of time. Tata Steel's study offers an ideal scenario; however, comparative sectoral analysis and size of the company can present how sectoral and organizational factors impact governance efficacy. Besides, further research might investigate how board diversity, including factors like sustainability competence, age diversity, and stakeholder representation influences ESG results. Another critical domain for future investigation is the evaluation of frameworks for measuring the impact of CSR. Notwithstanding the mandatory disclosures, existing models often lack the precision and standardization requisite for evaluating firms' long-term value creation. Investigating the convergence of CG, digital technology, and ESG more specifically how AI and big data improve accountability and transparency can add to the CG and sustainability discourse. Finally, more robust frameworks need to be developed to identify and alleviate greenwashing. Empirical research evaluating the substantive or symbolic nature of voluntary ESG disclosures may support both in policy formation and practice. These research avenues may further support in the development of a more flexible, stakeholder-focused, and resilient sustainability governance framework in emerging economies like India.

8 Policy Implications

The Companies Act, 2013, has been instrumental in the development of an important framework for bringing out sustainable business reforms in India and mandating CSR. Nevertheless, a polished policy is vital for obtaining more profound and pertinent results. Initially, industry-specific CSR and ESG guidelines must be developed which must acknowledge that sustainability priorities differ throughout sectors. This will enable firms to align CSR initiatives with pertinent operational issues in their operating environment. A systematic impact evaluation approach for CSR programs must be implemented to measure effectiveness beyond financial involvement. This would support policymakers and stakeholders to discern between genuine impact and just superficial compliance. Third is regulatory support for small and medium firms (SMEs) to adhere to CSR regulations via shared platforms, increased capacity, or financial support to minimize compliance cost constraints. The firms need to improve ESG practices by using independent third-party verification to minimize greenwashing risks and boost public confidence. Moreover, the government might foster gender and cognitive diversity on boards by designing and implementing appropriate training programs and incentives, therefore supporting inclusive governance. Ultimately, the promotion of digital transparency technologies like sustainability dashboards is indispensable for improving disclosure and stakeholder interaction. The new Policy should prioritize not just compliance but also the developing organizational capabilities for strategic, transparent, and inclusive sustainability practices.

9 Conclusion

The integration of CG and sustainability under the Companies Act, 2013, has revolutionized the landscape for responsible business in India. The Tata Steel case illustrates how legal compliance, coupled with corporate commitment and principles, may generate substantial social, environmental, and economic impacts. The incorporation of independent and gender-diverse directors, legislated CSR requirements under the act, and improved transparency through mandatory financial disclosures have together reinforced governance mechanism. These improvements have brought in concrete results such as increased brand value, effective risk management, increased stakeholder confidence, and measurable environmental benefits such as carbon reduction and water neutrality. However, challenges remain, particularly about the effectiveness of CSR project impact assessment, the potential for greenwashing, and the inconsistent adherence by the small enterprises. This shows that whereas law becomes the enabler for progress, sustainable success is rooted on the internalization of governance principles beyond simple legal compliance. Tata Steel's experience validates that integrating sustainability into corporate culture necessitates leadership commitment, stakeholder engagement, and continuous innovation. Eventually, CG must transition from a compliance-based framework to one that inculcates ethical leadership, accountability, and resilience. The Companies Act, 2013, serves both as a regulatory milestone and a driver for integrating sustainability into the core of business strategy of Indian firms—a vision that necessitates continued research, policy refinement, and industry commitment.

References

1. Adams, C. A., & Frost, G. R. (2008). Integrating sustainability reporting into management practices. *Accounting Forum, 32*(4), 288–302. https://doi.org/10.1016/j.accfor.2008.05.002
2. EY Global Board Report. (2023). *How can effective governance unlock value from sustainability?* https://www.ey.com/en_gl/insights/long-term-value/how-can-effective-governance-unlock-value-from-sustainability
3. Freeman, R. E., Harrison, J. S., & Wicks, A. C. (2007). *Managing for stakeholders: Survival, reputation, and success.* Yale University Press. https://www.jstor.org/stable/j.ctt1npxrg
4. Hristov, I., & Searcy, C. (2024). *Integrating sustainability with corporate governance: A framework to implement the corporate sustainability reporting directive through a balanced scorecard.* Advance online publication. https://doi.org/10.1108/MD-10-2023-1995
5. Khan, M., Serafeim, G., & Yoon, A. (2016). Corporate sustainability: First evidence on materiality. *The Accounting Review, 91*(6), 1697–1725. https://doi.org/10.2139/ssrn.2575912
6. KPMG. (2023). *Governance for sustainability: Building transparency and ethical practices.* KPMG Sustainability Report.
7. Mallin, C. A., & Michelon, G. (2011). Board reputation attributes and corporate social performance. *Corporate Governance: An International Review, Accounting and Business Research, 19*(5), 536–546. https://doi.org/10.1080/00014788.2011.550740
8. Michelon, G., Boesso, G., & Kumar, K. (2013). Examining the link between sustainability disclosure and sustainability performance. *Journal of Business Ethics, 116*(2), 297–309.
9. Ministry of Corporate Affairs. (2013). *The Companies Act, 2013.* Government of India.
10. Reliance Industries. (2022). *Annual report.* https://www.ril.com
11. Russell Reynolds Associates. (2024). *Global governance and ESG: Aligning corporate boards with sustainability.* Russell Reynolds Report.
12. SEBI. (2020). *Guidelines on corporate governance for listed entities.* Securities and Exchange Board of India.
13. Tata Steel. (2023). *Annual sustainability report.* https://www.tatasteel.com/media/18493/annual-report-2022-23.pdf

Financial Inclusion of the Slum Dwellers of Kolkata with Fintech Services: One More Step Towards SDG Goals

Swapnamoyee Palit, Atiba Batul, Ronismita Mishra, and Ananya Swain

Abstract

Financial Inclusion refers to the accessibility of economic entities to affordable and essential financial products and services, which meets their financial needs. This paper aims to find the factors contributing to FI of the women slum dwellers in Kolkata, through PMJDY, the factors contributing to influence these women slum dwellers in Kolkata, to have a bank account, and the impact of FI on the financial aspects of women slum dwellers, representing a very vulnerable section of the population. To achieve these objectives, it has used a structured questionnaire to collect primary data from the women slum dwellers of Kolkata. The stratified random sampling method was used. It has analysed the data in SPSS using techniques of logistic regression, descriptive statistics, etc. as is relevant. The findings shows that the PMJDY scheme has provided banking access to such deprived sections, has profound effect on their financial habits, gave access to several other governmental schemes, through their identification of the bank accounts, etc. It has particularly benefitted the women section, by mobilising their savings into channels of investment for their future. This has been a massive overcome from their habits of hoarding money,

S. Palit (✉)
KIIT Deemed to Be University, Bhubaneswar, India
e-mail: swapnamoyee.palit@gmail.com

A. Batul
School of Management Studies, Swami Vivekananda University, Barrackpore, Kolkata, India

R. Mishra
Centurion University of Technology and Management, Bhubaneswar, Odisha, India
e-mail: ronismita.mishra@cutm.ac.in

A. Swain
School of Economics and Commerce, KIIT Deemed to Be University, Bhubaneswar, India

S. Patnaik et al. (eds.), *People, Planet and Profit: Advancing Sustainable Business Practices*, Springer Proceedings in Geography and Sustainability,
https://doi.org/10.1007/978-3-032-11296-5_4

which ultimately finds its way to necessary or unnecessary consumptions, leaving them in financial stress. Thus, the findings have proved to some extent the beneficiary-centric theories of Financial Inclusion. It does find that FI benefits the financially deprived sections, by aligning them with the mainstream. It has impressed upon the idea of empowering the vulnerable sections in general and the women in these groups in particular, towards achievement of the SDG goals. It suggests the enhanced role of private–public partnership (PPP) in achieving a better result in these initiatives, through the Corporate Social Responsibility (CSR) policy involvement.

Keywords

Bank • CSR • Financial inclusion • Fintech • Pradhan Mantri Jan Dhan Yojana • Slum Dwellers

1 Introduction

Financial Inclusion (FI) refers to the accessibility of economic entities to affordable and essential financial products and services, which meets their financial needs of making transactions, saving, investing, lending and borrowing, credit, insurance, etc. The buzzword for these are the Fintech services, which refers to the use of Financial Technology (Fintech) for providing financial services. These include from the basic access to ATMs, debit, and credit cards, to complex applications like peer-to-peer (P2P) lending apps, different payment and investment apps, robo-advisors, crypto apps, etc. Again these aspects of FI should be safe, secure, and sustainable. FI enables these vulnerable groups to save, invest, transact, and expand businesses, in a nut-shell, making short- and long-term financial plans.

2 Conceptualising the Significance of Having a Bank Account

As Fig. 1 shows, having a bank transaction account is the gateway to other aspects of financial inclusion, like access to credit and insurance, capability to transact digitally, access to ATMs through credit and debit cards, accessibility to Internet banking, investing digitally in financial securities, etc. which can all be done only if the individual has a bank transaction account. So, having a bank account is the first gateway to the other aspects of financial inclusion, as it makes way to access the other fintech services. However, it is worth mentioning here that though it may provide the access to these fintech services, its actual utilisation would depend upon many beneficiary-centric factors, among which accessibility to the latest technologies is a prime factor.

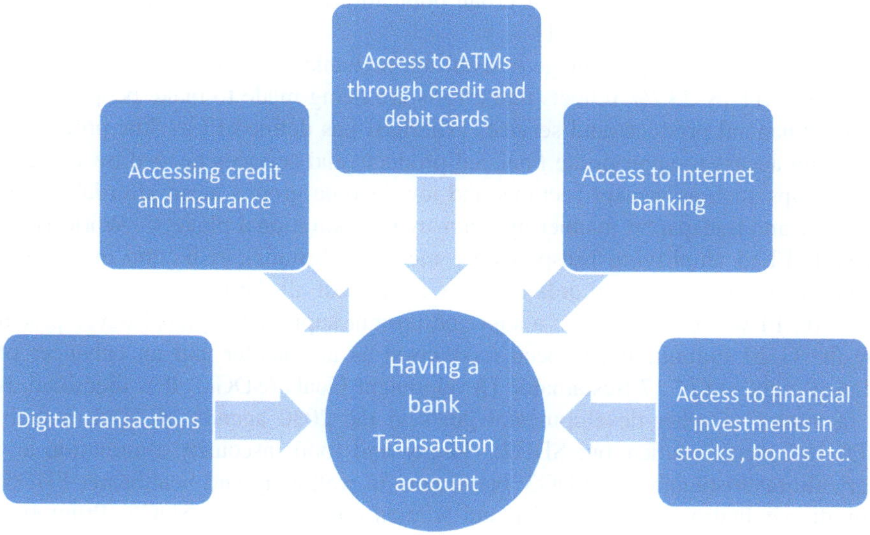

Fig. 1 Conceptualising the significance of financial inclusion and enhancement of fintech services through having a bank account. *Source* Authors' Own Conceptualisation

3 Theories of Financial Inclusion

The theories are very vital because they provide the link to the underlying literature about the concepts. The actual observations through data analysis thus help us to understand if they support the theories, and if not what are the deviations such that something new is observed. They thus help in developing new theories which can support the new found observations, and thus contribute further to the existing literature. There are several theories of financial inclusion which can be broadly divided under the following heads which are *Theories of financial inclusion beneficiaries* which explains about who can benefit from financial inclusion.

Theories of financial inclusion funding, which helps to explain how the funds needed to provide the essential fintech services can be raised. It highlights about the various stakeholders who can provide the required funds which can enhance provision of fintech services for financial inclusion.

Lastly, the *Theories about the Financial Inclusion delivery* points out the way in which these services have to be actually delivered, taking into account various types of beneficiaries.

4 Financial Inclusion's Present Status Analysis

As per Census 2001 and 2011 (consider the following Fig. 1), banking services availed by total population was 35.5% (constituting 30.1% of rural population and 49.5% of urban population), whereas in 2011, it was 58.7% (constituting 54.4%

of rural population and 67.8% of urban population), showing a significant rise by 23.2% in a span of a decade [1].

The change in the percentage of availing the banking services by citizens from 2001 to 2011 by 23.2% reflects that efforts are being made to make people aware of the financial products and services. The RBI has defined FI as "the process of ensuring access to appropriate financial products and services needed by vulnerable groups such as weaker sections and low-income groups at an affordable cost in a fair and transparent manner by mainstream institutional players". World Bank regards FI to amplifying prosperity by means of bringing reduction in poverty, which now has become an area of focus for almost all the countries [2]. In fact, in 2010, FI was considered as a supreme upright support for global development by the G-20 summit. It has been recognised as an enabler and an enhancer of almost eight out of 17 Sustainable Development Goals (SDGs). It is identified as a facilitator of other developmental goals in the 2030 agenda of SDG, namely SDG1: Poverty eradication; SDG2: Hunger and food insecurity elimination and agricultural sustainability; SDG3: betterment in wellbeing and healthcare; SDG5: non-discrimination among gender and women empowerment; SDG8: Promotion of economic growth and employment; SDG9: Boost in industry, innovation and infrastructure; SDG10: Diminishing inequality; and SDG17: Strengthening the means of implementation and global partnership for sustainable development [3]. Providing a wide range of financial products and services, like no-frills account and access to capital markets and credit, debit, insurance, pension, overdraft, etc. facilities, to all the people in the economy, is the basic subject matter of FI. However, merely provision of such products and services will not bring a fruitful picture of inclusive growth. One should also consider the concept of accessibility and affordability by all individuals apart from availability concept by mainstream financial role players. Therefore, any nation that wants to achieve inclusivity in terms of its growth requires to look after the cruciality of FI in achieving such a goal.

5 Presently Operative Major Financial Inclusion Scheme—*Pradhan Mantri Jan Dhan Yojana*

> In the words of the Prime Minister of India, Shri Narendra Modi—"To eradicate poverty, we must end financial untouchability".

As India too, like any other developing nation, wants to achieve inclusive growth, it has to constantly work upon removing barriers from the path of gaining complete FI. One such important effort brought in this context was the declaration of the National Financial Inclusion Mission, with the motive to make the provision of bank accounts for all the citizens of the country. For this, on 15 August 2014, GOI introduced a policy called *Pradhan Mantri Jan Dhan Yojana* (PMJDY). The then Finance Minister addressed PMJDY as "the foundation stone for government's people-centric economic initiatives". The facilities provided to people under this PMJDY scheme are as follows:

- Zero-balance savings account.
- India's domestic debit card, known as RuPay cards to be given to every account holder, which can also be used for e-commerce transactions.
- Liability to get Aadhar number.
- Accident insurance cover of up to Rs. 1,00,000, to the holder's family.
- Life cover of Rs. 30,000 (if account opened by 26 January 2015).
- Overdraft (loan) facility of Rs. 5000, if the holder has kept account active for 6 months.
- To avail overdraft facility, only Aadhar linked accounts are eligible (stopping overdraft frauds).
- Mobile financial facility.
- Interest on deposit and access to pension and insurance products.
- Since all the accounts opened will be under core banking solution, transfer of account from one branch to another is permitted.

PMJDY facilitated FI on a large scale by utilising these technologically backed products, such as RuPay debit cards, Aadhar linked accounts, electronic payment transfers, etc. [4]. As of 29/11/2023, PMJDY has banked 51.04 crore beneficiaries, amounting to Rs 208,855.26 crore, since its introduction in the economy [5]. PMJDY accounts from rural region is 67% approximately, and women accounts under PMJDY is 56% approximately, clearly stating its efforts towards inclusiveness of rural and vulnerable population into formal financial system.

GOI has also made efforts in the translation from *Jan Dhan* (wealth of people) to *Jan Suraksha* (security of people) in terms of FI. On 9 May 2015, it announced three security schemes for the society:

- Pradhan Mantri Jeevan Jyoti Bima Yojana (PMJJBY)
- Pradhan Mantri Suraksha Bima Yojana (PMSBY)
- Atal Pension Yojana (APY).

The Ministry of Finance considers these pension and insurance sector schemes as an enabler of the goal to construct a global social security mechanism for every citizen, with emphasis on under-privileged and poor population. An overview of the said schemes is provided in the following Table 1.

These social security policies have motivated the vulnerable and marginalised population to open zero-parity ledger and enjoy the benefits of the protection schemes on availing them at affordable expense along with the expansion of 55% in female population share [6]. The number of bank account holders in India increased with the inclusion of 80% of adult population in formal banking [7]. As per RBI, 251 million account holders availed mobile banking, and its volume of transaction rose beyond 3,047 billion during 2018–19 span. Preference given to ladies of poor households under the above-mentioned FI schemes reflects one of the motives of promoting empowerment among women through them. The cumulative enrolment of WB under PMJDY stands at 4.63 crore and that under social

Table 1 Overview of the operative financial schemes by the GOI

Schemes	Description	Criteria
PMJDY (2014)	• RuPay ATM or debit card • Life cover of Rs 30,000 • Overdraft facility of Rs 10,000 preferably to a woman of the family • Accidental insurance cover of Rs 0.2 million	• All Indian citizens with valid identity proofs • Preference to be given to the lady of the family
PMJJBY (2015)	• Annual premium of Rs 350 • In case of death, risk coverage of Rs 0.2 million	• PMJDY bank account • Age range is 18–50 years
PMSBY (2015)	• Annual premium of Rs 12 • In case of accidental death and full disability, risk coverage of Rs 0.2 million • In case of partial disability, risk coverage of Rs 0.1 million	• PMJDY bank account • Age range is 18–70 years
APY (2015)	• Fixed pension of Rs 1,000–5,000 to subscribers	• PMJDY bank account • Age range is 18–40 years

Source Researcher's Compilation based on PMJDY online portal. https://pmjdy.gov.in/

security schemes like PMJJBY, PMSBY, and APY stands at 0.60 crore, 1.87 crore, and 0.35 crore respectively [8].

6 Review of Literature

It was seen that since 2011, 1.2 billion people had availed banking services, though 1.7 billion people had not, and these 1.7 billion counts consisted largely of unemployed, poor, or rural population [9]. In fact, still a huge part of Indian population lacks the access to financial products and services [10]. The recent figures are more regrettable as only 20% of Indian adults have operative savings account with formal financial institutions but merely 7% can avail credit facility from them [2]. On an average, 650,000 odd number of villages in India fall under the head of exclusion due to unavailability or inaccessibility of formal services and the population residing in these villages do not have the knowledge of ongoing transformation in banking services or awareness about their products [11]. Since banks are the primary provider of basic financial services, banking inclusion/exclusion is often considered as a synonymous term for financial inclusion/exclusion. The financially excluded people tend to rely on informal financial services offered by money lenders for their huge or petty requirement purposes. To cover the already taken borrowed amount, they tend to rely on new borrowings, which leads them to get trapped in vicious debt cycle. Therefore, financial exclusion has been a major matter of concern, especially for under-developed and less developed economies [12]. Even the developed countries have been unable to ensure complete or universal FI [13].

Providing FI to women, by targeting them specifically, appears to be fruitful not only for them but also for the entire household and, thus, community, as their income contribution is larger than that of men towards household consumption. However, the exclusion tends to be higher in case of women than that of men, as they are the ones without any bank account of their own and fall under the major category of financially excluded section [14]. Women have plenty of potentiality to contribute directly to the growth of the economy, still they are excluded financially [15]. Some of the literature have stressed upon the significance of having more financial access, if the family has female-headed structure of household [16, 17]. FI in the formal financial system, defined in terms of the ownership of account at a regulated institution, a bank, credit union, Microfinance Institutions, Post Office, or with a digital payment service provider, showed that a majority of population excluded or unbanked globally are women [18].

7 Objectives

The objectives of the present study are as follows:

i. To find the factors contributing to financial inclusion of the women slum dwellers in Kolkata, through PMJDY
ii. To find the factors contributing to influence these women slum dwellers in Kolkata, to have a bank account
iii. To find the impact of FI on the financial aspects of the women slum dwellers of Kolkata, representing a very vulnerable section of the population.

8 Research Methodology

The present study has used the method of Logistic regression[1] to find the factors influencing them into financial inclusion, proxied here, in terms of joining the PMJDY. It also analyses the factors influencing them to have a bank account, which is a major factor of financial inclusion. The sampled units are the slum dwellers of Kolkata. A random sample of 300 women slum dwellers has been collected through direct interactions with the respondents of the households, from the slums under study. This Primary data has been collected by the researcher herself using a structured questionnaire. The sample of the survey is drawn out of its universe on the basis of the gradation of extent of area of slums, since Kolkata Municipal Corporation (KMC) does not maintain data on the basis of population of the slums.

The data has been analysed using the SPSS software.

[1] Logistic Regression method is used to find the relation with various independent variables, of the dependent variable which is binary (0, 1).

Frequently referred to as '*the city of joy*', Kolkata (earlier known as Calcutta) is the capital of West Bengal state. This dominant urban centre of eastern India is the town for trade, commerce, and manufacture. As of the year 2020, around 1.49 crores of people live in Kolkata. Due to extreme crises of houses, about three-fourths of the housing units are utilised for dwelling requirement. These urban settlements are known as *bastis* or slums, where about one-third of the population of the city resides. They have very little open space and lack basic living amenities, including proper sanitary facilities, though the government takes care of such places through resettlement and welfare programmes.

The area of sampling consists of 4 biggest slums of Kolkata, namely 'Shahid Smriti Colony' (ward 109—off EM Bypass), 'Mia Bagan Bustee' (ward 35—Beliaghata), 'Belgachia Bustee' (ward 3—Belgachia and Duttabagan neighbourhoods), and 'Bhanga Mahalla Bustee' (ward 32—near Bagmari), as per the KMC.

9 Data Analysis

Thus, having a bank account is taken as a proxy for their economic empowerment. Now, the logistic regression is used to find the most relevant factors which can prompt them to have a bank account. It was done in two ways. First, the relevant factors for their joining the PMJDY were identified. Then, in a sequel the factors which contribute in prompting them to open a bank account were analysed.

9.1 Logistic Regression Analysis of the Factors Contributing in Making Them Participate in the PMJDY

The logistic regression analysis was done to find the factors which significantly determined their involvement in the PMJDY. Table 2 shows the percentage in terms of their participation in PMJDY at 67% (approx.), with no independent variables involved. The Wald test result of the variables in the equation shows that it is significant with a sig value of $0.001 < 0.05$ (Table 3).

The model fit test is also provided by the chi-square results shown in Table 4, where all the values are found to be significant, reflecting the soundness of the variables taken to explain the model. This is supported by both Cox and Snell R square and Nagelkerke R square values shown in Table 5. Model summary and the Hosmer and Lemeshow Test is shown in Table 6. However, it is worth mentioning here that these are pseudo R square values in logistic regression.

The Hosmer and Lemeshow Test results should be insignificant with a significance value greater than 0.05. Here we got the value as 0.2648 which is preferable.

Table 7 classification table indicates the improvement in the explained percentage to 71.3 when the selected predictor variables are included. Thus, this indicated

Table 2 Classification table (Model: Participate in PMJDY with no Independent variable)

	Observed		Predicted		
			Have_PMJDY_acc		Percentage correct
			No	Yes	
Step 0	Have_PMJDY_acc	No	0	100	0.0
		Yes	0	200	100.0
	Overall percentage				66.7

[a]Constant is included in the model
[b]The cut value is 0.500
Source Author's own compilation in SPSS

Table 3 Wald test result of the variables in the equation

		B	S.E	Wald	df	Sig	Exp (B)
Step 0	Constant	0.693	0.122	32.030	1	< 0.001	2.000

Source Author's own compilation in SPSS

Table 4 Omnibus tests of model coefficients (Model: Participate in PMJDY, with no independent variable)

		Chi-square	df	Sig
Step 1	Step	64.518	15	< 0.001
	Block	64.518	15	< 0.001
	Model	64.518	15	< 0.001

Source Author's own compilation in SPSS

Table 5 Model summary (Model: Participate in PMJDY, no independent variable)

Step	–2 Log likelihood	Cox & Snell R Square	Nagelkerke R square
1	317.391[a]	0.194	0.269

[a]Estimation terminated at iteration number 20 because maximum iterations have been reached. Final solution cannot be found
Source Author's own compilation in SPSS

that the independent variables are able to predict 71% of the variation in the dependent variable.

Table 6 Hosmer and Lemeshow test (Model: Participate in PMJDY, no independent variable)

Step	Chi-square	df	Sig
1	17.638	8	0.248

Source Author's own compilation in SPSS

Table 7 Classification table (Model: Participate in PMJDY, with predictor variables)

	Observed		Predicted		
			Have_PMJDY_acc		Percentage correct
			No	Yes	
Step 1	Have_PMJDY_acc	No	34	66	34.0
		Yes	20	180	90.0
	Overall percentage				71.3

[a]The cut value is 0.500
Source Author's own compilation in SPSS

Table 8 identifies all the relevant factors which significantly influence the slum dwellers under study in joining the PMJDY. It finds individual monthly income, education level, expenditure on basic necessities, and the economic status (belonging to the APL or BPL group) as well as their expenditure on health and education are found to be significant factors with significance value < 0.05 and their coefficients are positive. Thus, these factors can be taken as the criteria of these individuals getting selected for the PMJDY.

9.2 Logistic Regression Analysis of Factors Influencing Banking Habits

In this section, the various factors affecting their banking habit are reflected. After several attempts, the variables which were selected based on reliability testing of the models are shown in the categorical variables' codings Table 9.

Table 10 shows the logistic regression percentage of having a bank account or not, without the explanatory variables. This is found to be 67% approximately.

Similarly, the Wald test is found to be significant with value less than 0.05 (in Table 11) and all chi-square values also significant (as shown in Table 12).

The variations in the dependent variable, explained by the independent variables taken, are given by the Cox and Snell R square at 63% and the Nagelkerke R square at 88% shown in Table 13, which are other indicators of the model fit. However, this being a logistic regression, it is to be noted that these are pseudo R square values.

Another indicator is the Hosmer and Lemeshow Test which should be insignificant for a good model fit. This requirement is also satisfied as can be observed in Table 14, with a significance value > 0.05. This is also reflected in its contingency table in Table 15.

The different variables affecting the banking habits of these slum dwellers when included gives an explained score of 96%, as shown in the classification Table 16. This reflects upon the goodness of the selected independent variables.

The noteworthy finding of the entire logistic regression is reflect in Table 17 above which shows that among all the factors taken to find the influence on their

Table 8 Variables in the equation (factors influencing participation in PMJDY)

		B	S.E	Wald	df	Sig	Exp (B)	95% C.I. for EXP (B)	
								Lower	Upper
Step 1[a]	Individual_monthly_Inc			8.953	4	**0.042**			
	Individual_monthly_Inc(1)	18.336	14,917.259	.000	1	0.999	91,888,766.858	0.000	0
	Individual_monthly_Inc(2)	–0.815	1.094	0.555	1	0.456	0.443	0.052	3.781
	Individual_monthly_Inc(3)	16.156	17,116.256	0.000	1	0.999	10,386,683.148	0.000	0
	Individual_monthly_Inc(4)	0.186	1.123	0.027	1	0.868	1.204	0.133	10.870
	Education_level			11.248	5	**0.047**			
	Education_level(1)	2.532	1.410	3.226	1	0.072	12.576	0.794	199.217
	Education_level(2)	21.236	14,860.464	0.000	1	0.999	1,669,723,879.333	0.000	0
	Education_level(3)	0.947	0.460	4.241	1	**0.039**	2.577	1.047	6.344
	Education_level(4)	0.594	0.474	1.573	1	0.210	1.812	0.716	4.585
	Education_level(5)	–0.174	0.495	0.124	1	0.725	0.840	0.319	2.214
	Total_HH_expn	0.000	0.000	2.806	1	0.094	1.000	1.000	1.000
	Basic_necessities_expn	0.004	0.001	7.471	1	**0.006**	1.004	1.001	1.006
	Cast_vote(1)	0.015	0.372	0.002	1	0.968	1.015	0.490	2.103
	Health_expn	0.004	0.001	8.058	1	**0.005**	0.996	0.994	0.999
	Edu_expn	0.005	0.002	6.951	1	**0.004**	0.995	0.992	0.999

(continued)

Table 8 (continued)

		B	S.E	Wald	df	Sig	Exp (B)	95% C.I. for EXP (B)	
								Lower	Upper
	Economic_ status(1)	3.359	1.555	4.664	1	**0.031**	28.761	1.364	606.464
	Constant	–3.255	1.375	5.606	1	**0.018**	0.039		

[a]Variable(s) entered on Step 1: Individual_Annual_Inc, Education_level, Total_HH_expn, Basic_ necessities_expn, Cast_vote, Health_expn, Edu_expn, Economic_status
Source Author's own compilation in SPSS

Table 9 Categorical variables codings

		Frequency	Parameter coding				
			(1)	(2)	(3)	(4)	(5)
Education_level	gradua	11	1.000	0.000	0.000	0.000	0.000
	HS	7	0.000	1.000	0.000	0.000	0.000
	no edu	112	0.000	0.000	1.000	0.000	0.000
	primar	90	0.000	0.000	0.000	1.000	0.000
	second	46	0.000	0.000	0.000	0.000	1.000
	upper	34	0.000	0.000	0.000	0.000	0.000
Individual_Annual_ Inc	0	21	0.000	0.000	0.000	0.000	
	10,001–20	7	1.000	0.000	0.000	0.000	
	5001–100	186	0.000	1.000	0.000	0.000	
	above 20	3	0.000	0.000	1.000	0.000	
	up to 500	83	0.000	0.000	0.000	1.000	
Have_PMJDY_acc	no	100	0.000				
	yes	200	1.000				
Economic_status	APL	16	0.000				
	BPL	284	1.000				
Cast_vote	no	50	0.000				
	yes	250	1.000				

Source Author's own compilation in SPSS

having a bank account, only their participation in the PMJDY was found to be significant, as is highlighted above, with a significance value of < 0.001. Thus the crux of the overall finding shows the tremendous importance of this scheme of PMJDY in achieving the financial inclusion of these very downtrodden and deprived class. While we found many factors like their overall expenditure on health, education, expenditures on their basic necessities, education level, etc. to be significantly influencing their participation in the PMJDY scheme, hoping to

Table 10 Classification table[a,b]

	Observed		Predicted		
			Have_bank_acc		Percentage correct
			No	Yes	
Step 0	Have_bank_acc	No	0	100	0.0
		Yes	0	200	100.0
	Overall percentage				66.7

[a]Constant is included in the model
[b]The cut value is 0.500
Source Author's own compilation in SPSS

Table 11 Wald test results (factors influencing participation in PMJDY)

		B	S.E	Wald	df	Sig	Exp (B)
Step 0	Constant	0.693	0.122	32.030	1	< 0.001	2.000

Source Author's own compilation in SPSS

Table 12 Omnibus tests of model coefficients

		Chi-square	df	Sig
Step 1	Step	299.754	16	< 0.001
	Block	299.754	16	< 0.001
	Model	299.754	16	< 0.001

Source Author's own compilation in SPSS

Table 13 Model summary (Model: Factors influencing banking habits)

Step	–2 Log likelihood	Cox & Snell R Square	Nagelkerke R Square
1	82.155[a]	0.632	0.878

[a]Estimation terminated at iteration number 20 because maximum iterations have been reached. Final solution cannot be found
Source Author's own compilation in SPSS

get some benefits in their distress, the scheme itself is the only influencing factor which has prompted them to have a bank account.

Table 14 Hosmer and Lemeshow test (Model: Factors influencing banking habits)

Step	Chi-square	df	Sig
1	9.593	8	0.295

Source Author's own compilation in SPSS

Table 15 Contingency table for Hosmer and Lemeshow test

		Have_bank_acc = no		Have_bank_acc = yes		Total
		Observed	Expected	Observed	Expected	
Step 1	1	29	29.683	1	0.317	30
	2	30	30.871	2	1.129	32
	3	28	27.069	2	2.931	30
	4	9	9.745	22	21.255	31
	5	4	1.322	26	28.678	30
	6	0	0.535	30	29.465	30
	7	0	0.347	30	29.653	30
	8	0	0.268	32	31.732	32
	9	0	0.141	30	29.859	30
	10	0	0.018	25	24.982	25

Source Author's own compilation in SPSS

Table 16 Classification table (Model: Factors influencing banking habits)

	Observed		Predicted		
			Have_bank_acc		Percentage correct
			No	Yes	
Step 1	Have_bank_acc	no	94	6	94.0
		yes	6	194	97.0
	Overall percentage				96.0

[a]The cut value is 0.500
Source Author's own compilation in SPSS

10 Major Findings

The finding of this analysis as well as the overall observations shows the following:

i. These sections of the population are found to be in overall financial stress.
ii. They were found to participate in the PMJDY scheme with the objective of getting some financial benefits.
iii. The PMJDY scheme has to a large extent been successful in achieving its objectives of providing banking access to such a deprived section.
iv. The inclusion in the scheme has some profound effect on their financial habits in terms of saving, consumption, and investment. They do have access to several other governmental schemes, through their identification of the bank accounts under the PMJDY scheme. They find themselves better-off then their counterparts, who did not participate in the PMJDY scheme, for whatever reasons.

Table 17 Variables in the equation (factors influencing having a bank account)

		B	S.E	Wald	df	Sig	Exp (B)	95% C.I. for EXP (B)	
								Lower	Upper
Step 1	Individual_Annual_Inc			0.221	4	.994			
	Individual_Annual_Inc(1)	16.716	14,954.457	0.000	1	0.999	18,189,178.160	0.000	
	Individual_Annual_Inc(2)	–0.284	2.204	0.017	1	0.897	0.753	0.010	56.543
	Individual_Annual_Inc(3)	16.015	18,570.693	0.000	1	0.999	9,021,758.211	0.000	
	Individual_Annual_Inc(4)	0.048	2.256	0.000	1	0.983	1.050	0.013	87.332
	Education_level			8.352	5	0.138			
	Education_level(1)	2.260	3.428	0.435	1	0.510	9.587	0.012	7939.583
	Education_level(2)	–1.603	1.252	1.640	1	0.200	0.201	0.017	2.341
	Education_level(3)	1.830	1.193	2.353	1	0.125	6.236	0.602	64.639
	Education_level(4)	0.168	1.163	0.021	1	0.885	1.183	0.121	11.551
	Education_level(5)	1.559	1.327	1.381	1	0.240	4.754	0.353	64.040
	Total_HH_expn	0.000	0.000	0.124	1	0.724	1.000	1.000	1.000
	Basic_necessities_expn	0.003	0.003	1.236	1	0.266	1.003	0.997	1.010
	Cast_vote(1)	–1.331	1.027	1.678	1	0.195	0.264	0.035	1.979
	Health_expn	–0.003	0.003	1.152	1	0.283	0.997	0.991	1.003
	Edu_expn	–0.004	0.004	0.895	1	0.344	0.996	0.989	1.004
	Economic_status(1)	1.619	3.275	0.244	1	0.621	5.049	0.008	3096.439

(continued)

Table 17 (continued)

		B	S.E	Wald	df	Sig	Exp (B)	95% C.I. for EXP (B)	
								Lower	Upper
	Have_PMJDY_acc(1)	**7.083**	**0.903**	**61.564**	**1**	**< 0.001**	**1191.312**	**203.070**	**6988.837**
	Constant	–4.786	2.849	2.822	1	.093	0.008		

[a]Variable(s) entered on Step 1: Individual_Annual_Inc, Education_level, Total_HH_expn, Basic_necessities_expn, Cast_vote, Health_expn, Edu_expn, Economic_status, Have_PMJDY_acc

Source Author's own compilation in SPSS

v. The scheme has particularly benefitted the women section, by mobilising their savings into channels of investment for their future. This has been a massive overcome from their habits of hoarding money, which ultimately finds its way to necessary or unnecessary consumptions, leaving them in financial stress.

Thus, the findings have proved to some extent the beneficiary-centric theories of Financial Inclusion. It does find that FI benefits the financially deprived sections, by aligning them with the mainstream.

11 Recommendations

This study's findings point to the existing potentialities, which can be tapped to enhance the achievements of financial inclusion schemes to a higher level. Besides many factors which can contribute towards higher FI objectives fulfilment of the government, this study most prominently suggests the involvement of the private–public partnership (PPP) through the Corporate Social Responsibility (CSR) way. As part of the CSR policy, these mega corporates can play a role in spreading the awareness towards financial inclusion, volunteer to make these vulnerable sections utilise the various schemes available for digitalisation of their finance in terms of saving, investment, and modes of payments, among others. These can require contributions in terms of giving them access to smart technologies which is vital to ensure actual accessibility and utilisation of these available FI technologies. Such pro-social initiatives will fuel the positive impacts of the various ongoing FI schemes. To give some instances of the CSR activities of corporates, Indian Mega corporates like Reliance Industries Limited (RIL) has spent Rs. 1186 crores on its CSR initiatives in the financial year (FY) 2021–22, Tata Consultancy Services (TCS) a whooping Rs. 716 crores and HDFC Bank limited about Rs. 736 crores, the ICICI Bank limited about Rs.266 crores, Infosys about Rs.344 crores, in the same FY, to name a few, towards their CSR initiatives which includes environmental issues, education, and skill development of vulnerable sections especially the women and youth for employability, health and poverty issues, etc. Including FI as part of such huge financial support through these CSR of corporates will make an incredible impact towards accelerating the FI process and thus help in fulfilling the needed SDG-related targets.

12 Conclusions

Financial inclusion is indispensable for economic progress of the economy. It is vital for the 'leave no one behind' concept of the Sustainable Development goals (SDGs) of the United Nations (UN), to be achieved by 2030. Without deliberate and wholesome efforts to target the vulnerable groups, its achievement would remain a daydream, with fading shifting milestones. This paper has significant contribution in highlighting a group which is vulnerable and overlooked, the slum dwellers. The unprecedented COVID-19 pandemic, which mauled every section under its evil feet financially, imposed the indispensability of digitalisation of financial transactions. The scenario of severe economic suffocation for these groups is not hard to imagine. Studies have highlighted how the financial inclusibility has to a large extent able to cushion them to an extent, in tiding over the sudden upheaval [19]. It has impressed upon the idea of empowering the vulnerable sections in general and the women in these groups in particular, towards achievement of the SDG goals. Streamlining a part of the potentialities of the CSR towards FI initiatives will enhance its reach and capabilities and help in economic progress of these vulnerable sections.

Declaration We declare that this article is an original manuscript written by us, and there are no conflicts of interest with anyone.

It has also not received any funding. it is part of the author's overall thesis, which has been submitted.

References

1. Rajeev, M. (2015). Financial inclusion and disparity: A case of India. *Global Labour University Working Paper No. 37,* International Labour Office.
2. Bhatia, S., & Singh, S. (2019). Empowering women through financial inclusion: A study of urban slum. *VIKALPA, The Journal for Decision Makers, 44*(4), 182–197.
3. Department of Economic and Social Affairs, United Nations. https://www.un.org/development/desa/en/news/2015
4. Vijay, R. (2019). Role of technology in fostering financial inclusion in India. *International Journal of Scientific & Technology Research, 8*(10)
5. Pradhan Mantri Jan Dhan Yojana (PMJDY) Progress Report. (2023). Department of Financial Services, *Ministry of Finance*, Government of India. https://pmjdy.gov.in/scheme
6. Barik, R., & Sharma, P. (2019). Analysing the progress and prospects of financial inclusion in India. *Journal of Public Affairs*, e1948, 1–6
7. The Global Findex Database. (2017). Measuring Financial Inclusion and the Fintech Revolution. https://documents.worldbank.org/en/publication/documents-reports/documentdetail/332881525873182837/the-global-findex-database-2017
8. NABARD Annual Report 2022–23. https://www.nabard.org/nabard-annual-report-2022-23.aspx
9. Financial Inclusion is a Key Enabler to Reducing Poverty and Boosting Prosperity. (2018). *The World Bank.* https://documents1.worldbank.org/curated/en/630671538158537244/pdf/The-World-Bank-Annual-Report-2018.pdf
10. Chaia, A., Dalal, A., Goland, T., Gonzalez, M. J., Morduch, J., & Schiff, S. (2009). Half the world is unbanked: State of financial inclusion policy in developing countries AFI. *Financial Access Initiative, Framing Note.* http://citeseerx.ist.psu.edu/viewdoc/download?doi=10.1.1.401.5164&rep=rep1&type=pdf
11. Bhattacharya, R. (2020). Financial Inclusion in India: An Assessment. In: Management Association (Eds.), *Wealth Creation and Poverty Reduction: Breakthroughs in Research and Practice* (pp. 559–568). IGI Global. https://doi.org/10.4018/978-1-7998-1207-4.ch033
12. Dymski, G. A. (2005). Financial globalization, social exclusion and financial crisis. *International Review of Applied Economics, 19*(4), 439–457.

13. Lenka, S. K., & Barik, R. (2018). A discourse analysis of financial inclusion: Post-liberalization mapping in rural and urban India. *Journal of Financial Economic Policy, 10*(3), 406–425.
14. Maurya, P. (2015). Financial inclusion and women empowerment in India. *ABS International Journal of Management, IV*(2), 81.
15. Capraro, V., & Sippel, J. (2017). Gender differences in moral judgment and the evaluation of gender-specified moral agents. *Cognitive Processing, 18*, 399–405. https://doi.org/10.1007/s10339-017-0822-9
16. Fletschner, D. (2008). Women's access to credit: Does it matter for household efficiency? *American Journal of Agricultural Economics, 90*(3), 669–683.
17. Hazarika, G., & Guha-Khasnobis, B. (2008). Household access to microcredit and children's food security in rural Malawi: A gender perspective (Discussion Paper Series IZA DP No 3793). Bonn, Germany: *Institute for the Study of Labor (IZA)*. https://www.worldbank.org/en/topic/financialinclusion/overview
18. The Global Findex Database. (2021). Financial Inclusion, Digital Payments, and Resilience in the Age of COVID 19. https://www.worldbank.org/en/publication/globalfindex
19. Batul, A., Ghosh, K. D., & Palit, S. P. (2023). Digital technologies, sustainable development goals and the grand societal challenges in the context of slum dwellers of Kolkata, India. In *Fostering Sustainable Development in the Age of Technologies*. ISBN: 978-1-83753-061-8, eISBN: 978-1-83753-060-1. Publication date: 13 December 2023.

Customer Retention: PCA-Enhanced AI Model

Rayaguru A. K. Das, Kumar Janardan Patra,
Rajendra Prasad Panigrahi, and R. Anshuman Dash

Abstract

This paper provides a comprehensive review of predictive models of machine learning and its application to forecast customer retention, a critical factor in ensuring the long-term viability and profitability of businesses across diverse sectors. The foundation of successful customer retention prediction rests upon the completeness, comprehensiveness, integrity, and overall quality of the dataset which is utilized as training dataset for the predictive models under consideration. The rapid evolution in the market place demands advanced analytics to optimize campaign strategies. This study explores an AI-driven framework using an 8000-sample dataset to enhance campaign performance. The approach begins with Principal Component Analysis (PCA) for feature selection, effectively simplifying high-dimensional data while preserving key information. An AI model Support Vector Machine (SVM), recognized for its robustness is employed for predictive analytics. The model achieves an outstanding accuracy of 92% with minimal error. These insights empower marketers to predict campaign outcomes with high precision, enabling data-driven optimizations. To evaluate the significance of variations among different strategies and audience segments, Analysis of Variance (ANOVA) is applied. This statistical method

R. A. K. Das
Department of CSE, Synergy Institute of Engineering and Technology, Dhenkanal, India

K. J. Patra (✉)
Department of CSE, OUTR, Bhubaneswar, India

R. P. Panigrahi
Department of CSE, IMIT, Cuttack, Odisha, India
e-mail: rpanigrahi@imit.ac.in

R. Anshuman Dash
Axis Computers, Cuttack, India

S. Patnaik et al. (eds.), *People, Planet and Profit: Advancing Sustainable Business Practices*, Springer Proceedings in Geography and Sustainability,
https://doi.org/10.1007/978-3-032-11296-5_5

validates the findings by identifying meaningful differences in campaign performance across groups, aiding strategic decision-making and resource allocation. The integration of PCA, SVM, and ANOVA creates a cohesive framework for driving impactful marketing decisions. Together, these techniques enable a scalable and practical solution for optimizing digital marketing efforts. The results of this paper emphasize the latent power of AI-powered methodologies to revolutionize marketing analytics by enhancing efficiency and accuracy, which leads to right decision-making. This study provides a roadmap for leveraging data-driven insights to achieve superior marketing outcomes.

Keywords

Digital campaign • Artificial intelligence • SVM • ANOVA • Customer retention • PCA

1 Introduction

In today's completive business environment, customer churn possess a great threat for survival of any business. For any company, the customers are the basis generating revenue which leads to its success. This is the reason the companies give due diligence to customer satisfaction. The customer's satisfaction can reflect in the Customer Relationship Management (CRM). Marketing department can take advantage of the CRM to select the target consumers for a cost-effective relationship. This leads to customer-oriented marketing campaign targeted at specific strata of customers.

High customer churn rates, representing the loss of customers to competitors or due to dissatisfaction, translate into substantial revenue losses and significantly inflate customer acquisition costs [1, 2]. The cost associated with acquiring a new customer often far exceeds the cost associated with retaining an existing customer [2, 3]. This economic reality underscores the urgent need for effective customer retention strategies. The companies come up with innovative schemes for customer retention. However, predicting the parameters for customer retention has become a Herculean task. Predictive modeling predominates the capability and potential of machine learning. This, in turn, poses a data-driven approach to proactively identify high-risk customers considering churning and implement targeted interventions to mitigate these losses [4, 5, This approach moves beyond reactive measures, allowing businesses to anticipate and address potential churn before it impacts their bottom line 6].

The subsequent sections of this paper are divided into the intricate related work, methodology, feature extraction, AI model, and ethical considerations associated with this crucial application of machine learning (ML).

AI and ML have turned into a vital tool in marketing analytics, particularly in predicting consumer behavior and optimizing advertisement campaigns. The integration of ML models allows businesses to forecast customer lifetime value

(CLV) and tailor marketing strategies to retain valuable customers [7]. By leveraging historical datasets, companies can make informed decisions about advertising expenditures and target specific customer segments with customized campaigns [7].

In the contemporary landscape of digital marketing, businesses must navigate vast amounts of data generated from CRM, various campaigns to refine strategies and boost performance. Leveraging innovative data analysis and machine learning tools and techniques can pointedly improve decision-making and campaign outcomes. This paper introduces a systematic approach for customer retention and thereby evaluating the cost-effectiveness of marketing campaigns using AI methods, for dimensionality reduction, feature extraction, and ML models for the purpose. This will help to make data-driven decisions. The prime goal of the current study is aimed at applying ML algorithms to digital marketing datasets, in order to predict customer retention thereby evaluating the success of the ad-campaign. The dataset used for this study includes a diverse range of variables such as advertising expenditure, click-through rates, customer demographics, and website visits. Data collection involves gathering these key metrics from several campaigns to establish a foundation for analysis. However, raw data might contain inconsistencies, missing values, and noise. This is why data cleaning is essential. Cleaning the dataset ensures the accuracy and reliability of the analysis by resolving these issues.

Once the data is cleaned, the next step is dimensionality reduction. For this purpose, Principal Component Analysis (PCA) is used. The main reasons of using PCA are five-fold: Dimensionality Reduction, Improvement in Model Performance, Noise Reduction, Visualization, and Feature Extraction. PCA is used to reduce the dimensionality of given data without losing the essence of it. The reduction of dimensionality is widely used in machine learning to simplify a big dataset into a user-friendly, small set to works further without losing the hidden patterns. PCA is used to convert the original dataset into a set of smaller uncorrelated variables. These are known as Principal Components. The Principal Components record the most significant information while reducing the dimensional complexity. This step allows for faster processing and prevents over-fitting by eliminating redundant or irrelevant features. After dimensionality reduction, feature extraction is performed to identify the most influential variables that contribute to campaign success. This involves selecting features such as ad-spend, click-through rates, and demographic data that significantly affect the outcomes of marketing efforts. By focusing on these essential features, the machine learning models are more likely to provide accurate predictions.

To further validate the results, the Analysis of Variance (ANOVA) test is conducted for the purpose of assessing the success of advertisement campaign across different marketing strategies and audience segments. ANOVA helps determine which factors or strategies most influence performance variations, offering valuable insights into effective resource allocation and optimization. The core of the methodology involves applying ML models for prediction. In the current study, three models of ML such as DT, LR, and SVM are employed and their results are compared and tested. These models are combined using a voting method to improve prediction accuracy by aggregating the strengths of each algorithm. The DT model in our case shows very impressive accuracy of 85%. It has Root Mean Square Error (RMSE) value 0.37 and Mean Absolute Error (MAE) value equal to 0.15. The SVM model performs slightly better, achieving an accuracy of 88%, an RMSE of 0.33, and an MAE of 0.12, indicating its excellent capability to successfully predict campaign outcomes. The Logistic Regression model provides an accuracy of 87%, an RMSE of 0.34, and an MAE of 0.13, showing its robustness in comparison to the other models. Through feature importance analysis, the study identifies key predictors of campaign success, with variables such as ad-spend, click-through rate, income, age, and website visits emerging as significant contributors. Ad-spend and click-through rate stand out as the most important features, with respective importance scores of 0.243 and 0.236, highlighting their central role in driving marketing performance.

This study demonstrates the applicable capability of AI and ML in the field of digital marketing by offering a structured and data-driven methodology for predicting and optimizing campaign strategies with an aim of retaining the customers. By utilizing PCA for dimensionality reduction, extracting key features, and applying statistical testing and machine learning models, marketers can make informed decisions that maximize ROI and campaign effectiveness. The findings underscore how AI-powered analytics can transform the marketing industry, leading to more efficient and impactful strategies.

2 Literature Review

Machine learning tools have emerged as pivotal in predicting customer retention. These tools enable businesses to analyze customer behavior and implement targeted strategies to mitigate churn. The following are some of the key aspects of how machine learning models contribute to customer retention. ML can utilize heuristic learning, mathematical models and knowledge acquisitions.

In a specific study [8], ML models like Random Forest and SVM are used to predict the customer churn in occasional high accuracy of 95%. The predictive analytical tools enable businesses to identify at-risk customers, facilitating timely interventions that can significantly reduce churn rates [9]. The ML models are employed to predict consumer behavior post-ad-exposure. Various classification models, such as LR, DT and RF, have been used to analyze demographic and socio-economic factors influencing purchase decisions. Notably, a Random Forest model achieved a 95% accuracy rate in predicting purchase likelihood, highlighting its effectiveness in enhancing marketing strategies [10]. Various algorithms, including DT, LR, and ensemble methods like XGBoost, have been effectively utilized for churn prediction [11, 12]. Ensemble methods, particularly Gradient Boosting and XGBoost, have shown superior performance in handling imbalanced datasets, which is crucial for accurate churn predictions [13]. DL frameworks, such as LSTM networks, were developed to model user interests and predict ad-responses. These models consider temporal sequences of user interactions, offering improvements over static feature-based approaches. By capturing the dynamic nature of user behavior, these models enhance the accuracy of ad-response predictions and campaign-specific click predictions [14]. The ML models no doubt offer robust predictive capabilities, but they are not transparent, which can hamper their adoption. To address this, explainable AI approaches, such as Local Interpretable Model-Agnostic Explanations (LIME) together with Contextual Importance and Utility (CIU), have been proposed. These methods provide both local and global explanations of model predictions, improving the interpretability of advertisement effectiveness predictions [15]. Predictive modeling helps in estimating CLV, so that the businesses can prioritize impotent and high-value customers which leads to formulation of appropriate marketing strategies [7, 16]. Understanding customer segment transitions enables advertisers to optimize resource allocation and improve return on investment through targeted campaigns [16].

ML algorithms can be instrumental in identifying user behavior trends, which help in optimizing advertising services and customizing promotional offers. By analyzing user interactions on digital platforms, businesses can enhance customer engagement and increase sales. The application of ML in advertising has shown to significantly boost revenue and brand publicity [10].

3 Methodology

a. Flow of work

The proposed work process begins with data collection, after which data cleaning is undertaken to improve data quality. Then, feature extraction is carried out, and PCA is applied to reduce dimensionality. An ANOVA test is used to determine significant features, which are applied in AI model implementation. Finally, model performance is evaluated, and results are analyzed for market insights (Fig. 1).

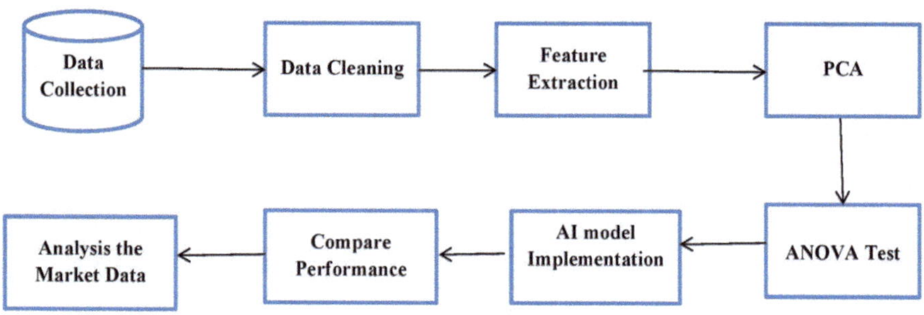

Fig. 1 Flow of work

b. Data collection

For the purpose of this study, the dataset was chosen from Kaggle for the purpose of modeling customer retention. The original dataset consists of 8000 records and 16 different characteristics.

c. Pre-processing and data cleaning

One of the critical steps in processing is pre-processing the dataset. The dataset may contain missing values. These missing values are substituted with the preliminary technique like mean, median, and mode. The outliers were handled by PCA. The scaling of variables is done using the ANOVA technique. In order to retain the quality and reliability of the dataset, first, we applied data cleaning techniques. After that missing values are handled using mean, median, and mode imputation methods. This approach allowed us to fill in incomplete data points based on statistical measures, ensuring consistency across the dataset. Additionally, rows with extensive missing information that could not be reliably imputed were removed to maintain the integrity of the analysis.

d. PCA

PCA is considered as one of the most popular multivariate techniques. It can analyze several inter-correlated dependent variables. The main aim of this is to get the information. Its goal is to obtain the critical and covert information from the dataset, to a set of new orthogonal variables known as principal components. It is capable of displaying similarity of patterns of the observations as points in the map (Table 1).

Further enhancement of the dataset is done by the PCA. In our case, the application of PCA is done for dimensionality reduction. PCA preserves the most important information while reducing the high-dimensional data to a smaller set of uncorrelated components (Fig. 2).

This process effectively reduced redundancy and eliminated less significant features, streamlining the dataset for analysis. By minimizing noise and focusing on the most influential variables, PCA refined the dataset and improved its overall quality (Fig. 3).

Table 1 PCA

Feature	PC1	PC2
Age	−0.418	0.501167
Income	−0.10929	0.659164
AdSpend	0.575849	−0.00073
ClickThroughRate	−0.52658	−0.05647
WebsiteVisits	0.452153	0.557804

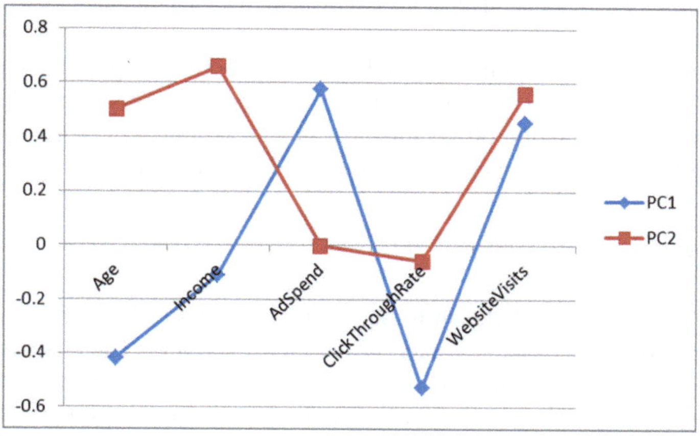

Fig. 2 PCA

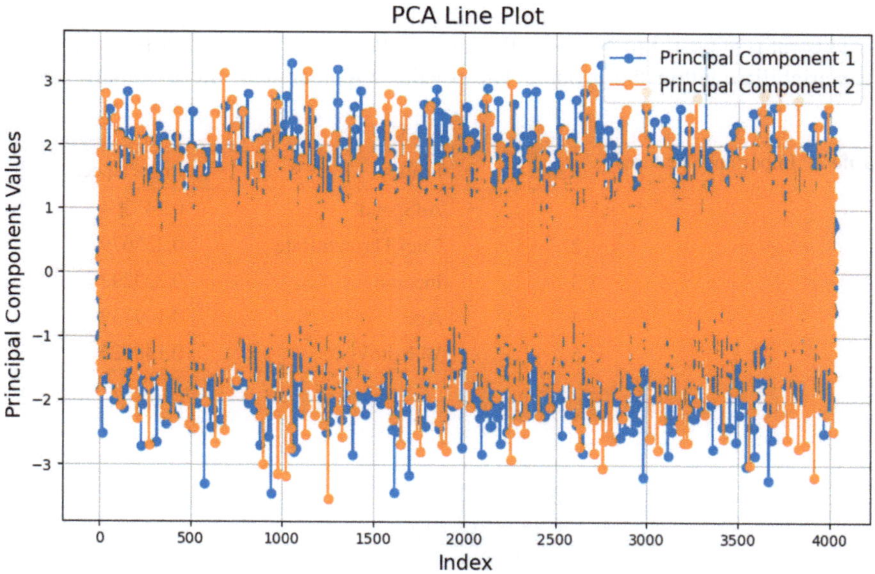

Fig. 3 PCA implementation

e. Feature extraction

After applying PCA and refining the dataset, we identified and extracted the most significant features critical to the success of the analysis. These features, such as ad-spending, click-through rates, and audience demographics, were selected based on their contribution to explaining the variability in the data. By isolating these key attributes, we reduced data complexity and focused on the most impactful variables (Table 2).

The extracted features thus achieved were then used as inputs for AI models, ensuring that the analysis was both efficient and precise. This targeted approach enabled the implementation of machine learning algorithms to generate actionable insights and optimize campaign performance (Figs. 4 and 5).

f. ANOVA Test

The ANOVA test was undertaken to determine the statistical significance of each feature's contribution to the variance in the target variable. The F statistic along with the p-value was taken into account for determining the statistical significance. The F statistic is the division between mean square between groups (MSB) and the mean square within groups (MSW).

$$Fstatic = \frac{MSB}{MSW}$$

This ratio is useful to compare the variability among groups that is greater than the variability within the groups. The p-value together with the F statistic shows

Table 2 Importance features

Rank	Feature	Importance
1	AdSpend	0.243264
2	ClickThroughRate	0.235679
3	Income	0.223292
4	Age	0.149739
5	WebsiteVisits	0.148025

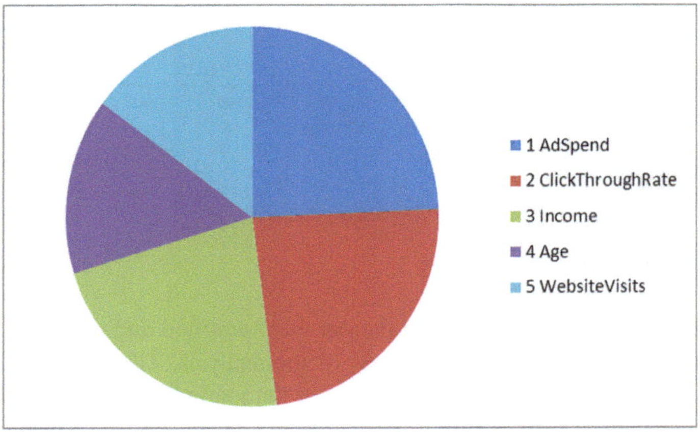

Fig. 4 Graphical presentation of importance features

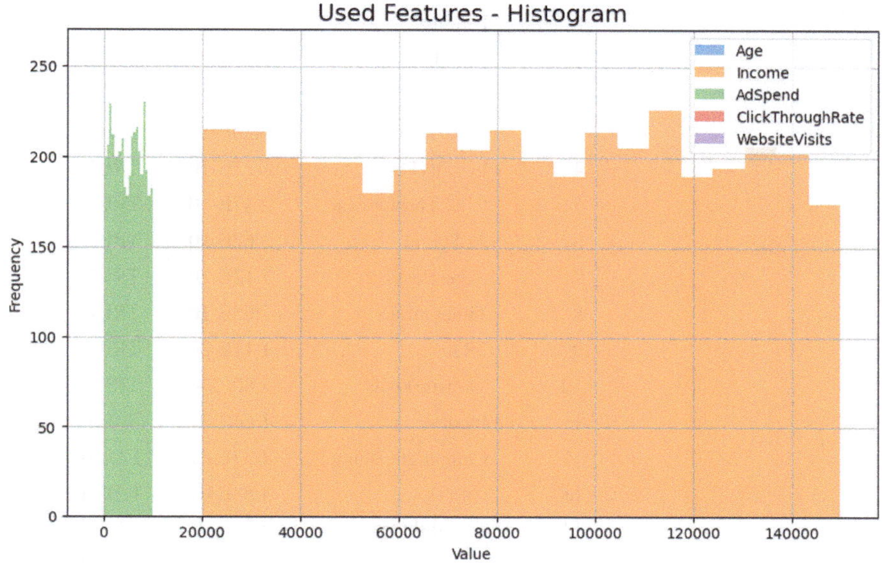

Fig. 5 Histogram depicting different significant features

the probability of observed data being True. A low p-value in this case less than 0.05 is used to determine the model selection.

The p-values of different features less than 0.05 were considered significant, indicating a strong influence on the target variable. The results revealed that several features, such as Clicks ($p = 1.97E\text{-}15$), Opens ($p = 9.76E\text{-}13$), and Previous Purchases ($p = 6.30E\text{-}12$), significantly impacted the outcome. Other notable features, including Ad Spend, Click-Through Rate, and Time on Site, also demonstrated statistical significance with p-values well below the threshold. Conversely, features with a p-value greater than or equal to 0.05, such as Income ($p = 1.64E\text{-}01$), Campaign Channel ($p = 4.57E\text{-}01$), and Age ($p = 4.88E\text{-}01$), did not contribute significantly to the variance. These findings suggest that while some demographic and behavioral factors, like gender and social shares, had minimal impact on the target variable, engagement-related metrics like clicks, opens, and previous purchases played a pivotal role.

The ANOVA results provided a clear understanding of which features were most influential, enabling their prioritization in the subsequent stages of analysis and modeling. This step ensured that only the most impactful features were incorporated into the AI framework, enhancing both the efficiency and accuracy of the predictive model (Table 3 and Fig. 6).

Interpreting Results: (ANOVA).

Low P-Value (< 0.05): The feature contributes significantly to the variance in the target variable.

High P-Value (≥ 0.05): The feature does not contribute significantly.

Table 3 ANOVA test

Rank	Feature	P-Value	Significant
1	2Clicks	1.97E-15	TRUE
2	2Opens	9.76E-13	TRUE
3	PreviousPurchases	6.30E-12	TRUE
4	LoyaltyPoints	1.23E-10	TRUE
5	ClickThroughRate	2.47E-10	TRUE
6	AdSpend	3.62E-10	TRUE
7	PagesPerVisit	7.12E-10	TRUE
8	TimeOnSite	7.65E-10	TRUE
9	3Rate	1.11E-06	TRUE
10	WebsiteVisits	4.80E-06	TRUE
11	Income	1.64E-01	FALSE
12	CampaignChannel	4.57E-01	FALSE
13	Age	4.88E-01	FALSE
14	SocialShares	5.76E-01	FALSE
15	Gender	7.28E-01	FALSE

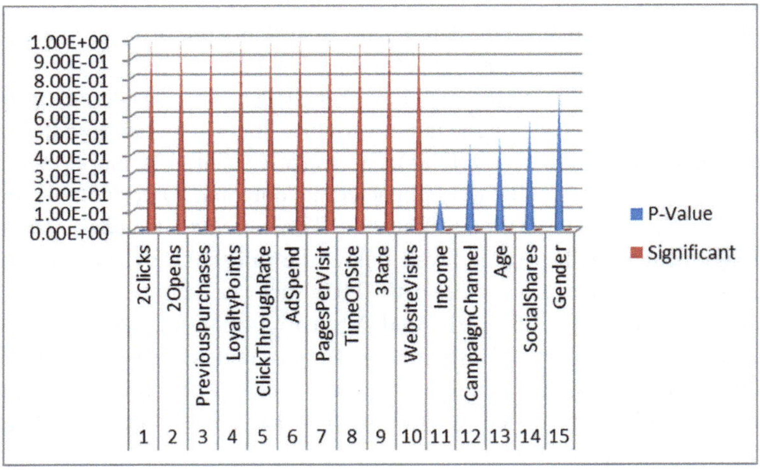

Fig. 6 ANOVA

4 AI Model

For this study, three AI models namely DT, SVM, and LR are considered to analyze and predict campaign outcomes. Each model has distinct capabilities in processing data and predicting results. The relative performance of these considered models was evaluated with the help of accuracy metrics such as RMSE, MAE, and F1 Score.

Decision Tree (DT):

The DT algorithm splits the dataset into tree like subsets on the basis of feature values. Such a tree-like structure makes interpreting and implementing easy. In this case an accuracy of 0.87, RMSE of 0.37, MAE of 0.15, and F1 Score of 0.87 were achieved. DT is easy to understand and implement, and is usually used effectively in ML algorithms. DT is versatile, interpretable, and easy to predict and understand complex relationships within data [17]. DT is also capable of modeling non-linear data. DT models are quick to construct. DT only shows the predictors that are crucial to building the DT. DT is well suited for large datasets. While using DT, it must be kept in mind that the DT is more sensitive to outliers and missing values thus makes it susceptible to over-fitting [18, 19].

SVM:

SVM is considered as one of the popular supervised learning technique. SVM is very reliable and works well in small and medium-sized datasets. It is also a very common model for classifying text. The SVM is used for its effective performance. The attractive two significant features of SVM are the use of kernel function which represents the classification function and the dot product and finding a suitable hyperplane which in one hand maximizes separation between classes and minimizes the over-fitting of data.

The expression $^{\rightarrow}w\ ^{\rightarrow}x + b = 0$ represents a hyperplane in a feature space. In the context of a Linear SVM, this equation defines the separating hyperplane that separates two classes of sets of data points $^{\rightarrow}x$. The vector $^{\rightarrow}w$ is a normal vector to hyperplane, b is the bias which shifts the hyperplane along normal vector, and "." is the dot product. SVM uses "regularization parameter" which controls the trade-off between experimental error and complexity of assumption space used [17]. The predictions of KNN and SVM were compared in [20], and showed that SVM can provide a higher accuracy, especially large dataset than KNN. In our case it accomplishes data processing and analysis with the intent of identifying hidden patterns. It works by identifying an optimal hyperplane to separate data points, making it highly effective for complex and high-dimensional data. Out of these three models, the SVM delivered the comparatively best results with an impressive accuracy of 0.92, RMSE of 0.33, and MAE of 0.12, with F1 Score of 0.91.

LR:

LR is used for binary classification tasks, estimating the probability of a target variable. LR is a powerful and well-known way for control classification [21]. This provides binary classification suitable for data processing. This statistical classification model according to the probability shows fraud with logistics curves. The value of this logistics curve varies from 0 to 1, so it can be used to interpret the members' probability of the classroom. The dataset submitted by the input data in the model is classified for model training and model testing. The models are tested for minimum thresholds for prediction. Depending on the probability

Table 4 Comparison table

Model	Accuracy	RMSE	MAE	F1 Score
DT	0.87	0.37	0.15	0.87
SVM	0.92	0.33	0.12	0.91
LR	0.90	0.34	0.13	0.89

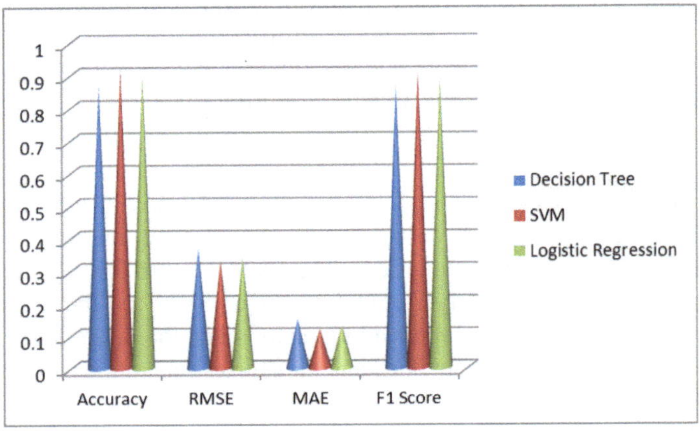

Fig. 7 Presentation of comparison

of some thresholds, the logistics return can be divided using one line and the dataset can be divided into exactly two areas. This model too performed well, with impressive accuracy of 0.90, RMSE of 0.34, MAE of 0.13, and 0.89 as F1 Score (Table 4).

Based on the performance metrics, SVM model appeared as the top-performer. The SVM emerged as the most accurate model. It could achieve the highest accuracy (0.92) and simultaneously the lowest error values (RMSE: 0.33; MAE: 0.12). Logistic Regression demonstrated strong performance, closely following SVM, while Decision Tree, though effective, showed slightly lower accuracy with slightly higher error values. These results show the importance and significance of selecting the appropriate AI model to ensure accurate, efficient, and reliable predictions for analyzing campaign performance (Fig. 7).

5 Comparison of Model Strength and Weakness

The strength and weakness are given in the following tabular form.

Algorithms	Strengths	Weakness
DT	Helps in visual analysis. It can be used in both regression and classification problems	Problem of over-fitting. Due care has to be taken to deal with the outliers and missing values. Handling these are bit tricky in DT
SVM	Strong generalization ability; small size dataset; multi-dimensional	Sensitive to kernel and outlier
LR	Less calculation cost; easy to understand	Under-fitting; less accurate

6 Ethical Considerations: Responsible Use of Customer Data

Predictive modeling for customer retention involves the use of sensitive customer data. Due consideration may be attributed to ethical implications [4, 22]. Transparency and data privacy are paramount [23]. Businesses must confirm to relevant rules and regulations for data usage [4]. The data privacy can be enforced in the following manner keeping in mind Indian laws such as Digital Personal Data Protection Act, 2023 (DPDP Act), and the Consumer Protection Act, 2019; the guidelines of CCPA may be followed for the US and GDPR for EU. In order to comply with ethical guideline and to protect the privacy as demanded by Law, the following steps may be adopted:

- Data Minimization
- Anonymization and De-identification
- Ethical Guidelines and Training
- Privacy-Enhancing Technologies (PETs): Explore and adopt privacy-enhancing technologies like encryption
- Ethical Review Boards
- Role-based access (RBAC) to the data may be granted to restrict the data access.

The bias mitigation can be done by fair use of data and the dataset which includes data from all the strata of customers. Transparency in the use of AI is attained by providing clear usage notices and explanations in non-technical language. The stakeholders must know that the result comes from AI processing and through proper audit trails. Strict data security measures must be implemented with an aim to protect customer dataset from unauthorized access or misuse. It is crucial to avoiding biases in data collection and training the model as it may result in bigoted outcomes [24, 22]. For example, some models trained on biased datasets may unfairly target specific demographic groups for retention interventions. Fairness and accountability are essential aspects of responsible AI development. Developing trust with customers is vital for the continuing success of any customer retention strategy. Open communication about data usage and model transparency can foster trust and ensure ethical AI practices. Commitment to responsible data handling is fundamental for maintaining customer relationships and building a sustainable business.

7 Constraint of the Model and Steps for Improving Scalability

One of the primary obstacles in expanding AI capabilities is handling the vast amount and different levels of data complexity that AI systems need to be trained for seamless functioning. As companies advance their AI initiatives, they frequently encounter a wider array of datasets that differ in structure, format, and quality. The sheer amount of data increases the computing power necessary to process, cleanse, and store this information grows dramatically. For instance, training a machine learning model on a small dataset may take just a few hours on one machine, but scaling that model to accommodate terabytes of data can necessitate distributed computing networks, cloud services, and considerable financial resources. These factors can be major constrains for scalability. However the following steps may be taken to improve the scalability:

- Parallel Processing
- Cloud Computing
- Containerization of Microservices.

8 Challenges in Deploying the Framework and Recommendation

The real-world deployment of a framework developed in a controlled environment can be very challenging. The challenges of deployment and the recommendation in overcoming the challenges are mentioned below.

Challenges	Recommendation
Poor, incomplete, or biased data	Synthetic data generation or data augmentation techniques may be used
Changes in data distribution over time which is known as concept drift may degrade performance	Advised to monitor model performance for drift detection. Further adversarial training may be implemented to improve robustness
Computational costs and training like high computational costs	Cloud-based scalable solution may be used
Existing system may not support the AI frameworks. There may be API compatibility issues	Containerization like Docker and Kubernetes is recommended for smooth deployment. Furthermore, microservice architecture is recommended for better AI integration

(continued)

(continued)

Challenges	Recommendation
AI models can be vulnerable to attacks which may lead to data breach and model theft	Model encryption techniques is recommended to protect the API. Recommended to detect anomaly for better performance. Periodically penetration testing can be done on the system

9 Conclusion and Future Work

Here, we focused on the significant capability of leveraging Artificial Intelligence (AI) and advanced statistical techniques for optimizing digital marketing campaigns. By applying PCA for dimensionality reduction, feature extraction, and predictive modeling using DT, SVM, and LR, we have demonstrated a robust approach for analyzing marketing data and predicting campaign outcomes with respect to customer retention. The integration of PCA allowed for efficient data processing, while feature extraction focused on the most impactful variables, such as ad-spend and click-through rates, enhancing the performance of our models. Among the tested algorithms, SVM outperformed others with the highest accuracy, confirming its effectiveness in predicting campaign success. The use of Analysis of Variance (ANOVA) further enriched the analysis by validating the significance of different campaign strategies, ensuring that decisions are backed by statistical rigor. The results indicate that a data-driven approach, combining dimensionality reduction, feature selection, and machine learning, can substantially enhance decision-making, allowing marketers to optimize their strategies for higher engagement and improved ROI. Further, integrating real-time data streams into the model could enable continuous optimization and adjustments to campaigns based on up-to-the-minute results. This framework offers valuable insights and a scalable methodology for businesses seeking to improve their digital marketing efforts. By focusing on the most influential factors and using predictive modeling, organizations can make informed decisions, reduce inefficiencies, and adapt to the dynamic nature of digital marketing.

References

1. Bahnsen, A. C., Aouada, D., & Ottersten, B. (2015). A novel cost-sensitive framework for customer churn predictive modeling. Springer Nature.
2. Das, L., Salman, R., Sabeer, S., Ansari, S. K., Aarif, M., & Rana, A customer retention using machine learning. In *IEEE Uttar Pradesh Section International Conference on Electrical, Computer and Electronics Engineering*.
3. Aramola, J. O. D., Ladipupo, O. O. O., & Usa, A. G. M. A. Process 11. Framework for subscriber anagement and retention in Nigerian telecommunication industry.
4. Babadoan, B. (2024).Harnessing AI and predictive analytics to revolutionize customer retention strategies.

5. Patil, K. (2024). Customer churn prediction using machine learning. *Interantional Journal of Scientific Research in Engineering and Management.*
6. Sai, C. A. L. V. S., Kumari, K., & Gupta, B. A. (2024). Churn prediction based on customer segmentation in banking industry using machine learning techniques.
7. Jamalpur, B., Singh, D., Kumar, B., Nagpal, A., Pawar, S., & Banerjee, D. (2024). Applications of deep learning in marketing analytics: Predictive modeling and segmenting customers. In *2024 International Conference on Communication, Computer Sciences and Engineering.*
8. Ganesh, C., Haritha, R., Nirmal Kumar, M., Renuka, K., & Sakthivel, M. (2024). Machine learning based customer churn prediction on telecom industry.
9. Patel, A., & Kumar, A. G. (2023). Predicting customer churn in telecom industry: A machine learning approach for improving customer retention.
10. Malhi, A., Madhikermi, M., Maharjan, Y., & Främling, K. (2021). Online product advertisement prediction and explanation in large-scale social networks. In *2021 Eighth International Conference on Social Network Analysis, Management and Security (SNAMS).*
11. Kunekar, P., Usman, M., Veena, C., Singla, A., Anute, N., & Polke, N. (2024). Enhancing advertising initiatives: Using machine learning algorithms to engage targeted customer. In *2024 International Conference on Communication, Computer Sciences and Engineering (IC3SE)*, pp. 1433–1437.
12. Kandula, P. (2024). Decoding consumer behavior: Predictive analytics of social media advertising effectiveness. *Interantional Journal of Scientific Research in Engineering and Management.*
13. Altairey, H. A., & Al-Alawi, A. I. (2024). Customer churn prediction in telecommunication and banking using machine learning: A systematic literature review. https://doi.org/10.1109/icetsis61505.2024.10459439
14. Sonune, S., Mohite, A., Borhude, A., Patil, Y., & Prof. Thorat, Y. (2024). Customer churn prediction using machine learning algorithms. *International Journal of Scientific Research in Science and Technology.*
15. Das, R. A. K., & Khan, A. B. (2021). A study of agile iterative development methodology on web application quality. In *In digital democracy–IT for change* (pp. 19–26). Springer Singapore.
16. Gharibshah, Z., Zhu, X., Hainline, A., & Conway, M. (2020). Deep learning for user interest and response prediction in online display advertising. *Data Science and Engineering, 5*, 12–26.
17. Hervé, A., & Williams, L. J. (2010). Principal component analysis. *Wiley Interdisciplinary Reviews: Computational Statistics, 2*(4), 433–459.
18. Abidar, L., Zaidouni, D., Asri Ikram, E. L., & En-Nouaary, A. (2023). 5.Predicting customer segment changes to enhance customer retention: A case study for online retail using machine learning. *International Journal of Advanced Computer Science and Applications.*
19. Rudd, D. H., Huo, H., Islam, M. R., & Xu, G. (2023). Churn prediction via multimodal fusion learning: Integrating customer financial literacy, voice, and behavioral data. In *International-Conference on Behavioral, Economic, and Socio-Cultural Computing.*
20. Adeniran, I. A., Efunniyi, C. P., Osundare, O. S., & Abhulimen, A. O. (2024). Implementing machine learning techniques for customer retention and churn prediction in telecommunications.
21. Sikri, A., Jameel, R., Idrees, S. M., & Kaur, H. (2024). Enhancing customer retention in telecom industry with machine learning driven churn prediction. *Dental science reports.*
22. Evgeniou, T., & Pontil, M. (2001). *Support vector machines: Theory and applications* (Vol. *2049*, pp. 249–257).
23. Bansal, M., Goyal, A., & Choudhary, A. (2022). A comparative analysis of K-nearest neighbor, genetic, support vector machine, decision tree, and long short term memory algorithms in machine learning. *Decision Analytics Journal, 3*, 100071.
24. Hosmer, D. W., Jr., Lemeshow, S., & Sturdivant, R. X. (2013). *Applied logistic regression.* Wiley.
25. Gallo, A., (2014). The value of keeping the right customers, HBR,

26. Priyam, A., Abhijeeta, G. R., Rathee, A., & Srivastava, S. (2013). Comparative analysis of decision tree classification algorithms. *International Journal of current engineering and technology, 3*(2), 334–337.
27. Friedl, M. A., & Brodley, C. E. (1997). Decision tree classification of land cover from remotely sensed data. *Remote Sensing of Environment, 61*(3), 399–409.

Mapping the Landscape of Sustainable Tourism in India a Bibliometric Review and Systematic Synthesis

Sobhana Tripathy and Ananya Mitra

Abstract

The 2024 Travel and Tourism Development Index (TTDI) by the World Economic Forum ranks India 39th out of 119 countries, up from 54th in 2021. As a major global tourist destination, India faces the critical task of preserving its natural and cultural heritage through sustainable tourism practices. Sustainability aims to minimize negative impacts while maximizing positive outcomes, ensuring tourism benefits future generations and the planet. While global research on sustainable tourism has flourished since the 1990s, India's contribution has been relatively limited. In 2023, the Indian Ministry of Tourism launched the "Travel for LiFE" programme under Mission LiFE, a collaboration with the Ministry of Environment, UNWTO, and UNEP. The primary objective of the paper is to conduct a bibliometric analysis to examine global trends and India's role in sustainable tourism research. Using software tools, the study visualizes and analyses scientific publications, citations, co-authorship networks, and key research topics. It identifies trends, influential research areas, and emerging gaps, such as the role of technology in sustainability and the effectiveness of tourism policies. The findings highlight collaborative networks between Indian and international researchers, the institutions driving research, and potential future research directions, ultimately contributing to a more sustainable and equitable tourism industry. The paper also undertakes a secondary data analysis of domestic and foreign tourist inflow post COVID in India. Key findings reveal

S. Tripathy · A. Mitra (✉)
School of Economics and Commerce, Kalinga Institute of Industrial Technology (KIIT) Deemed to be University, Bhubaneswar, Odisha, India
e-mail: ananya.mitrafhu@kiit.ac.in

S. Tripathy
e-mail: 2481167@kiit.ac.in

S. Patnaik et al. (eds.), *People, Planet and Profit: Advancing Sustainable Business Practices*, Springer Proceedings in Geography and Sustainability,
https://doi.org/10.1007/978-3-032-11296-5_6

that India plays a growing role in shaping sustainable tourism research, particularly through its collaborative efforts with international researchers, while also highlighting critical gaps in areas such as the integration of technology and the assessment of tourism policy impacts.

Keywords

India • Sustainable tourism • VOSviewer • Visualization

1 Introduction

The current global challenges demand that all sectors of the economy innovate continuously, adapting to rapid changes while striving for sustainability. Many destinations, heavily dependent on tourism for their survival, are facing growing degradation [1, 2]. Consequently, there is an urgent need to implement more sustainable tourism practices, which has become a key priority for governments, businesses, and travellers around the world [3]. In 2015, the United Nations General Assembly established a clear set of objectives through Agenda 2030, which outlined the Sustainable Development Goals. UN tourism formerly known as UNWTO is one of the leading international organizations to promote tourism as a catalyst for sustainable development and inclusive economic growth globally. It provides direction and assistance in the formulation of tourism policies synergized with the United Nation Sustainable Development Goals [2]. World Tourism Organization (UNWTO) describes sustainable tourism as an approach that addresses the prerequisites of present travellers and local communities while ensuring that future generations can also benefit from this approach, and it involves the careful management of resources to balance economic, social, and cultural demands, while preserving key ecological functions, biodiversity, and essential life-support systems [4]. India's commitment to sustainable tourism is reflected in its alignment with the UN's Sustainable Development Goals, particularly through initiatives like the "Travel for LiFE" programme, which focuses on promoting sustainable tourism practices while addressing environmental and socio-economic challenges. "Sustainable tourism" does not refer to a particular form of tourism; rather, it encompasses any form of tourism that operates within the structure of sustainability and functions in harmony with the doctrine of sustainable development. It can take various forms, such as eco-tourism, cultural tourism, community-based tourism, rural tourism, and other emerging fields like accessible tourism, responsible tourism, wildlife tourism, and geo tourism [3].

1.1 Travel for LiFE

India, as the largest nation with enhanced connectivity, tourism infrastructure, and a variety of destinations and offerings, has risen to prominence as one of the foremost global tourism hubs in recent years [5]. The G20 Summit 2023 notably highlights the theme "One Earth, One Family, One Future," promoting a collective pledge towards responsible living to ensure a cleaner and more sustainable future. The Travel for LiFE initiative, part of Mission LiFE, strives to promote awareness about sustainable tourism while motivating both travellers and tourism enterprises to adopt environmentally friendly practices. The TFL programme identifies eight themes to address various aspects of sustainability such as (i) Save Water, (ii) Save Energy, (iii) Say No to Single-Use Plastic, (iv) Respect Local Culture and Heritage, (v) Empower Local Businesses and Communities, (vi) Reduce Waste, (vii) Consume Local Foods, and (viii) Conserve Natural Heritage. As per Ministry of Tourism, Government of India, Sep. 2023 data, the programme strives to integrate sustainability into the tourism sector, fostering the development of a responsible, resilient, and eco-friendly tourism industry. Also, the Ministry of Tourism, Government of India, took part in the first Working Tourism Group (WTG) meeting of 2024, held virtually during February 28–29, under Brazil's G20 Presidency. At this meeting, the Brazilian G20 Presidency advanced the agenda of the G20 Tourism Working Group. The group's activities will be aligned with Brazil's priorities and the theme of its G20 presidency, "Building a Just World and a Sustainable Planet." Based on Sustainable Tourism India Newsletter Issue 13, Ministry of Tourism, Government of India data, the session aimed to build on the initiatives and discussions from previous G20 presidencies, focusing on promoting global tourism cooperation with an emphasis on sustainability and equity. The emphasis lies in proactive measures and fostering sustainable future collaborations among public, private, and voluntary sectors, aiming for well-coordinated planning, decisive actions, and meaningful outcomes in preserving the global heritage landmarks.

1.2 Sustainable Tourism

Sustainable tourism is a holistic approach that seeks to balance environmental stewardship, socio-cultural integrity, and economic feasibility. In the last few years, the expanding focus on sustainable tourism research has highlighted the need for a thorough review of existing knowledge in the field. The bibliometric approach is especially valuable in dynamic fields such as sustainable tourism, where analysing research trends and their effects is crucial for fostering future progress. While there seems to be a growing interest in the topic, bibliographic studies on sustainability in tourism remain relatively scarce [6]. Even though sustainable tourism is gaining popularity, there is a lack of extensive bibliometric research on the topic [7]. However, more extensive research is required to gain a deeper understanding of the factors that lead to and foster successful sustainable tourism [8]. The purpose of the present study is to bring together or to organize the sustainable tourism literature which are fragmented and lack clear framework, to identifying key themes, concepts, and research gaps in this field thereby providing guidance and support to academic practitioners conducting research in sustainable tourism, contributing to the development of knowledge and practical strategies for the effective management of tourist destinations.

2 Research Questions

This study seeks to fill these knowledge gaps by exploring four key research questions (RQs):

RQ1. What is the co-authorship network in sustainable tourism, at the level of the top 10 countries actively publishing on the subject, organizational collaboration, and at the individual author level?

RQ2. What are the major thematic areas of sustainable tourism research?

RQ3. What are the future research prospects in the sustainable tourism research that would be advisable for researchers exploring in this field?

RQ4. What are the emerging tourism trends in the post-COVID period?

By addressing these questions, the study aims to identify patterns of collaboration, knowledge dissemination, and emerging trends that will inform and guide future research efforts, thereby contributing to the advancement of sustainable tourism scholarship and practices.

3 Methodology

This study uses a bibliometric approach to analyse sustainable tourism research, leveraging Scopus as the primary data source. Scopus was selected for its comprehensive and authoritative coverage of peer-reviewed literature in sustainable tourism, ensuring high-quality data for bibliometric analysis, while VOSviewer was used to visualize and analyse citation networks, co-authorship, and key research trends effectively. A total of 9357 documents were retrieved using the search term "sustainable tourism," encompassing journal articles, conference papers, and book chapters. The analysis focuses on co-authorship networks at both the country and individual author levels using VOSviewer. For RQ1, the study identifies co-authorship relationships between the top 10 countries in sustainable tourism, and also examines the collaboration patterns among individual authors. Additionally, keyword co-occurrence analysis is performed to address RQ2 by identifying the main thematic areas in sustainable tourism research. To answer RQ3, the study explores emerging research trends and gaps to propose future directions for sustainable tourism research. Visualization tools like network maps, heatmaps, and cluster diagrams will be used to present the co-authorship and thematic networks, with centrality measures applied to identify influential authors, countries, and themes. RQ4 analyses secondary data collected from states/UTs Tourism Department, Govt. of India. The time period is post-COVID years 2022–2023. To get an idea of the trend and distribution across 36 states and UTs of India, statistical tools like regression, AVOVA, Turkey's HSD, and Chi Square test have been conducted. The findings will be interpreted in the context of existing literature, providing insights into key trends and suggesting areas for future investigation. Although the study is limited to Scopus-indexed and English-language publications, it aims to offer a comprehensive view of the current state and future opportunities in sustainable tourism research.

4 Systematic Literature Review

The most prolific author in the field is C.M. Hall, with 52 publications, followed closely by S. Gössling (39) and X. Font (36). These authors are key figures in sustainable tourism research, with Hall contributing to a significant body of work in the field. Notably, Bramwell, B. (34), Lane, B. (32), and Boley, B.B. (30) also rank high, indicating their strong involvement in shaping the discourse on sustainable tourism. The institutions leading the research output include Griffith University, which supports two high-output authors, Hall and Lane, with a total of 139 articles. University of Johannesburg and University of Surrey are also prominent contributors, with 127 and 96 articles, respectively. These universities are likely to play an important role in advancing sustainable tourism knowledge globally (Table 1).

In terms of national contributions, the United Kingdom is the leading country with 843 articles, followed by the United States (830) and China (744). These countries have a strong presence in sustainable tourism research, with the UK and the US consistently producing high volumes of research. Other countries such as Spain (707), Australia (642), and Italy (635) also contribute significantly, suggesting that sustainable tourism is a topic of global importance across diverse regions. Notably, countries like Indonesia, India, Malaysia, and Portugal also appear in the data, indicating that research in sustainable tourism is increasingly being undertaken in both developed and developing nations.

In India, 29% of the research papers on sustainable tourism are published in the domain of business, management, and accounting because most often the main focus of business management research is on balancing profitability with sustainability by adopting eco-friendly methods and practices. The central themes of social science research are preserving the cultural heritage, impact of sustainable tourism on local communities, and equitable distribution of resources which make it the second largest domain in the area of sustainable tourism research contributing 20% of the total. Again 17% of the total research is represented by economics, econometrics, and finance due to the economic impact of tourism such as employment, major source of GDP, and foreign exchange earnings. The fourth largest category in the above chart is environmental science research as it plays

Table 1 Top 10 contributors based on number of publications

Author	Number of articles	Organization	Number of articles	Country	Number of articles
Hall, C.M.	52	Griffith university	139	UK	843
Gössling, S.	39	University of Johannesburg	127	United States	830
Font, X.	36	University of Surrey	96	China	744
Bramwell, B.	34	The University of Queensland	93	Spain	707
Lane, B.	32	Griffith business school	84	Australia	642
Boley, B.B.	30	Oulun Yliopisto	82	Italy	635
Saarinen, J.	27	University of Waterloo	69	Indonesia	532
Weaver, D.B.	27	The Hong Kong Polytechnic University	63	India	459
Dolnicar, S.	25	Universiti Sains Malaysia	61	Malaysia	368
Han, H.	23	Universidade do Algarve	61	Portugal	353

Source Scopus database

a crucial role in facing the challenges of global sustainability and development. The research in the other domains like computer science, engineering, agriculture hold less percentage due to the low significance in sustainability tourism research (Fig. 1).

Figure 2 depicts the number of research papers released in the area of sustainable tourism in India. From the year 1991 up to 2010, the number of publications in sustainable tourism prospects is almost negligible which indicate lack of awareness of the sustainability issues and challenges along with lesser academic interest in this area at that time. From the year 2011 gradually it started showing an upward trend due to the possible reasons of over tourism, pollution, and other environmental concern. Introduction of global sustainability frameworks like SDG is another reason for the upward trend from 2015. But there is a major shift in upward trend in the sustainable tourism research which is visible from 2020 due to the post-pandemic effect on tourism industry. In 2024, the research is in the peak due to the increased awareness of sustainability challenges through various government initiatives taken by the Government of India.

Graph 3 shows the leading organizations sponsoring the sustainable tourism research. University Grant Commission has contributed the highest funding by

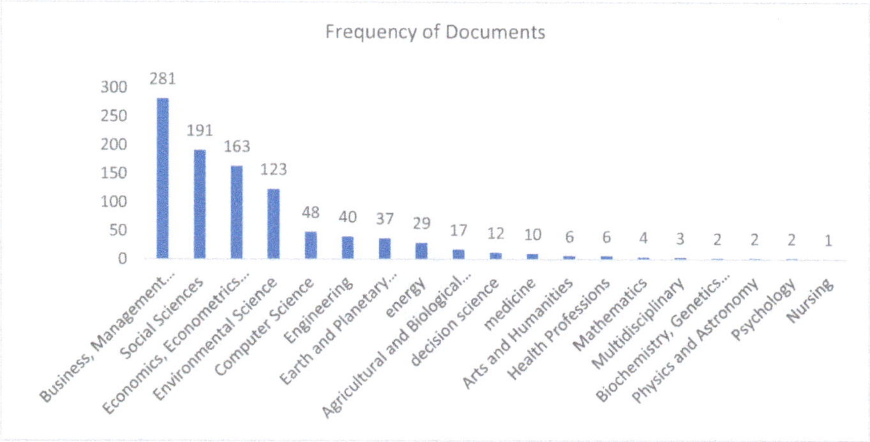

Fig. 1 Frequencies of documents based on research area

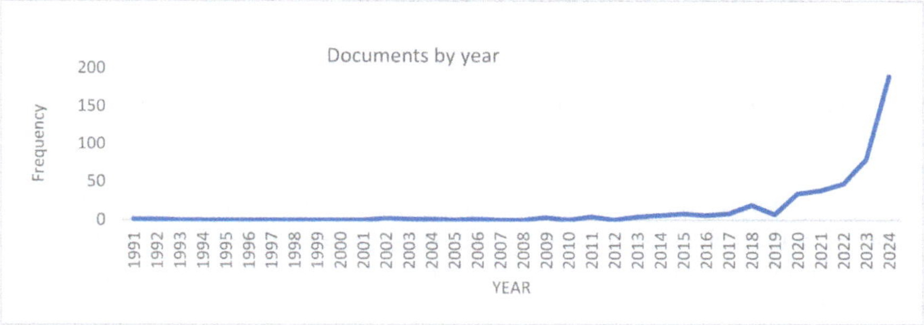

Fig. 2 Frequencies of documents based on years

sponsoring 5 research publications in the sustainable tourism research. Open Society Foundation, Ministry of Environment, Forest and Climate Change, Mekelle University, King Saud University, and Indian Council of Social Science Research also showcase their increasing interest in funding two publications each in the sustainable tourism research. The presence of international sponsoring organizations indicates the global interest of sustainable tourism research. The organizations like Bundesministerium für Bildung und Forschung (Germany) and AC3 have the smaller sponsorship funding (1 each) and are the less frequent contributors in the sustainable tourism research.

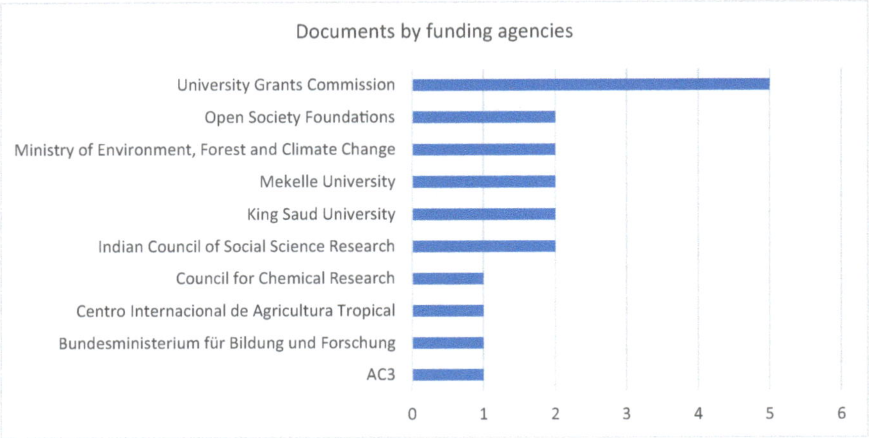

Fig. 3 Frequencies of documents based on funding agencies

4.1 Visualizing Bibliometric Networks

Figure 4 illustrates the co-authorship network among the top 10 countries actively publishing on sustainable tourism, as sourced from the Scopus database. The visualization reveals distinct clusters based on the strength of collaborative ties, with countries like the United States, China, and the United Kingdom serving as central hubs connecting other nations. Notably, there is a strong collaboration between countries in the same region, such as Malaysia, Indonesia, and India, forming one cluster, while European countries like Italy, Spain, and Portugal form another. This clustering likely reflects regional proximity, shared socio-economic contexts, and common sustainability challenges, which make collaboration more feasible and relevant. Additionally, countries with higher research funding, advanced academic infrastructure, and global outreach, such as the United States and the United Kingdom, emerge as global connectors, enabling cross-continental collaborations. These findings underline the importance of fostering cross-regional partnerships to bridge gaps between the Global North and South in sustainability discourse, as regions with fewer resources may rely on collaborative efforts to enhance their research contributions.

Figure 5 focuses on organizational collaborations among institutions publishing a minimum of 10 articles on sustainable tourism. The network highlights prominent academic institutions, such as the School of Hospitality and Tourism and the School of Business and Economics, which serve as key nodes in the research ecosystem. Collaborative ties between departments of management, environmental science, and geography research units suggest a multidisciplinary approach to addressing sustainability challenges. The dominance of certain organizations could be attributed to their specialized focus on hospitality and tourism, their access to funding, and their global academic partnerships. Furthermore, geographic location

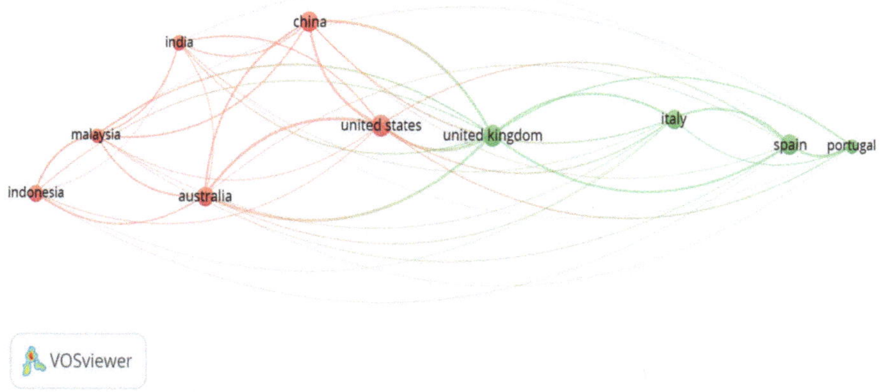

Fig. 4 Co-authorship analysis country-wise. *Source* Authors own. (Top 10 Countries as per scopus database and their networking)

and institutional reputation likely play roles in attracting international collaborations. For instance, institutions in regions heavily dependent on tourism economies, such as Southeast Asia or Southern Europe, may naturally prioritize research on sustainable tourism, aligning with local industry needs and policy agendas. These findings highlight the pivotal role of academic institutions in fostering knowledge sharing across diverse disciplines, emphasizing the need for interdepartmental synergy to generate holistic solutions for sustainable tourism.

Figure 6 depicts the co-authorship network among individual authors with at least five publications and five citations. The network is characterized by multiple clusters, each representing a community of researchers working closely on shared themes within sustainable tourism. Authors such as Hall C. Michael, Bramwell Bill, and Font Xavier emerge as central figures, demonstrating strong collaborative connections and significant contributions to the field. The prominence of these authors can be attributed to their extensive academic output, strong citation

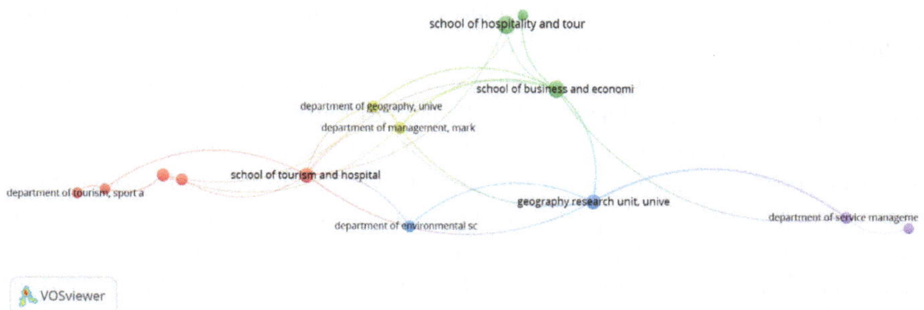

Fig. 5 Co-authorship analysis organization-wise publishing a minimum 10 articles on sustainable tourism

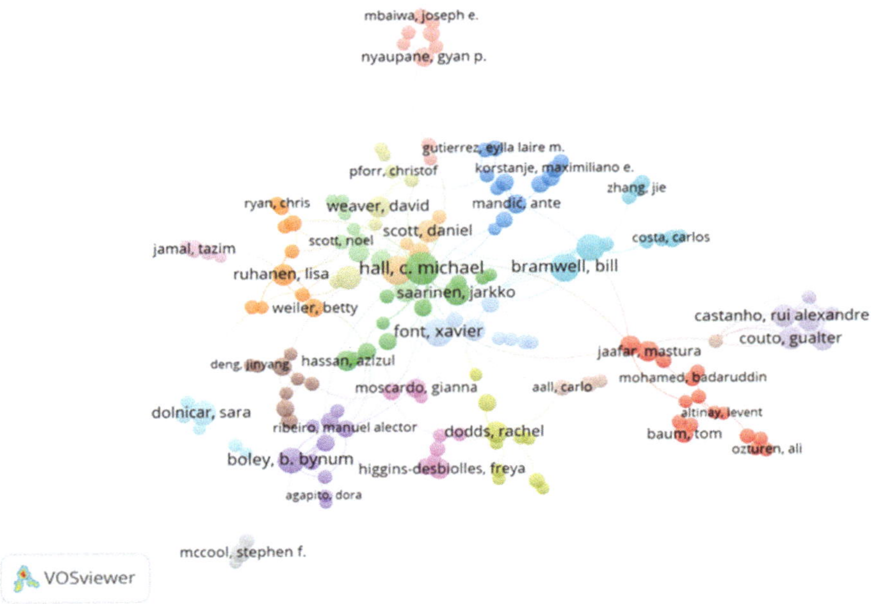

Fig. 6 Co-authorship analysis author-wise

impact, and ability to build global networks through conferences and collaborative projects. Conversely, the dispersed nature of some clusters indicates niche areas of research, where smaller groups of researchers focus on specific aspects of sustainability, such as ecotourism or community-based tourism. The results reflect the increasing need for interdisciplinary work, with central authors often acting as bridges between distinct research clusters. These findings highlight the pivotal role of prolific researchers in shaping the trajectory of sustainable tourism studies and underline the importance of fostering inclusive networks to ensure equitable dissemination of knowledge.

Figure 7 presents a co-occurrence analysis of author keywords with a minimum threshold of 30 occurrences, visualized using VOSviewer. The central theme identified in the visualization is "sustainable tourism," which dominates the network, connecting with keywords such as "tourism development," "climate change," "environment," and "resilience." This indicates that sustainable tourism is a highly significant and well-studied area within the field of tourism research. The thematic areas derived, such as tourism development and environmental conservation, reflect global priorities on balancing tourism growth with environmental sustainability. The inclusion of terms like "post-pandemic tourism" underscores the recent interest in understanding how the COVID-19 pandemic has reshaped tourism practices, with researchers emphasizing recovery strategies aligned with sustainability principles. The findings suggest an interdisciplinary approach to sustainable tourism, integrating economic, environmental, and social dimensions.

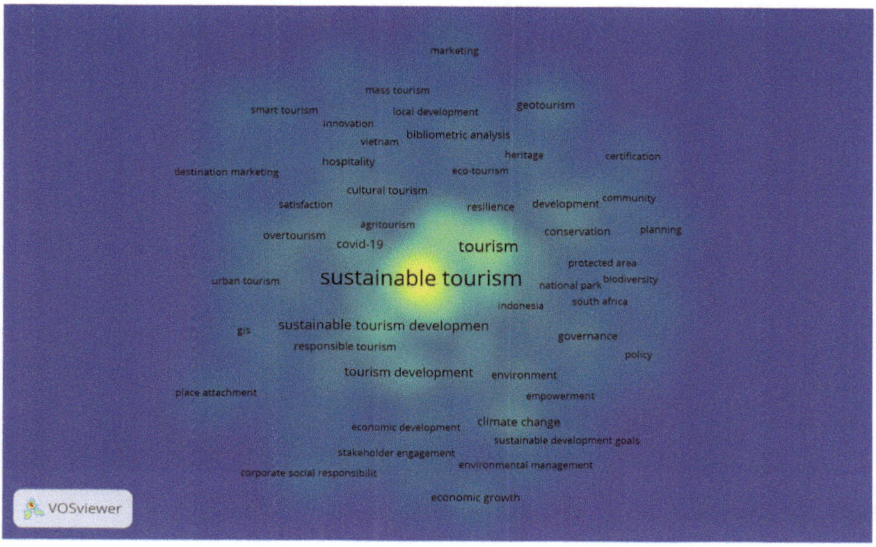

Fig. 7 Co-occurrence of author keyword

Figure 8 focuses on the co-occurrence analysis of indexed keywords, again with a minimum threshold of 30 occurrences. "Sustainable development" and "eco-tourism" emerge as focal points, closely tied to terms like "economic impact," "climate change," and "cultural heritage." This indicates a broader scope of research interests that transcend traditional tourism studies to include sustainability goals and the socio-economic impacts of tourism. The thematic areas emphasize critical concerns such as climate change and cultural preservation, reflecting an increasing acknowledgment of tourism's role in addressing global environmental challenges and conserving cultural assets. Moreover, the prominence of terms like "geographic focus" and "regional planning" highlights the importance of localized and context-specific strategies for tourism development. These insights suggest that tourism research is evolving towards a holistic framework that requires combined force of ecological stewardship, community involvement, and strategic policy-making.

5 Post-COVID Trend and Distribution of Tourists in India

Post-COVID India's economy demonstrated robust growth, with GDP expanding by 8.2% in 2023, up from 7.0% in 2022. This economic momentum was reflected in the tourism sector, where domestic tourist visits (DTV) surged by 44.9% and foreign tourist visits (FTV) increased by 163.7% in 2023 compared to the previous year. Secondary data shows major states like Uttar Pradesh, Tamil Nadu, and

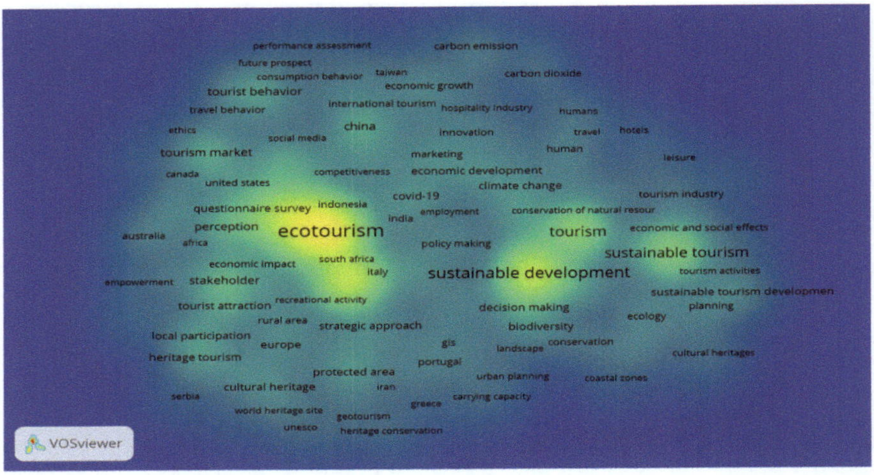

Fig. 8 Co-occurrence analysis of indexed keyword

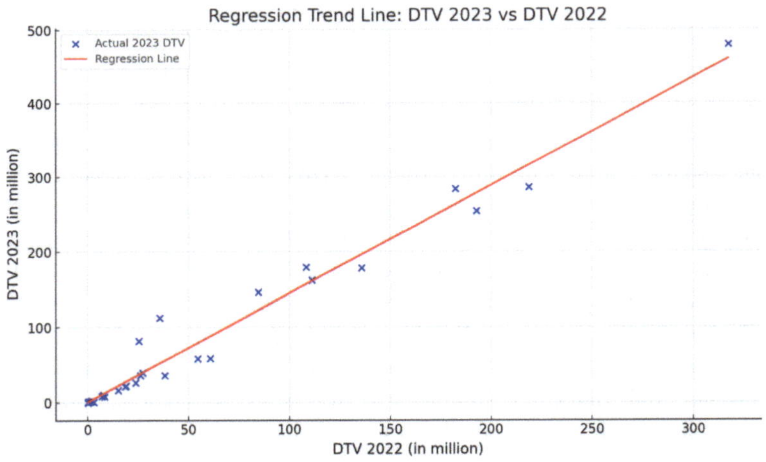

Fig. 9 Regression trend line of tourist visits in India post COVID

Karnataka as the top three states for domestic tourism in 2023, with Uttar Pradesh leading at 19.1%. These states, along with Maharashtra and Andhra Pradesh, dominate India's tourism landscape, highlighting regional popularity and infrastructure supporting high volumes of domestic travellers:

$$\text{Growth Rate} = 41.57 + 1.17\log(\text{DTV}_{2022}) \tag{1}$$

The regression results in Eq. (1) show no evidence of beta-convergence in domestic tourist visits across Indian states between 2022 and 2023, meaning states with higher tourist numbers in 2022 did not grow more slowly in 2023 as convergence theory would suggest. The parallel growth suggests a positive correlation between economic performance and tourism activity. The relationship was slightly positive but statistically insignificant, indicating that some already popular states (like Uttar Pradesh, Rajasthan, and Maharashtra) continued to grow at strong rates.

The regression analysis (2) reveals a strong positive linear relationship between DTV in 2022 and 2023 across Indian States and UTs, with an R^2 value of 0.974, indicating that 97.4% of the variation in 2023 DTV can be explained by the 2022 figures:

$$\mathrm{DTV}_{(2023)} = 8.6373 + 1.2936\mathrm{DTV}_{(2022)} \quad (2)$$

The regression equation suggests that, on average, each additional million domestic tourists in 2022 contributed to a 1.29 million increase in 2023.

The points in the trend line graph show a strong momentum effect where regions with high tourist numbers in one year tend to maintain or grow their appeal. States significantly above the trend line outperformed expectations, likely due to targeted tourism initiatives, while those below may need to reassess their strategies.

Domestic Tourist Visit Result Analysis: The ANOVA results indicate a statistically significant difference in domestic tourist visits across different regions in India (p-value = 0.003). Tukey's Honestly Significant Difference (HSD) test confirms that North-East, East, and South show notable variations compared to other regions, with North-East having the lowest tourist influx while South has one of the highest. The Chi-square test, however, suggests no significant difference in the distribution of domestic tourists across regions (p-value = 0.97). The variation in domestic tourism is likely due to regional economic disparity, accessibility, cultural attractions, and pilgrimage tourism. The South and North regions house famous religious and heritage sites (e.g., Tamil Nadu's temples, Rajasthan's forts), contributing to higher domestic tourist volume. Meanwhile, the North-East sees less domestic travel.

Foreign Tourist Visit Result Analysis: ANOVA reveals a statistically significant difference among regions for foreign tourists (p-value = 0.012). Tukey's HSD test shows that North-East and East have significantly fewer foreign tourists, while West and Central, and North host relatively higher numbers. The Chi-square test confirms that the distribution of foreign tourist visits across regions is statistically similar (p-value = 0.97). The discrepancy in foreign tourist arrivals arises from popular destinations like Delhi, Rajasthan, Maharashtra, and Goa attracting more international travellers due to historical landmarks, business hubs, and coastal tourism. South India also draws foreign visitors, particularly to Kerala, Karnataka, and Tamil Nadu, for wellness tourism, beaches, and cultural heritage. In contrast, North-East and some Eastern states struggle with low foreign tourist numbers.

6 Key Challenges in Fostering Sustainable Tourism Practices in India

A coordinated, prompt response is desperately needed in the current tourism environment, emphasizing strengthened capability, enhanced research activities, information mobilization, and better communication strategies. Effective promotion and implementation of sustainable tourism practices necessitates cooperation among stakeholders. In the current era of globalization and digitalization, Indian tourism sector lags significantly and is far behind in the use of AI and AR technology as compared to other advanced countries like the US, the UK, and China [9] due to a variety of factors, including the need for data protection, a lack of robust AI technological infrastructure, a lack of awareness among tourism industry stakeholders, and the difficulty of integrating AI with existing systems.

In addition, many locations in India, which are filled with natural beauty and scenic landscapes like the north-eastern region of India, remain unexplored due to inadequate infrastructure and connections. Besides that, improper trash management, severe environmental deterioration, and a lax attitude among residents towards eco-friendly practices all impede sustainable tourism development in India [10]. Environmental awareness is essential for understanding sustainable activities. When the quantity of visitors surpasses the environment's ability to cope, a natural hazard arises [11]. Natural habitats, both tangible and intangible cultural history, and mutually beneficial economic advancements would all be well preserved by doing this [12].

The planning approach for historical tourism urgently needs to focus on really integrating the communities. An awareness campaign is required to inform the public about the value of heritage, the effects of tourism, and their part in managing their resources. Building trust and communicating with the local communities are also necessary to reassure them that their issues and recommendations will be taken into consideration. The tourism sector will develop when people begin to coordinate and are willing to keep participating in the relevant policy-making process [13]. The Indian government is working diligently to implement sustainable tourism practices for the country's overall tourism industry development under the Travel for LiFE programme in partnership with the Ministry of Tourism. To demonstrate the North Eastern States of India's tourism potential both domestically and internationally, the Ministry of Tourism hosted the 11th edition of the International Tourism Mart in the NER in Shillong, Meghalaya. There is still more work to be done in this area to overcome the underlying challenges and for the successful implementation of the eight policies under the mission LiFE programme to address the sustainability challenges.

7 Future Scope

The research on sustainable tourism has notable gaps. Despite the increasing global interest in this domain, bibliometric analyses reveal that certain areas, such as the role of technology in enhancing sustainability and the effectiveness of tourism policies, remain underexplored. The fragmented nature of existing literature and the lack of a cohesive framework also indicate areas where more systematic investigation is needed. These gaps highlight the need for a multidisciplinary approach and greater collaboration between stakeholders from different regions. Contradictions in the research often stem from regional differences, the scope of studies, and the varied methodologies employed. In the current context, while some research highlights the economic benefits of sustainable tourism, others highlight challenges such as unequal resource distribution and the strain on local communities. Additionally, differing interpretations of "sustainability" can lead to varying outcomes in studies, complicating the establishment of universally applicable best practices. Future research should explore the intersection of sustainable tourism with emerging fields such as artificial intelligence, digital marketing, and smart tourism [14]. Additionally, longitudinal studies on the long-term impacts of tourism policies and practices can provide deeper insights. Addressing region-specific challenges and expanding research beyond English-language publications will also enhance inclusivity and relevance in this domain. More extensive work needs to be done in the research area converging various disciplinary knowledge.

8 Conclusion

The findings effectively address the research questions by mapping global and individual co-authorship networks, identifying thematic areas, and suggesting future research directions. The co-authorship analysis highlights significant collaborative clusters, indicating active research hubs. The thematic analysis identifies critical areas such as climate change, resilience, and community involvement, aligning with global sustainability goals. These findings provide a clear direction for researchers to focus on underexplored areas while leveraging existing collaborations. Key recommendations include fostering cross-regional and interdisciplinary collaborations to bridge research gaps and ensure equitable knowledge dissemination. Increasing the focus on technology integration, policy effectiveness, and local community empowerment is also advised. Encouraging more targeted funding and sponsorship, especially from regions with limited research output, can further enhance the global research landscape. Standardizing methodologies and definitions in sustainable tourism studies could address contradictions and promote consistency. Best practices that could be recommended include adopting a holistic approach that integrates economic, environmental, and social dimensions [15]. Encouraging participatory research involving local communities and stakeholders ensures practical applicability and inclusivity. Promoting the standardization of key indicators for sustainability and leveraging bibliometric tools for trend analysis can guide strategic decision-making. Learning from successful models in regions with robust sustainable tourism frameworks can also serve as a valuable guide.

References

1. Rhonda, D. B., Malatesta, S., & di Friedberg, M. S. (2022). Promoting education for sustainable development in the Maldives: Exploring the link between theory and practice. *Prospects, 52*(3), 529–544.
2. HeSong, G., Zhu, D., & Waqas, M. (2024). How to decouple tourism growth from carbon emission? A panel data from China and tourist nations. *Heliyon, 10*(15), e35030.
3. Lina, P., Grigaliunaite, V., & Bogoyavlenska, Y. (2024). A bibliometric review of innovations in sustainable tourism research: Current trends and future research agenda. *Sustainability, 16*(16), 7124.
4. Huamanchumo, E., Marie, R., Villalba-Condori, K. O., Botezan, I., & Jimenez, R. S. (2024). Ecotourism, sustainable tourism and nature based tourism: An analysis of emerging fields in tourism scientific literature. *Geo Journal of Tourism & Geosites, 54*, 953–966.
5. Snigdha, K., Chowdhary, N., Singh Kaurav, R. P., & Tiwari, P. (2024). Motivations of urban millennials for spiritual travel in India. *Tourism Recreation Research, 49*(2), 410–425.
6. Angels, N., Sanchez-Rebull, M. V., & Lara, A. B. H. (2019). Tourism research on sustainability: A bibliometric analysis. *Sustainability, 11*(5), 1377.
7. Weaver, D. B., & Lawton, L. J. (2007). Twenty years on: The state of contemporary ecotourism research. *Tourism Management, 28*(5), 1168–1179.
8. Samuele, M., Butturi, M. A., Gamberini, R., & Martini, U. (2023). Indicators for sustainable touristic destinations: A critical review. *Journal of Environmental Planning & Management, 66*, 1–30.
9. Reejo, R. J. (2023). A review on marketing tourism in the era of AI: Future scope for sustainable tourism in India. In *AI and emotional intelligence for modern business management* (pp. 321–347).
10. Joseph, E. K., Varghese, B., Kallarakal, T. K., & Antony, J. K. (2021). Sustainable tourism practices: A perception of backwater tourism destinations in South Kerala India. *Geo Journal of Tourism & Geosites, 38*(4), 1232–1238.
11. Naik, S., & Chanda, R. S. (2025). Examining the impact of factors on sustainable tourism practices: An empirical study in the Indian context. *Cogent Social Sciences, 11*(1).
12. Chopra, A., & Pandey, A. (2025). I, me, and the outsider—An exploratory study to identify antecedents of sustainable tourism ecosystem in India. *Cogent Social Sciences, 11*(1).
13. Chauhan, E. (2020). Challenges to community participation in heritage tourism development: Case studies of Shahjahanabad and Nizamuddin Basti in New Delhi, India. *WIT Transactions on Ecology & the Environment, 1*, 225–233.
14. Upadhyay, S. K., & Upadhaya, G. (2024). Empowering travelers: Enhancing consumer awareness and understanding of travel insurance benefits. *Journal of Tourism & Hospitality, 18*.
15. Parashar, A., Kumar, N., & Agarwal, P. (2025). Role and importance of sustainable bleisure travel in India: With special reference to nature-based health tourism. In *Smart travel and sustainable innovations in Bleisure tourism* (pp. 29–44). IGI Global.

Chemical Fertilizer Usage in Indian Agriculture: Trade-Off Between Carbon Emissions and Food Security

Smrutirekha Mohanty and Ananya Mitra

Abstract

Chemical fertilizer is a two-edged sword for the world's highest populous country, India. On one side, it can cut down the hunger statistics (SDG2) and on the other CO_2 present in the environment (SDG13). To yield the sword, one must understand its existence's mechanism and purpose. The current paper tries to explore the relationship between chemical fertilizer, food grain production, hunger statistics, and annual average temperature from 1981 to 2023. The paper uses robust econometric time-series tools like Pairwise Granger Causality Test, Johansen Cointegration Test, and Vector Error Correction Model. The result showed that there exists a short-run and long-run relation between the variables, but the extent of causality is limited.

Keywords

NPK led CO_2 emission • Food grains • Hunger statistics • Temperature • Econometrics • India

S. Mohanty · A. Mitra (✉)
School of Economics and Commerce, Kalinga Institute of Industrial Technology (KIIT) Deemed to be University, Bhubaneswar, Odisha, India
e-mail: ananya.mitrafhu@kiit.ac.in

S. Mohanty
e-mail: smrutirekha.mohantyfhu@kiit.ac.in

S. Patnaik et al. (eds.), *People, Planet and Profit: Advancing Sustainable Business Practices*, Springer Proceedings in Geography and Sustainability,
https://doi.org/10.1007/978-3-032-11296-5_7

1 Introduction

Chemical fertilizers have been an integral component of Indian agriculture, supporting unprecedented increases in food grain production and serving as one of several global initiatives to eliminate hunger and achieve food security [1]. Chemical fertilizers are central to satisfying the nutritional requirements of an ever-growing population in one of the world's largest agrarian economies, India. But there's a complicated trade-off in relying on chemical fertilizers. These inputs increase agricultural productivity but disproportionately increase the greenhouse gas (GHG) emissions of CO_2 and N_2O, which adds to global warming and climate change [2]. Additionally, excessive fertilizer use impacts soil health, water resources, and environmental sustainability [3]. Climate change is deeply tied to the challenges of India's agriculture because increasing temperatures impact the amount of crops that can be produced [4]. This interplay between chemical fertilizer use and its environmental impacts is made more complicated by the nation's commitments to fulfill the Sustainable Development Goals (SDGs). For instance, Zero Hunger (SDG 2), Climate Action (SDG 13), and Responsible Consumption and Production (SDG 12) explicitly require strategies to increase food grain production in such a way as not to increase fertilizer-driven emissions excessively [5]. India faces critical agricultural problems such as eroding soil quality alongside its dependence on water supplies and requirements for sustainable practices so one can understand the complete effects of chemical fertilizers better. The national response to rising food requirements along with environmental stressors makes the fertilizer impact both soil quality and water reserve conditions and biological diversity stronger than ever. The effects of fertilizers on farming practices have impacts far wider than those related to carbon dioxide. Based on information from the Central Pollution Control Board (CPCB), too many fertilizers are used in India, which has spoiled much of the water, especially in areas like Punjab, as nitrogen-rich fertilizers have resulted in unsafe levels of nitrate in the groundwater. According to the Ministry of Agriculture and Farmers Welfare, applying excessive fertilizers to the land in Uttar Pradesh and Bihar has caused the soil to deteriorate due to unbalanced nutrients. According to the Indian Council of Agricultural Research (ICAR), using excessive fertilizers has decreased organic carbon in the soil in many locations across the country, contributing to further issues with soil quality. According to a study by the National Institute of Hydrology, fertilizers from land reach the Ganges River and increase eutrophication in aquatic habitats, and reduce the quality of water. As a result, we should also monitor soil health, the quality of our water, and losses in biodiversity, which already measure CO_2 emissions, to improve sustainable farming practices. A comprehensive assessment of how fertilizers relate to the Sustainable Development Goals (SDGs) depends on analyzing these issues. Better evaluation of contemporary agricultural sustainability requires an intensive investigation of identified challenges. Environment and community concerns need to influence decisions about agricultural productivity for sustainable results.

However, there is very little research on trade-offs between temperature change, food grain production, hunger alleviation, and carbon emissions, even less so within the Indian context. "This gap is particularly concerning given the urgency of addressing these challenges in a country where agriculture employs a large portion of the workforce and sustains millions of livelihoods" [6] and millions of livelihoods. For its part, India must understand the role of chemical fertilizers in a multi-dimensional equation that determines agricultural productivity and its environmental and broader development dimensions.

This study investigates the multiple and very complex trade-offs between chemical fertilizers used in Indian agriculture. In our analysis, we aim to elucidate the connections among fertilizer-induced productivity gains, fossil CO_2 emissions, climate change, and food security, as related to the SDGs. Therefore, the research provides a complete structure for policymakers, researchers, and professionals to address the twin challenge of increasing both food production to meet population increase while keeping within environmental confines.

2 Literature Gap

Scopus database was used to access the literature on the dual impact of chemical fertilizer usage and the leading trade-off between two SDGs of zero hunger and climate action. By synthesizing existing literature, it tries to demonstrate the trade-offs between food security and climate change mitigation.

It illustrates in Fig. 1 an increase in chemical fertilizer research and implications in the last five years (2023–2024). The trend can be explained by research priorities evolving, policy interventions, and awareness of trade-offs between agricultural productivity and environmental sustainability. Studies that link chemical fertilizers to hunger also show a spike in 2024, with increased emphasis on their contribution to resolving world hunger as sustainable development goals and sustainable agriculture.

It is evident that attention to research of chemical fertilizers and food grains is given consistently, peaking in 2023 and showing the significance of fertilizers in crop production. Advancements in technology, policy changes, or agricultural crises that increased the need for productivity-improving inputs may be associated with this consistency.

While studies on chemical fertilizers and CO_2 emissions have been less frequent, they gained momentum in 2023 as the first indications of the growing awareness regarding the environmental trade-offs. This implies that such planet and socio-economic considerations are of growing concern, as sustainability and climate change mitigation become prime objectives in national and international research agendas.

The chemical fertilizer uses and CO_2 with temperature combination are still relatively unexplored, and this may be an area of research in terms of the combined impacts of climate and fertilizer use. They could be a result of the complexity of

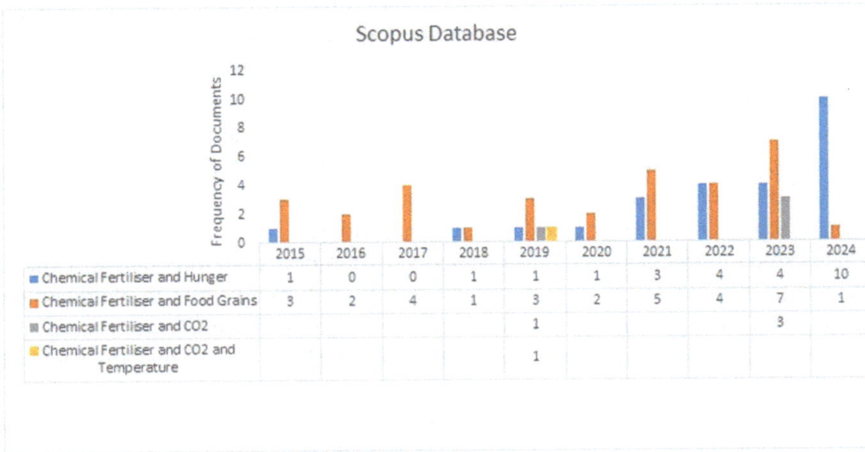

Fig. 1 Chemical fertilizer trade-off

these topics, which require an interdisciplinary approach and refined climate modeling tools. In addition, funding and institutional focus tend to focus on quick agricultural productivity solutions while leaving out long-term environmental impacts.

3 Objectives

The above trend highlights an urgent need for interdisciplinary research exploring the trade-off between agricultural productivity and environmental sustainability, especially in the context of global climate change and food security. Aligning with the research gap, the objectives of the study are.

1. To examine the dual impact of chemical fertilizers on enhancing food grain production (SDG 2) while contributing to greenhouse gas emissions and climate change (SDG 13).
2. To investigate the causal relationships between fertilizer usage, food grain production, hunger alleviation, and carbon emissions, focusing on India's agricultural context using the Granger Causality Test.
3. To investigate the short- and long-run relationships between chemical fertilizer usage, food grain production, hunger statistics, and carbon emissions in the context of Indian agriculture, using the Vector Error Correction Model (VECM) and co-integration analysis.

4 Methodology

The study aims to analyze secondary data from 1981 through 2023 to examine the interaction between chemical fertilizer usage, food grain production, hunger statistics, and carbon emissions in the Indian agricultural scenario. The secondary data have been collected from government publications and reports [7]. Trends and key indicators are summarized by descriptive statistical tools, while graphical methods are employed to visualize these relationships and patterns over the study period. Econometric techniques, such as the Granger Causality Test, are applied to the study of causal relationships between variables, while co-integration analysis and the VECM are used to investigate the short- and long-run relationships on the dual impacts of fertilizers on food security (SDG 2) and environmental sustainability (SDG 13).

5 Role of Chemical Fertilizers in Food Grain Production

The soil, types of crops grown, and how they are grown influence fertilizer use in India, leading to unequal application of nutrients and fertilizers across the country. The Fertilizer Association of India (FAI) states that Punjab and Haryana, where active wheat and rice cultivation takes place, consume much fertilizers, and Punjab alone accounts for almost 10% of India's total fertilizer use, though it grows food on only 1.5% of the nation's total land. ICAR has found that using too many fertilizers in India causes significant nitrous oxide emissions. Yet, portions of Rajasthan and the Northeast part of India tend to use fewer fertilizers, according to the Ministry of Agriculture and Farmers Welfare, with Nagaland and Mizoram each having fertilizer consumption below 15 kg/ha, less than the national average of 140 kg/ha. Because these areas use their lands less, the National Food Security Mission (NFSM) points out that there may not be enough nutrients for crops to thrive, gaining food security. Differing ways of using fertilizers result in effects on both food and the sustainability of the climate. Excessive use of land and resources in some places leads to damage to the environment and more greenhouse gas, while underuse in other areas jeopardizes crops and food supply. Peninsular Southeast Asian farmers should use fertilizers carefully and in a way that maintains the health of their land and protects the environment.

In particular, Nitrogen (N), Phosphorus (P), and Potassium (K) chemical fertilizers have made major contributions to boosting the yields for staple crops, such as rice, wheat, and maize, and preventing starvation. Fertilizer consumption shot up in India between 1960 and 2020, making the country self-sufficient in food grains. However, they have produced diminishing returns from fertilizer use and heavy dependence on specific forms, like urea, causing nutrient imbalances and soil health breakdown [7].

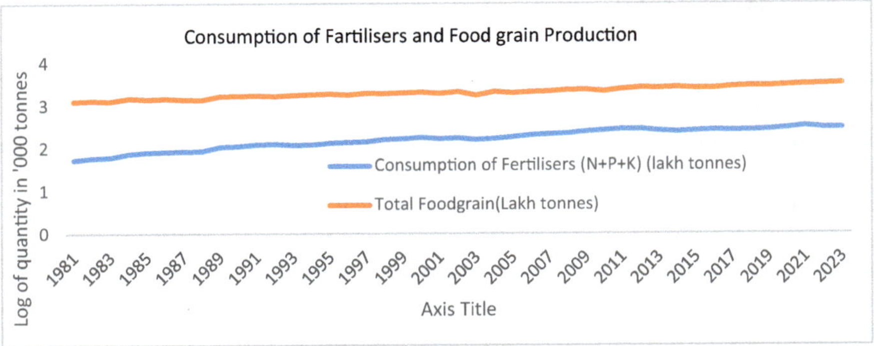

Fig. 2 Consumption of fertilizers and food grain production

Shown in Fig. 2 are trends in fertilizer (N + P + K) consumption (in lakh tonnes) and food grain production over a period. An increase in food grain output is likely to be depicted to go hand in hand with an increase in fertilizer usage. If this is true, fertilizers have made much of agricultural production possible, although when overused, it might slow production or cause environmental problems.

For food grains and NPK, we run Pairwise Granger Causality Tests (Table 1), following which the F-statistic of 0.00510 with a very high p-value of 0.9435 means that changes in food grain production do not lead to any significant changes in NPK fertilizer usage. The non-causality of fertilizer application relative to changes in food grain production may be explained by the fact that fertilizer application is conditioned by broader agricultural practice and policy or availability of fertilizers. Moreover, fertilizer use is probably adjusted by soil quality or crop-specific requirements rather than in response to food grain production. Having an F statistic of 10.2782 and a low p-value of 0.0027 strongly illustrates that a large amount of NPK fertilizer contributes to many landscapes on food grain yields. So it's most likely that returning to this use of fertilizers for crop productivity is supposed to be a highly input-intensive system. The research outcome demonstrates that NPK fertilizer application enables crops to improve their growth while generating superior yield outputs. The application of NPK fertilizers supplements essential soil nutrients such as nitrogen, phosphorus together with potassium. The impact of fertilizers becomes stronger in healthy and fertile soils. Farmers achieve higher food production rates, together with larger harvest yields, due to this outcome.

6 Hunger and Food Security

On the largest scale, chemical fertilizers made hunger and food security accessible and affordable across India. Therefore, fertilizers have increased agricultural productivity and helped to reduce malnutrition and hunger in line with SDG 2.

Table 1 Pairwise Granger Causality Tests on food grains and NPK

Pairwise Granger Causality Tests			
Sample: 1981 2024			
Lags: 1			
Null hypothesis:	Obs	F-statistic	Prob
FOODGRAINS does not Granger Cause NPK	41	0.00510	0.9435
NPK does not Granger Cause FOODGRAINS		10.2782	0.0027

Source Author's own calculation

However, fertilizers are not evenly distributed by region, and uneven benefits result from improper application. A large proportion of India's agricultural workforce are small-holder farmers, who often lack adequate knowledge, have low access to resources, and lack the ability to balance fertilization. The importance of moving from chemical to sustainable inputs for Indian agriculture is not comprehensible to them. The authors observe that while small farmers may choose chemical inputs because of their cheaper availability, sustainable input utilization is considered more expensive. The equitable distribution mechanisms and targeted interventions are the tools that are needed to bridge these gaps in ways that will ensure holistic and sustainable food security [8].

Pairwise Granger Causality Tests on food grains and hunger statistics are represented in Table 2. The results indicate an insignificant causality relationship between food grain production to hunger as indicated by the F-statistic of 0.64805 and p-value of 0.4319. The complexity and multinationalism of hunger may be responsible for this lack of causality: hunger is not simply a matter of the availability of food grains. Besides a degree of income distribution, food accessibility, poverty, and political instability, hunger levels are also influenced by other factors. Food grain production may not always be synchronized with the regional food distribution and local demand; an additional increase in production may not always alleviate hunger in some areas.

The result of hunger to food grain production is not significant based on the F statistics = 0.23440, and p-value = 0.6345. Food insecurity, marked by hunger,

Table 2 Pairwise Granger Causality Tests on food grains and hunger statistics

Pairwise Granger Causality Tests			
Sample: 2001 2021			
Lags: 1			
Null hypothesis:	Obs	F-statistic	Prob
FOODGRAINS does not Granger Cause HUNGER	20	0.64805	0.4319
HUNGER does not Granger Cause FOODGRAINS		0.23440	0.6345

Source Authors own calculation

does not directly affect agricultural output. Instead, weather conditions, technological advancements in farming, and government agricultural policy are more determinant factors of food grain production. In addition, it might be that there was such a lack of a causal link because food grain production decisions are largely determined by market conditions and the incentives associated with that area, rather than by the immediate hunger level.

7 Role of Chemical Fertilizers in Carbon Emission

Chemical fertilizers put us on a road to climate change in several ways: one is through nitrous oxide (N_2O) emissions, a powerful greenhouse gas. Fertilizer application emissions represent a large proportion of agricultural emissions in India. It has also been shown that local nitrogen cycles are altered by over-fertilization, leading to localized increases in temperature, impeding crop resilience and ecosystem stability [9]. Production and application of chemical fertilizers are energy-intensive processes that enormously contribute to carbon emissions. The gas released via the Haber–Bosch process, which is the most common method for synthesizing ammonia for nitrogen fertilizers, is usually in large amounts of CO_2. Nitrous oxide (N_2O), which is estimated to be about 300 times more potent than CO_2 as a greenhouse gas, is emitted whenever nitrogen fertilizers are applied.

CO_2 emission trend over the use of chemical fertilizers (NPK) shown in Fig. 3 is likely to be trending steadily or exponentially in terms of agricultural practices which is a key indicator for an increase in CO_2 emission over time because it is displaying a steady increase and also because it indicates it is a result of soar environmental concern relating to climate change. They may fluctuate and suggest either interventions or changes in policies or technologies that temporarily reduce emissions. First, we calculate the CO_2 equivalent emission of NPK fertilizer to show the trend.

To calculate CO_2-equivalent emissions for 1 lakh tons (0.1 million tons) of NPK fertilizer with a ratio of 4:2:1 (N: P_2O_5:K_2O):

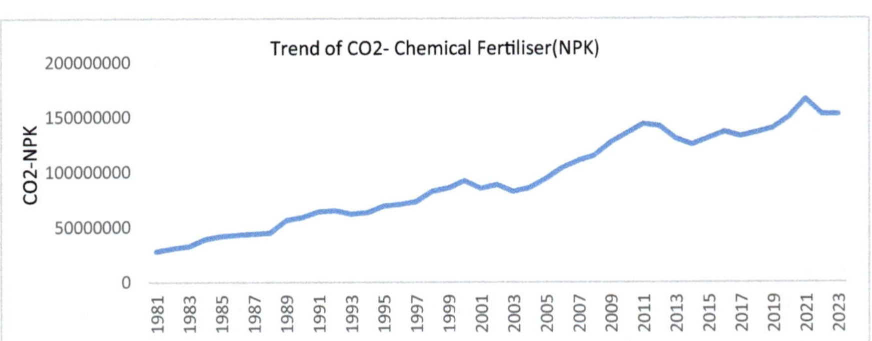

Fig. 3 CO_2-Chemical Fertilizer (NPK)

Calculation of NPK Content.
Total fertilizer used: 0.1 million tons (1 lakh tons).

(i) NPK composition: (4:2:1)
Nitrogen (N): 4/7 X 0.1 = 0.05714million tons N = 57,143 tons N.
Phosphorus (P_2O: 2/7 × 0.1 = 0.02857 million tons P_2O = 28,571 tons P_2O.
Potassium (K_2O): 1/7 × 0.1 = 0.01429 million tons K_2O = 14,286 tons K_2O.

(ii) Calculation of Production Emissions
Nitrogen: 57,143 tons N × 5 kg CO_2/kg = 8285,715 tons CO_2.
Phosphorus (P_2O_5): 28,571 tons P_2O_5 × 0.3 kg CO_2/kg = 8,571 tons CO_2.
Potassium (K_2O): 14,286 tons K_2O × 0.3 kg CO_2/kg = 4,286 tons CO_2.
Total production emissions = 285,715 + 8,571 + 4,286 = 298,572 tons CO_2.

(iii) Calculation Field Emissions
Nitrogen (from N_2O emissions).
= 57,143 tons N × 3.7 kg CO_2-eq/kg = 211,429 tons CO_2.

(iv) **Total Emissions** = Production emissions + Field emissions
= 298,572 + 211,429 = 510,001 tons CO_2-equivalent.

Using 1 lakh tons of NPK fertilizer with an NPK ratio of 4:2:1 will generate approximately 5,10,001 tons of CO_2-equivalent emissions.

8 NPK-Led Carbon Emission and Annual Temperature

The destabilization of ecosystems via local microclimate changes imposed through fertilizer-induced temperature variations impacts species composition and ecological balance. The excessive use of fertilizers in India has been associated with environmental instability in agricultural regions to the extent of disturbing the system of ecosystem health [10].

In Fig. 4, the graph comparing carbon emissions from NPK fertilizers and average temperatures seems to signal a possible correlation. Such a trend might occur for both the average temperature trend and the amount of carbon emissions as a result of greenhouse gas, and would be consistent with the idea that higher carbon emissions lead to global warming. The rise of both variables is likely a result of the effect of agricultural emissions on climate change.

Table 3 represents Pairwise Granger Causality Tests on average annual IMD temperature and NPK-led CO_2. A p-value of 0.1292 and an F statistic of 2.40264 suggest there is no significant causality from average IMD temperature to CO_2 NPK levels. The absence of this relationship could simply be because fertilizer use

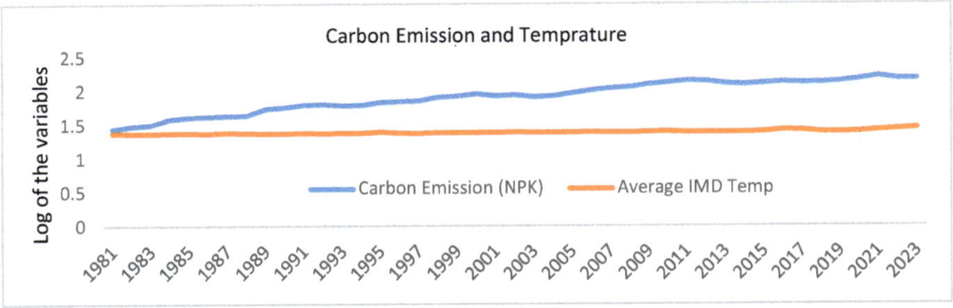

Fig. 4 Carbon emission and temperature

Table 3 Pairwise Granger Causality Tests on average annual IMD temperature and NPK-led CO_2

Pairwise Granger Causality Tests			
Sample: 1981 2024			
Lags: 1			
Null hypothesis	Obs	F-Statistic	Prob
AVG_IMD_TEMP does not Granger Cause CO_2_NPK	41	2.40264	0.1292
CO_2_NPK does not Granger Cause AVG_IMD_TEMP		4.31260	0.0445

Source Author's own calculation

and CO_2 emissions follow more closely with industrial and agricultural production processes as opposed to temperature changes. Temperature, however, might not have an immediate effect on agricultural practices so fertilizer use is not causal in the short run—that's what the Granger causality test does measure. Some adaptation to changing temperatures may be possible by the agricultural sector over longer times than this analysis covers. A significant relationship exists, with CO_2 Emissions and NPK Fertilizer Use affecting average IMD temperature with an F statistic of 4.31260 and p-value of 0.0445. That implies that human activities, such as those in agriculture and industry, are causing the warming of the climate. The NPK fertilizers increase greenhouse gas emissions, causing the global temperature to escalate. This statistical evidence demonstrates the role that agricultural inputs have in affecting the climate patterns, which were already linked to human-induced emissions and global warming.

9 Trade-Offs: Carbon Emission Versus Zero Hunger

Chemical fertilizers have significantly enhanced agricultural output since they provide essential nutrients to the crop. Nevertheless, their over-application can cause the soil to become acidic and sick, leaving little nutrient supply and straining the soil structure [11]. The soil microbial communities are also negatively affected by this acidification: the abundance of these beneficial microorganisms decreased whose presence is important for nutrient cycling and plant health [12]. However, as a consequence these changes cause soil degradation, meaning that such soil is less fertile and has poor agricultural productivity. Furthermore, when soils are over-fertilized, nitrogen can leach through the soil into waterways and contaminate soil to the point of eutrophication [11]. To counteract such adverse effects, organic amendments and agrochemical applications must be practiced sustainably and the results should be managed based on their nutrient requirements. Practices, such as applying these principles, can maintain soil health, assure long-term agricultural productivity, and help protect the environment quality.

Increasing fertilizer use is often part of the domain of food security, as people strive to raise average crop yields. This practice conflicts with climate action goals; however, as important as global greenhouse gas emissions, nitrogen-based fertilizer production and application make up about 5% of global greenhouse gas emissions [13]. This creates a huge challenge between the imperative to reduce emissions and the need for enhanced agricultural productivity for India. While the agricultural sector is fundamentally important to food security, it also plays a key role in the country's greenhouse gas emissions [14]. To align with 12 SDGs, focusing on responsible consumption and production, it is necessary to start transitioning to sustainable agricultural practices. Practices that not only enhance productivity but also mitigate environmental impacts and maintain long-lasting sustainability [15]. Such ecological changes restrict efforts to ensure stable food production, an essential objective of SDG-2, which aims to "End hunger, achieve food security and improved nutrition and promote sustainable agriculture" [16]. Studies globally have shown that unsustainable food production practices like poor fertilizer use promote environmental degradation and further climate change, threatening food security [17].

10 Short- and Long-Run Relationship

To investigate long-run relationships with NPK-led CO_2 emissions, NPK fertilizer use, food grains production, hunger, and annual average temperature of India, during 1981–2023 Johansen cointegration test was used. Five model configurations ranging from trend (linear, quadratic) to intercepts (none, linear, quadratic) were considered, and two test methods (Trace, Maximum Eigenvalue) were used to detect the number of cointegrating relations under each of them. The long-run equilibrium relationships among the variables were assessed to determine the rank of cointegration. Furthermore, information criteria (Log Likelihood, Akaike, and Schwarz) were used to guide model selection as well as to select the best model, knowing the degree of fit.

When a quadratic trend is assumed, results indicate a 1 cointegrating relationship, and both the Trace and Maximum Eigenvalue tests confirm this conclusion. As the information criteria validate further, the Akaike (82.51) and Schwarz (84.50) values of the quadratic trend model are the best ones. Since a quadratic trend, compared to a linear or no trend, provides a closer explanation of the data's long-term dynamics, this inclusion makes sense. Simpler models exhibit more than one cointegrating relation in the Trace test, but the quadratic trend simplifies these into one significant equation. Under this trend, the Maximum Eigenvalue test also implies a single cointegrating relation, in agreement with the finding that a single long-run relationship is most appropriate for this dataset (Table 4).

We reject the null hypothesis of no cointegration at the 0.05 significance level and also accept the existence of one cointegrating equation from the Trace test with a test statistic of 79.62, which surpasses the critical value, i.e., 55.25. The Maximum Eigenvalue test illustrates the same result as this statistic of 53.03 for no cointegration is greater than the critical value of 30.82. As the p-values for the first tests (no cointegration) of the null hypotheses are very low (<0.001 corresponding to 0.0001 and 0.0000 respectively for the Trace and Maximum Eigenvalue tests), these results indicate strong statistical significance and suggest the possibility of the existence of one long-run relationship between the variables. However, the

Table 4 Johansen cointegration test results

Sample (adjusted): 2003 2023				
Included observations: 21 after adjustments				
Trend assumption: Quadratic deterministic trend				
Series: CO_2_NPK FOODGRAINS HUNGER IMD_TEMP				
Lags interval (in first differences): 1 to 1				
Unrestricted Cointegration Rank Test (Trace)				
Hypothesized no. of CE(s)	Eigenvalue	Trace statistic	0.05 critical value	Prob.**
None *	0.919950	79.62027	55.24578	0.0001
At most 1	0.593140	26.59313	35.01090	0.2962
At most 2	0.300473	7.708103	18.39771	0.7126
At most 3	0.009655	0.203733	3.841466	0.6517
Trace test indicates 1 cointegrating eqn(s) at the 0.05 level				
*Denotes rejection of the hypothesis at the 0.05 level				
**MacKinnon-Haug-Michelis (1999) p-values				
Unrestricted Cointegration Rank Test (Maximum Eigenvalue)				
Hypothesized no. of CE(s)	Eigenvalue	Max-Eigen statistic	0.05 critical value	Prob.**
None *	0.919950	53.02713	30.81507	0.0000
At most 1	0.593140	18.88503	24.25202	0.2189
At most 2	0.300473	7.504370	17.14769	0.6576
At most 3	0.009655	0.203733	3.841466	0.6517
Max-eigenvalue test indicates 1 cointegrating eqn(s) at the 0.05 level				
*Denotes rejection of the hypothesis at the 0.05 level				
**MacKinnon-Haug-Michelis (1999) p-values				

Source Authors own calculation.

following hypotheses (up to 1, 2, and 3 cointegrating equations) cannot reject null and hence there are no further cointegrating equations. Because of this, the model appears to detect a single cointegrating relationship characterized by a strong long-run equilibrium among the variables and short-run dynamics exhibit no further cointegration.

The short- and long-run relationships of variables are studied using the Vector Error Correction Model (VECM). The model has lagged values of the variables included to capture their time dependency; and it is assumed that the variables are cointegrated, i.e., they have a long-term equilibrium relationship with each other. The CointEquation (CointEq1) describes long-run dynamics while the Error Correction Term (ECT) quantifies the rate at which short-run disequilibria are corrected to place them back on long-run equilibrium. Also included in the VECM are the first and second lags of the variables to examine the immediate and delayed effects. Finally, the model performance is assessed with various statistics such as R-squared, Akaike information criterion (AIC), Schwarz criterion (SC), and the F statistic, to test for goodness of fit and significance of relationships. The p-values of the error correction terms (CointEq1) and coefficients imply the power of the variables to adjust to long-term equilibrium (Table 5).

Results for the model show mixed instantaneous and instantaneous relationships between the variables. The coefficient of t-statistic = −9.19 in the cointegrating equation of the above model shows that in the long run, food grain production has a strong negative impact on the NPK-led CO_2 emissions. This implies that higher food grain production is linked with lower CO_2 emissions, as perhaps better farming techniques or better efficiency. But, the temperature in its favor has a positive association with CO_2 emissions, with a t-statistic of 2.89, meaning that the increase in emissions can be associated with the changes in agricultural practices or more use of energy to adapt to climate changes. Statistically, we find that the error correction term of average annual temperature is significant (t stat = −4.04), while the correction of NPK-led CO_2 emissions and food grain production is not significant, implying that changes in CO_2 emissions and food production do not adjust directly to the long-run equilibrium. The strong effect of lagged food grain production and lagged temperature (t statistics = –3.30 and 3.12, respectively) on their own change suggesting that past values of these variables strongly influence their future dynamics.

Model fit statistics (R-squared and AIC) suggest that food grain production and temperature explain relatively more variation in the dependent variables than CO_2 emissions (food grains: R^2 = 0.537; temperature: R^2 = 0.434; CO_2 emissions: R^2 = 0.205). This lower R^2 for CO_2 emissions indicates that other factors (not included in the model) may be affecting the dynamics of CO_2 emissions. Furthermore, the F statistics indicate that food grain production and temperature each have statistically significant explanatory power for their respective dependent variables; for CO_2 emissions, however, there is weaker explanatory power. Hence, Akaike and Schwarz criteria suggest that the food grain production equation (AIC = 12.38) fits the data better than other equations, that is, the proposed model is likely to predict better food grain production. The residual sum of squares for CO_2

Table 5 Vector error correction estimates

Sample (adjusted): 1984 2023			
Included observations: 40 after adjustments			
Standard errors in () and t-statistics in []			
Cointegrating Eq.	CointEq1		
CO_2_NPK(−1)	1.000000		
FOODGRAINS(−1)	−125,530.7		
	(13,659.3)		
	[−9.19013]		
IMD_TEMP(−1)	42,010,603		
	(1.5E + 07)		
	[2.88775]		
C	−8.71E + 08		
Error correction	D(CO_2_NPK)	D(FOODGRAINS)	D(IMD_TEMP)
CointEq1	−0.070512	−1.97E-07	−2.22E-08
	(0.07566)	(1.3E-06)	(5.5E-09)
	[−0.93201]	[−0.14794]	[−4.03944]
D(CO_2_NPK(−1))	0.199603	−9.53E-07	1.22E-08
	(0.18162)	(3.2E-06)	(1.3E-08)
	[1.09899]	[−0.29835]	[0.92475]
D(CO_2_NPK(−2))	0.129034	1.35E-05	2.18E-08
	(0.21777)	(3.8E-06)	(1.6E-08)
	[0.59252]	[3.52346]	[1.37822]
D(FOODGRAINS(−1))	−18,678.57	−0.623073	−0.001934
	(10,739.4)	(0.18878)	(0.00078)
	[−1.73925]	[−3.30054]	[−2.47723]
D(FOODGRAINS(−2))	−16,727.85	−0.297725	−0.000664
	(9693.52)	(0.17039)	(0.00070)
	[−1.72567]	[−1.74727]	[−0.94214]
D(IMD_TEMP(−1))	−2,896,215	−51.15292	0.537142
	(2,367,943)	(41.6240)	(0.17211)
	[−1.22309]	[−1.22893]	[3.12098]
D(IMD_TEMP(−2))	−648,551.6	13.61676	0.114587
	(2,661,327)	(46.7812)	(0.19343)
	[−0.24369]	[0.29107]	[0.59240]
C	3,788,530	51.36885	0.061964
	(1,332,026)	(23.4145)	(0.09681)
	[2.84419]	[2.19389]	[0.64003]

(continued)

Table 5 (continued)

R-squared	0.205253	0.537437	0.433642
Adj. R-squared	0.031402	0.436251	0.309752
Sum sq. resids	1.21E + 15	372,861.3	6.374626
S.E. equation	6,140,808	107.9440	0.446326
F-statistic	1.180626	5.311390	3.500199
Log likelihood	−677.5133	−239.5592	−20.02646
Akaike AIC	34.27567	12.37796	1.401323
Schwarz SC	34.61344	12.71574	1.739099
Mean dependent	2,985,036	50.04750	0.100750
S.D. dependent	6,239,557	143.7659	0.537217
Determinant resid covariance (dof adj.)	8.08E + 16		
Determinant resid covariance	4.14E + 16		
Log likelihood	−935.5070		
Akaike information criterion	48.12535		
Schwarz criterion	49.26534		
Number of coefficients	27		

Source Author's own calculation

emissions is especially high (1.21E + 15) indicating that the model does not fully explain variability in CO_2 emissions, because the emissions relate to factors that are more complex than just NPK fertilizer use. This is because CO_2 emission due to NPK is less than ten percent of the total CO_2 emission from the main sources.

11 Conclusion

The result revealed the use of chemical fertilizer in Indian agriculture, in the context of the dual-edge implications of this important input in agriculture. However, NPK fertilizers on the one hand have contributed significantly in achieving self-sufficiency in food grain production, reducing hunger and achievement of Sustainable Development Goal 2. On the other side, however, overuse of their application has led to heightened environmental problems like carbon emissions and temperature fluctuation, which are serious hurdles to Sustainable Development Goal 13.

The research advocates ways to deal with these challenges while transitioning toward sustainable agricultural practices. Some of these include ensuring optimal fertilizer application, finding organic alternatives, and ensuring that we find equitable ways through which we distribute fertilizers to smallholder farmers. Food production objectives are, of course, important to policymakers, but they should be driven by an integrated approach that takes into account environmental sustainability goals. The solutions to reconcile ecological preservation with productivity require interdisciplinary research; targeted interventions are also needed.

The Indian government needs to put in place policies supporting Integrated Nutrient Management (INM) with financial benefits for organic and bio-based fertilizers to minimize chemical fertilizer use. The government should restructure fertilizer subsidies to benefit effective products while providing technology incentives to enhance precise agricultural techniques for better fertilization practices. The combination of improved soil health surveillance systems with time-sensitive farming services and wider agricultural advisory services will enable farmers to decrease their use of chemical products. R&D investment for sustainable fertilizer types must happen simultaneously with efforts to boost water efficiency standards by implementing fertilization accompanied by drip irrigation systems. Government intervention should focus on two aspects: safeguarding environmental outcomes from fertilizer overuse and implementing practices to preserve soil health together with water systems and eco-systems. Public campaigns aimed at farmers should demonstrate both environmental and economic advantages of sustainable fertilizer practices for creating sustainable production while safeguarding environmental health and food security. This study reinforces a more comprehensive assessment of trade-offs set into motion through chemical fertilizer application by taking into account the multifaceted impacts. The call is to re-imagine agricultural systems in India, one that is resilient, sustainable, and equitable in response to growing global challenges.

References

1. Food and Agriculture Organization of the United Nations (2021). *The state of food and agriculture 2021: Making agrifood systems more resilient to shocks and stresses*. FAO.
2. Lassaletta, L., Billen, G., Garnier, J., & Bouwman, L. (2014). How changes in nitrogen inputs to terrestrial ecosystems have reshaped nitrogen fluxes to the ocean. *Biogeochemistry, 118*(1–3), 225–249.
3. Zhang, X., Davidson, E. A., Mauzerall, D. L., Searchinger, T. D., Dumas, P., & Shen, Y. (2015). Managing nitrogen for sustainable development. *Nature, 528*(7580), 51–59.
4. Gupta, R. K., Sharma, N., & Singh, P. (2017). Climate change impacts on Indian agriculture and adaptation strategies. *Indian Journal of Agricultural Economics, 72*(3), 319–332.
5. United Nations (2015). *Transforming our world: The 2030 agenda for sustainable development*. United Nations.
6. World Bank (2020). Employment in agriculture (% of total employment)—India, (accessed 2020). https://data.worldbank.org
7. Bhan, S. (2021). The impact of fertilizer use on agricultural productivity in India: Trends and challenges. *Agricultural Economics Research Review, 34*(1), 101–113.
8. Paul, K., Murari, K., & Sasidharan, S. (2023). Sustainability transition for Indian agriculture. *Science and Reports, 13*, 7290.
9. Sahil, F. M., Narayanan, M., & Ilampooranan, I. (2023). Nitrous oxide emissions dynamics in Indian agricultural landscapes: Revised emission rates and new insights. *Agriculture Ecosystems and Environment, 356*, 108639

10. Rana, V., Singh, S., & Gupta, S. (2023). Impact of fertilizer-induced environmental changes on Indian agriculture. *Journal of Environmental Sustainability, 12*(3), 456–470.
11. Zhou, M., Brandt, P., Pelster, D., Rufino, M. C., Robinson, T., & Butterbach-Bahl, K. (2020). Regional nitrogen budget of the Lake Victoria Basin, East Africa: Syntheses, insights, and future directions. *Agriculture, Ecosystems and Environment, 295*, Article 106890.
12. Guo, J. H., Liu, X. J., Zhang, Y., Shen, J. L., Han, W. X., & Zhang, W. F., et al. (2010). Significant acidification in major Chinese croplands. *Science, 327*(5968), 1008.
13. Menegat, S., Ledo, A., & Tirado, R. (2022). Greenhouse gas emissions from global production and use of nitrogen synthetic fertilisers in agriculture. *Scientific Reports, 12*
14. Sharma, A. K. (2023). Sustainable pathways toward reimagining India's agricultural systems. *Communications Earth and Environment, 4*, Article 902
15. Philip, B., & Suresh, G. (2024). India's agriculture sector's journey towards sustainable development goals: assessing governmental interventions and sectoral outcomes. *Universal Journal of Agricultural Research, 12*(2), 332–341.
16. Department of Economic and Social Affairs, Goal 2, accessed 2015. https://sdgs.un.org/goals/goal2
17. Smith, J. A., & Turner, K. (2022). Global food systems and the environmental consequences of fertilizer use. *Environmental Science and Policy, 128*, 89–99.

The Impact of CSR Perception on Employee Performance: Exploring the Mediating Role of Person-Organization Fit in Public Sector Organizations

Abhijit Mohanty, Amita Panda, Surama Mohanty, and Sonam Subhadarshini

Abstract

This research explores the connections between employees' views on Corporate Social Responsibility (CSR), the alignment between individual's and the organization, and employee performance in workplace-environments. It examines how perception toward CSR influences employee performance, with person-organization fit serving as a mediating variable. The increasing focus on CSR as a strategic tool for organizational success necessitates understanding its impact on employee-level outcomes. Data were gathered from employees through structured questionnaires designed on three key constructs: perceptions of CSR, person-organization fit, and employee performance. Responses were collected from 341 employees working in public sector units (PSUs) using purposive sampling. Hypothesis testing and model evaluation were performed utilizing Partial Least Squares Structural Equation Modeling (PLS-SEM). The findings demonstrate that perceptions of corporate social responsibility (CSR) significantly affect person-organization fit, which subsequently enhances employee performance, underscoring its mediating role in the CSR-performance relationship. These results emphasize the importance of CSR initiatives in promoting

A. Mohanty (✉)
Department of Business Administration, Utkal University, Vanivihar, Bhubaneswar, Odisha, India
e-mail: abhijitmohanty.1986@gmail.com

A. Panda
Institute of Management Bhubaneswar (IMB), Bhubaneswar, Odisha, India

S. Mohanty
Bhubaneswar, Odisha, India

S. Subhadarshini
TACT, Chandrasekharpur, Bhubaneswar, Odisha, India

S. Patnaik et al. (eds.), *People, Planet and Profit: Advancing Sustainable Business Practices*, Springer Proceedings in Geography and Sustainability,
https://doi.org/10.1007/978-3-032-11296-5_8

alignment between employees and organizational values, thereby improving performance outcomes. By exploring the mediating role of person-organization fit, this study adds to the growing body of research on CSR and its effects on employee performance.

Keywords

Corporate social responsibility (CSR) • Person-organization fit • Employee performance • PLS-SEM analysis • PSUs

1 Introduction

Corporate Social Responsibility (CSR) has a longstanding history in India, evolving significantly from its informal inception to its current formalized and mandatory structure. During the pre-independence era, CSR existed in an unorganized format, yet Indian companies demonstrated awareness of their social responsibilities by contributing to education, healthcare, and other social services. These initiatives were largely inspired by Mahatma Gandhi, who advocated for the fulfilment of societal responsibilities by businesses. Over time, the concept of CSR in India has undergone substantial transformation, becoming more structured and integrated into corporate governance. What were once isolated philanthropic activities have now become an essential element of corporate strategy, as reflected in the CSR pyramid's highest level of responsibility.

The contributions of corporate entities such as TATA and Birla highlight the historical importance of CSR in India. These companies, recognized as pioneers of CSR in the country, have consistently participated in societal improvement since a period when the concept was minimally acknowledged. In contemporary India, corporate social responsibility has progressed beyond conventional philanthropy to include a variety of initiatives aimed at sustainable development and societal well-being. The Government of India's mandate for companies to allocate a minimum of 2% of their average net profits from the preceding three years to CSR activities has further institutionalized this practice. Companies must now document their CSR initiatives in specified formats within annual reports, thereby ensuring accountability and transparency.

The scope of CSR encompasses the recognition of workers' fundamental human rights, the promotion of work-life balance, and the implementation of policies aimed at enhancing employee well-being [1]. Empirical evidence indicates that CSR activities have a positive effect on organizational-level outcomes, such as financial performance, market share, stock prices, and sustained competitive advantage [2]. Early research primarily concentrated on macro-level outcomes, particularly financial performance, while offering insufficient consideration of the micro-foundations of CSR.

Recent studies have focused on the effects of corporate social responsibility (CSR) on individual outcomes, especially regarding employee attitudes and behaviors. Employee perceptions of corporate social responsibility significantly influence various factors, including organizational commitment, turnover intentions, burnout, organizational citizenship behaviors, job satisfaction, task performance, creativity, work engagement, organizational identification, and voicing behavior [3–5]. Studies have investigated the influence of CSR on employee engagement and performance, asserting that CSR exerts a more significant impact on employee engagement [6], organizational citizenship behavior [7], pro-environmental behaviors [8], and job satisfaction [9]. Conversely, inconsistent CSR strategies may adversely affect employee morale and trust [10]. Although research on CSR is expanding, its impact on individual outcomes, especially employee performance, is still inadequately examined.

A significant gap remains in comprehending the impact of CSR initiatives on employee performance via mediating mechanisms such as Person-Organization Fit (P-O Fit). P-O Fit, denoting the congruence between employee values and organizational practices, is essential for cultivating a sense of belonging and coherence. Although prior research has recognized this association, its empirical confirmation in Indian public sector units (PSUs) remains scarce, limiting contextual understanding. This limited examination of the mediating function of P-O Fit restricts understanding of how CSR could be strategically used in PSUs to enhance employee performance and organizational efficacy.

This study addresses this gap by examining the impact of CSR activities on employee performance within the unexplored context of PSUs in India, highlighting the mediating role of P-O Fit. This study examines the alignment of CSR initiatives with employee values and expectations, aiming to offer practical insights for developing CSR strategies that enhance motivation and performance within the public sector workforce. The study enhances theoretical understanding of CSR by applying the Stimulus-Organism-Response (S-O-R) framework more systematically to analyze individual-level outcomes. In this model, CSR initiatives act as the stimulus (S), affecting employee perceptions and alignment (organism, O) via P-O Fit, which in turn enhances employee performance as the response (R). There are many evidences where this model has been used to assess the employee engagement. Hasan et al. [11] has utilized it to study the impact of organizational culture on employee engagement and employee performance, whereas the theory was used to explore the impact of perceived CSR on psychological meaningfulness and employee engagement [12]. This study utilizes the S-O-R framework to systematically examine the psychological and behavioral mechanisms that underlie the impacts of CSR, thereby expanding the literature within the public sector context. This study addresses a significant gap in CSR research and emphasizes the role of employee alignment as a mediating mechanism. The study establishes a solid basis for improving the strategic execution of CSR initiatives, thereby promoting both individual and organizational effectiveness within public sector organizations.

2 Literature Review

2.1 Employee Perception of CSR and Person-Organization Fit

The concept of Person-Organization Fit (P-O fit) is well-established in organizational behavior literature as a crucial factor influencing employees' workplace attitudes and behaviors. Research indicates that employees who recognize alignment between their personal values and organizational values exhibit more positive attitudes and enhanced efficiency in their roles [13]. Employee perceptions of person-organization fit are notably influenced by corporate social responsibility initiatives, which frequently enhance the sense of alignment between employees and their organizations [14].

Corporate Social Responsibility (CSR) initiatives enhance job fit among employees by aligning organizational values with their expectations, resulting in increased work commitment [15]. Environmental CSR policies and practices enhance perceived P-O fit by clarifying organizational values, goals, and attributes, which fosters a shared sense of purpose [16, 17]. Employees who perceive alignment with their organization are more likely to experience heightened belongingness, thereby reinforcing the organization as a significant referent for their self-definition [18].

Employees with a strong identification with their organization are more inclined to internalize its ideals, beliefs, and objectives, thereby motivating them to engage in behaviors that align with organizational values [19, 20]. This alignment enhances employee attitudes and positively influences customer-oriented behaviors. Reference [21] demonstrated that in China's hotel industry, elevated perceptions of internal and external CSR, along with a stronger P-O fit, contributed to enhanced work engagement and subsequently improved customer orientation. The findings highlight the significant role of CSR in promoting P-O fit and its wider implications for organizational success.

H1: Employee Perceptions of Corporate Social Responsibility Positively Influence Their Sense of Person-Organization Fit

2.2 Person-Organization Fit and Employee Performance

Person-Organization Fit (P-O fit) denotes the degree of compatibility between an individual and their organization, encompassing the alignment of values, objectives, and organizational culture. This alignment is essential for enhancing employee performance and overall organizational results. Jufrizen et al. [22] emphasized that P-O fit denotes the alignment between the fundamental values of employees and organizations, asserting that this congruence enhances mutual attraction and facilitates organizational harmony. Lary et al. [23] introduced a conceptual framework highlighting the essential role of person-organization fit in influencing employee behavior and performance, thereby reinforcing its importance in organizational settings. Rajper et al. [24] indicated that organizational citizenship behavior serves as an intervening variable in the relationship between person-job fit, P-O fit, and employee performance, illustrating the complex influence of fit on employee outcomes, similarly, a positive correlation among P-O fit, job satisfaction, and employee performance, indicating that job satisfaction mediates the relationship between fit and performance outcomes [25]. It was also demonstrated that person-job fit significantly influences employee performance, with greater alignment resulting in improved performance outcomes [26]. A lack of alignment leads to reduced employee performance, highlighting the importance of ensuring congruence between employees and their organizations.

H2: Person-Organization Fit Positively Influences Employee Performance

2.3 Employee Perception of CSR and Employee Performance

The perception of corporate social responsibility (CSR) among employees has garnered increasing attention in organizational research, with various studies investigating its impact on key outcomes including employee performance, innovative behavior, ethical conduct, and organizational citizenship behaviors. Kowalczyk et al. [26] highlighted the influence of stakeholder pressure, cultural factors, and CSR practices on brand performance, corporate reputation, and employee identification in various cultural contexts. The beneficial effects of corporate social responsibility (CSR) on employee engagement and commitment have been seen within the banking sector, and it was found very important for fostering internal organizational outcomes [27]. A significant positive relationship has been established between employees' perceptions of CSR and their innovative behavior, with organizational identification and employee well-being playing crucial mediating roles [28]. The role of CSR in shaping employees' ethical behavior has also been recognized, with organizational commitment serving as a mediating factor in this relationship [29]. The importance of internal CSR practices in encouraging intrapreneurial behavior and enhancing social performance, thereby contributing to organizational success, has been emphasized in prior studies [30, 31]. Furthermore, CSR has been identified as a catalyst for fostering creativity among employees, while also ensuring alignment between employee awareness, experiences, and organizational objectives, ultimately leading to improved performance [32, 33]. At last, a positive and significant impact of CSR on employee performance was observed among the employees of public sector banks from Indonesia [34]. Although substantial evidence connects CSR to employee performance, a significant gap exists in comprehending the mechanisms by which CSR perceptions affect performance in particular contexts, such as the public sector in India. The mediating role of Person-Organization Fit (P-O Fit) is underexplored, indicating a need for further investigation to address this gap.

H3: Employee Perceptions of CSR Positively Influence Their Performance

2.4 Person-Organizational Fit as a Mediator

A multitude of studies have investigated the mediating role of person-organization fit (P-O fit) in influencing various organizational and individual outcomes in different contexts. Siyal et al. [35] investigated the mediating role of person-organization fit in the relationship between high-performance human resource practices and negative employee outcomes in public sector universities, highlighting its significance in mitigating adverse effects. The role of person-organization (P-O) fit and employee trust as mediators is well established, particularly in connecting employees' perceptions of corporate social responsibility (CSR) to increased organizational loyalty [36]. The mediating roles of job satisfaction and person-organization fit are acknowledged in elucidating the connection between public service motivation and job performance, encompassing both task and contextual performance dimensions among public servants in Ho Chi Minh City, Vietnam [37]. Research has confirmed the beneficial effect of employees' perceptions of environmental corporate social responsibility (CSR) practices on pro-environmental behaviors (PEBs) in organizations, with person-organization fit and organizational identification serving as sequential mediators in this relationship [38]. Therefore, we propose

H4: Person-Organization Fit Mediates the Relationship Between Employees' Perceptions of Corporate Social Responsibility and Their Job Performance

3 Research Methodology

This study employs a quantitative research design to examine how employees' perceptions of Corporate Social Responsibility (CSR) affect employee performance, with Person-Organization Fit (P-O Fit) acting as a mediating variable (refer Fig. 1). Quantitative methods effectively analyze relationships between variables and provide statistical evidence, which guarantees the reliability and generalizability of findings [39]. Data were collected using a structured survey questionnaire distributed to employees in chosen public sector units (PSUs) in India. The online data collection utilized Google Forms, a platform recognized for its efficiency and user-friendliness in survey distribution, which facilitates a wider reach and greater convenience [40]. The survey link was disseminated through social media channels like LinkedIn and WhatsApp to strategically reach respondents and promote prompt engagement. Online surveys facilitate quick data collection from a dispersed workforce, overcoming logistical barriers associated with face-to-face data collection in large or multi-location organizations. Moreover, e-questionnaires allow respondents the flexibility to complete the survey at their convenience, which has been shown to improve response rates and data accuracy [41]. The method

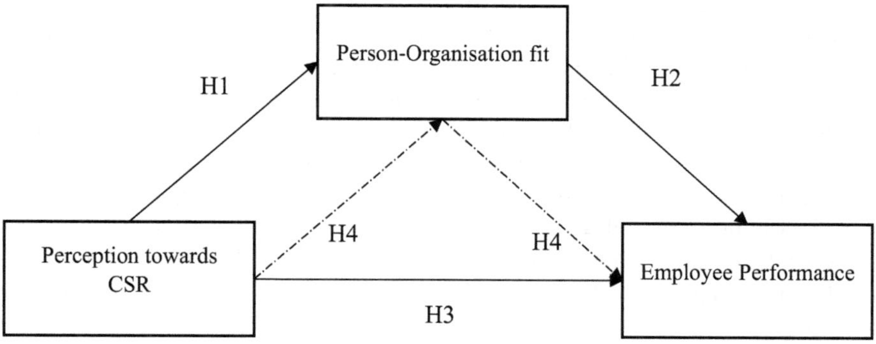

Fig. 1 Conceptual model

is also aligned with previous CSR-related studies that have successfully used e-surveys to collect data on employee attitudes, perceptions, and behaviors [41].

The sample size for this study was established by employing Daniel Soper's approach for determining the minimum sample size necessary for Structural Equation Modeling (SEM). With an expected effect size of 0.3, a statistical power of 0.8, three latent variables, 20 observed variables, and a significance level of 0.05, the minimum sample size required was determined to be 323. To improve the reliability and generalizability of the results, the study utilized a slightly larger sample size of 341 respondents. Employing a larger sample size guarantees a more rigorous statistical analysis and reduces the likelihood of sampling errors, as indicated in previous studies [42]. A purposive sampling method was employed to specifically focus on employees who are actively engaged in or well-informed about their organization's CSR initiatives. This approach guarantees the participation of individuals with pertinent experience and understanding, thereby improving the credibility of the results [43]. While this sampling method increases content validity, it may limit generalizability [44].

The survey questionnaire consisted of two distinct sections. The initial segment collected information on demographics and job profiles, encompassing factors like age, gender, education, years of experience, and organizational role. The second section assessed three critical variables—CSR perception, P-O Fit, and employee performance—utilizing validated scales derived from previous research. The measurement of CSR utilized eight items sourced from Turker [45] and Lee et al. [46] and used them to study the influence of hotel employees' perception of CSR on organizational commitment [47]. A sample item is "The company demonstrates commitment to ethical business practices." The evaluation of P-O fit was conducted utilizing six items created by Afsar and Badir [48] and was used to assess the impact of CSR on employee loyalty [49]. The items include "I feel that my skills and abilities are a good match for the needs of this organization." Finally, the evaluation of employee performance was conducted using items modified from Podsakoff et al. [50], and it was used by Silva et al. [51] to evaluate the influence of CSR on employee performance in construction industry. A sample item is “I met the established objectives and goals because what was expected from me was clearly explained to me.” These scales were modified to fit the context of PSUs while preserving their original psychometric properties [52]. All items were assessed using a 1–5 Likert scale, where 1 represents strongly disagree and 5 indicates strongly agree. The analysis employed Partial Least Squares Structural Equation Modeling (PLS-SEM), a robust method particularly effective for intricate models that encompass various constructs and accommodate small to medium sample sizes. PLS-SEM facilitates the examination of both direct and indirect relationships among variables, rendering it especially effective for exploring the mediating role of P-O Fit in the connection between CSR and employee performance [42].

To address potential common method variance (CMV), the Harman single-factor test was conducted. The findings indicated that no individual factor was responsible for the majority of the variance, suggesting that common method variance is not a significant concern in this study [53]. The SPSS results indicate that the single factor accounted for merely 36% of the total variance, falling significantly short of the 50% threshold. Consequently, this study shows no indication of bias.

4 Data Analysis

Descriptive study was employed to analyze profiles of the respondents, providing a detailed overview of the sample characteristics. Subsequently, PLS-SEM was utilized as the primary analytical technique to evaluate the relationships among the variables.

4.1 Demographic and Professional Profile

The demographic profile of the 341 respondents given in Table 1 offers a valuable snapshot of the workforce characteristics within the selected public sector units (PSUs). The sample comprised predominantly male participants (69.7%), while females accounted for 30.3%.

The age distribution indicates that maximum 40.6% of respondents were under 30 years of age. Regarding educational status, 52.3% of the sample were graduates, while 28.3% possessed postgraduate or higher qualifications. The distribution of job departments comprised production/technical (32.2%), finance (28.2%), HR (24.4%), and marketing/sales (15.4%). 77 percent of respondents occupied non-executive roles, whereas 23 percent held managerial positions. In terms of work experience, maximum 30.9 percent had 10–15 years of service, followed by 28.7 percent who had an experience of 5–10 years. This distribution ensures the inclusion of insights from both long-tenured and early-career employees, which is crucial for assessing alignment with organizational values and performance-related outcomes.

Table 1 Demographic and professional profile

Variable	Category	Frequency	Percent
Gender	Male	238	69.7
	Female	103	30.3
Age	Less than 30	138	40.6
	30–40 years	98	28.7
	40–50 years	69	20.1
	50 and above	36	10.7
Education	Under 10th	16	4.7
	Intermediate	50	14.8
	Graduate	178	52.3
	PG or above	97	28.3
Job department	Production/Technical	110	32.2
	Finance	97	28.3
	HR	83	24.2
	Marketing/Sales	53	15.4
Job position	Executive/Managerial	78	23
	Non-executive/non-managerial	263	77
Work experience	Less than 5 year	66	19.5
	5–10 years	98	28.7
	10–15 years	105	30.9
	Above 15 years	71	20.9
Total		341	100

4.2 Assessment of Measurement Model

In PLS-SEM, evaluating the measurement model requires an examination of the reliability and validity of the constructs to ensure that the indicators accurately measure the latent variables. The reliability of the indicators was confirmed by analyzing the outer loadings, all of which surpassed the suggested threshold of 0.7, with values ranging from 0.883 to 0.962 (see Table 2). This indicates that the indicators serve as robust and dependable measures of their respective constructs [42]. The high outer loadings further indicate that each item makes a significant contribution to its corresponding construct.

Reliability of internal consistency was evaluated through Cronbach's alpha and composite reliability measures (see Table 2). Cronbach's alpha values were 0.982 for "Perception toward CSR," 0.97 for "Employee Performance," and 0.963 for "Person-Organization Fit," all of which surpass the acceptable threshold of 0.7. Composite reliability values were high, recorded at 0.985 for "Perception toward CSR," 0.976 for "Employee Performance," and 0.97 for "Person-Organization Fit,"

Table 2 Assessment of measurement model

Constructs	Item	Loadings	Alpha	C.R	AVE
Perception toward CSR	CSRQ1	0.950	0.982	0.985	0.891
	CSRQ2	0.930			
	CSRQ3	0.942			
	CSRQ4	0.933			
	CSRQ5	0.945			
	CSRQ6	0.962			
	CSRQ7	0.942			
	CSRQ8	0.946			
Employee performance	EPQ1	0.913	0.97	0.976	0.87
	EPQ2	0.939			
	EPQ3	0.944			
	EPQ4	0.956			
	EPQ5	0.936			
	EPQ6	0.910			
Person-organization fit	POQ1	0.883	0.963	0.97	0.846
	POQ2	0.884			
	POQ3	0.937			
	POQ4	0.944			
	POQ5	0.929			
	POQ6	0.938			

thereby confirming the internal consistency of the constructs. The results validate that the items within each construct consistently assess the same fundamental concept. Convergent validity was evaluated using the Average Variance Extracted (AVE). The AVE values for "Perception toward CSR" (0.891), "Employee Performance" (0.87), and "Person-Organization Fit" (0.846) surpassed the minimum recommended threshold of 0.5 [54]. This suggests that the constructs explain a substantial portion of the variance in their indicators, providing strong evidence of convergent validity.

The Fornell and Larcker criterion given in Table 3 was applied to evaluate discriminant validity by comparing the square root of the Average Variance Extracted (AVE) for each construct with the correlation values between constructs [54]. The analysis revealed that all constructs had square root AVE values higher than their corresponding correlations with other constructs. These findings confirm that each construct is distinct and captures a unique dimension of the conceptual model, thereby establishing discriminant validity.

Table 3 Discriminant validity (Fornell and Larcker criterion)

Constructs	Employee performance	Perception toward CSR	Person-organization fit
Employee performance	0.933		
Perception toward CSR	0.041	0.944	
Person-organization Fit	–0.525	0.132	0.92

4.3 Assessment of Structural Model

The evaluation of the structural model in Partial Least Squares Structural Equation Modeling (PLS-SEM) as shown in Fig. 2 and Table 4 is essential for assessing the relationships among latent variables and testing the proposed hypotheses [42]. This process entails evaluating the model's predictive capability and path coefficients to confirm that the theoretical relationships are substantiated by the data. Key assessments involve evaluating the R^2 values to quantify the variance explained by the independent variables. Path coefficients were analyzed to assess the strength and significance of the relationships between constructs in the model, along with effect sizes (f^2) to evaluate the relative impact of predictors on the dependent variables [55]. The predictive relevance (Q^2) of the model was evaluated through the Stone-Geisser test, which assesses the model's ability to accurately predict data points [42]. These evaluations offer a thorough understanding of the structural relationships and the overall robustness of the model, establishing a basis for significant conclusions and actionable insights.

The assessment of structural models focuses on the relationships among constructs and their capacity for explanation. The R^2 values indicate that the model

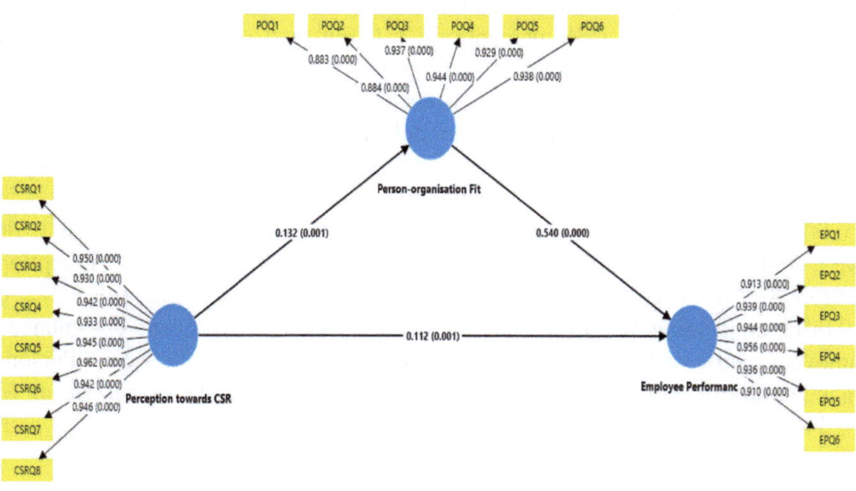

Fig. 2 Structural model with factor loadings, beta coefficients

Table 4 Path analysis and hypotheses testing

Path	Beta	T stats	Sign	Results
Perception toward CSR -> Employee performance	0.112	3.300	0.001	H1 supported
Perception toward CSR -> Person-organization fit	0.132	3.272	0.001	H2 supported
Person-organization Fit ->Employee performance mediation effect	0.540	17.735	0.000	H3 supported
Perception toward CSR ->Person-organization fit ->Employee performance	0.171	3.055	0.002	H4 supported

explains 28.8% of the variance in "Employee Performance" and 11.7% of the variance in "Person-Organization Fit," where a R2 value above 0.10 is considered acceptable in social sciences [56]. Therefore, the R-square values indicate that the predictors significantly contribute to the dependent variables [42]. The f^2 effect sizes offer insights into the relative contributions of each predictor to the dependent variables. The f2 values assess the impact of each predictor construct on the dependent construct, with values of 0.02, 0.15, and 0.35 indicating small, medium, and large effects, respectively [57] and the size of the effect with a value less than 0.02 indicates that the variable has no effect [42]. The perception of CSR exhibits a small influence on both employee performance ($f^2 = 0.047$) and person-organization fit ($f^2 = 0.088$) exhibits a medium effect. In contrast, person-organization fit demonstrates a large impact on employee performance ($f^2 = 0.403$), highlighting its essential mediating function. The Q^2 predictive relevance values are 0.019 for "Employee Performance" and 0.044 for "Person-Organization Fit," suggesting a small yet significant predictive relevance of the model [55]. The results validate the structural model's adequacy in explaining and predicting the relationships among CSR perceptions, P-O Fit, and employee performance, highlighting the significant mediating role of P-O Fit.

The structural path analysis results offer a detailed understanding of the relationships among the variables in the model and confirm the acceptance of the proposed hypotheses. The direct effect of "Perception toward CSR" on "Employee Performance" was found to be statistically significant ($\beta = 0.112$, $t = 3.3$, $p = 0.001$), supporting H3, which suggests that employees' perceptions of CSR have a positive impact on their performance. Similarly, the path from "Perception toward CSR" to "Person-Organization Fit" is significant ($\beta = 0.132$, $t = 3.272$, $p = 0.001$), confirming H1, which suggests that CSR initiatives positively enhance employees' sense of fit with their organization. The relationship between "Person-Organization Fit" and "Employee Performance" is highly significant and strong ($\beta = 0.54$, $t = 17.735$, $p < 0.001$), providing robust evidence to accept H2, which states that P-O Fit positively influences employee performance.

The indirect effect of "Perception toward CSR" on "Employee Performance" through "Person-Organization Fit" was found to be statistically significant ($\beta =$

0.171, t = 3.055, p = 0.002), confirming H4, which states that P-O Fit mediates the relationship between employees' perceptions of CSR and their job performance. The Variance Accounted For (VAF) method was applied to identify the type of mediation effect present in the relationship among CSR perceptions, Person-Organization Fit (P-O Fit), and employee performance. This method calculates the proportion of the total effect that is mediated by the indirect effect [42]. The indirect effect, derived from the product of the path coefficients for "Perception toward CSR → P-O Fit" ($\beta = 0.132$) and "P-O Fit → Employee Performance" ($\beta = 0.54$), was calculated as 0.071. The direct effect of "Perception toward CSR → Employee Performance" was 0.112, resulting in a total effect of 0.183. Using the formula, the mediation effect value was determined to be 38.9%, indicating partial mediation [42, 58]. Partial mediation implies that CSR perceptions influence employee performance both directly and indirectly through P-O Fit, with the indirect pathway contributing significantly to the overall relationship.

5 Discussion

The results demonstrate that employee perceptions regarding CSR have a favorable impact on their sense of Person-Organization Fit (H1), as CSR programs assist to match company values with those of workers. This alignment improves employee performance (H2) by promoting engagement, commitment, and a feeling of belonging. Furthermore, CSR views have a direct influence on employee performance (H3), indicating that workers are driven to repay socially responsible activities with better job results.

The mediation study confirms H4, implying that CSR impacts performance both directly and indirectly via P-O Fit. This dual approach highlights CSR's strategic role in molding employee behavior via value alignment. These findings are similar with previous research, such as Zhao et al. [14] and Sarfo et al. [29], which discovered that CSR improves congruence between employee expectations and corporate ideals, resulting in better outcomes such as ethical conduct and organizational commitment. Farooquia and Nagendra [25] found that CSR improves work happiness and performance, whereas Kowalczyk and Kucharska [25] discovered that a good P-O Fit greatly improves job performance. When employees think that CSR projects are in line with their values (good P-O Fit), they are more likely to do a better job. This shows how important it is for CSR to be in line with employees' values in order for it to improve employee performance.

6 Managerial Implications and Theoretical Implications

This research provides practical insights for public-sector management. The direct and indirect impacts of CSR views on employee performance highlight the need for developing CSR efforts that reflect employee values. Effective communication about these activities may promote Person-Organization Fit (P-O Fit), which improves engagement and performance. Integrating CSR into business culture and engaging workers in CSR initiatives helps boost their feeling of belonging and motivation. Thus, CSR should be considered as a strategic instrument for workforce development rather than just a compliance necessity. Consistent internal communication campaigns, newsletters, and success stories regarding PSUs' CSR actions may raise awareness and deepen value alignment, improving perceived Person-Organization Fit [59]. The research adds to the increasing CSR literature by concentrating on the understudied public sector environment. It experimentally confirms the two pathways—direct and mediated by P-O Fit—that relate CSR perceptions to employee performance. The Stimulus-Organism-Response (S-O-R) paradigm provides a structured lens through which to explore the underlying psychological and behavioral factors. These results support and expand on previous research, identifying employee alignment as an important mediator and laying the groundwork for future studies in both the public and commercial sectors.

7 Conclusion

This research emphasizes the critical significance of Corporate Social Responsibility (CSR) views in improving employee performance, with Person-Organization Fit (P-O Fit) acting as a crucial mediating element in public sector firms. The results demonstrate that CSR improves performance in both direct and indirect ways, with alignment between employee values and CSR programs improving engagement, commitment, and outcomes. The found partial mediation effect stresses the need for building P-O Fit to optimize the impact of CSR on employee engagement in PSUs, which supports prior research on staff alignment and organizational success [14, 19]. The Stimulus-Organism-Response (S-O-R) model provides a systematic framework for understanding the psychological and behavioral processes that drive these interactions. The findings show that public sector managers should see CSR as a strategic lever for staff alignment and productivity, rather than just a compliance instrument. Communicating CSR activities properly and engaging workers may improve P-O Fit and lead to beneficial results. This research has limitations, purposive sampling and online data collecting may have an impact on generalizability, and the lack of organizational-level controls restricts the scope of application. Future research should use a variety of sample strategies, incorporate control variables, investigate longitudinal or mixed-methods approaches to improve causal findings, and can use this approach in private-sector companies. Finally, CSR may be used as a transformational approach by aligning company principles with employee goals, resulting in long-term individual and organizational success.

References

1. Střihavková, E., Svobodová, J., & Vysloužilová, D. (2021). Corporate social responsibility of organizations as part of a quality management system. *Production Engineering Archives, 27*
2. Chaudhary, R. (2019). Corporate social responsibility and employee performance: A study among Indian business professionals. *Journal of Business Ethics, 159*(2), 473–489. https://doi.org/10.1007/s10551-017-3780-4
3. Donia, M. B., Ronen, S., Tetrault Sirsly, C. A., & Bonaccio, S. (2019). CSR by any other name? The differential impact of substantive and symbolic CSR attributions on employee outcomes. *Journal of Business Ethics, 157*, 503–523
4. Ilkhanizadeh, S., & Karatepe, O. M. (2017). An examination of the consequences of corporate social responsibility in the airline industry: Work engagement, career satisfaction, and voice behavior. *Journal of Air Transport Management, 59*, 8–17.
5. Mory, L., Wirtz, B. W., & Göttel, V. (2017). Corporate social responsibility: The organizational view. *Journal of Management and Governance, 21*, 145–179.
6. Gupta, N., & Sharma, V. (2016). The relationship between corporate social responsibility and employee engagement and its linkage to organizational performance: A conceptual model. *IUP Journal of Organizational Behavior, 15*(3), 59.
7. Tourigny, L., Han, J., Baba, V. V., & Pan, P. (2019). Ethical leadership and corporate social responsibility in China: A multilevel study of their effects on trust and organizational citizenship behavior. *Journal of Business Ethics, 158*(2), 427–440. https://doi.org/10.1007/s10551-017-3745-6
8. Afsar, B., Cheema, S., & Javed, F. (2018). Activating employee's pro-environmental behaviors: The role of CSR, organizational identification, and environmentally specific servant leadership. *Corporate Social Responsibility and Environmental Management, 25*(5), 904–911. https://doi.org/10.1002/csr.1506
9. Sanusi, F. A., & Johl, S. K. (2020). A proposed framework for assessing the influence of internal corporate social responsibility belief on employee intention to job continuity. *Corporate Social Responsibility and Environmental Management, 27*(6), 2437–2449. https://doi.org/10.1002/csr.2025
10. Scheidler, S., Edinger-Schons, L. M., Spanjol, J., & Wieseke, J. (2019). Scrooge posing as Mother Teresa: How hypocritical social responsibility strategies hurt employees and firms. *Journal of Business Ethics, 157*, 339–358.
11. Hasan, H., Astuti, E. S., Afrianty, T. W., & Iqbal, M. (2020). Impact of organizational culture on employee engagement and employee performance: A stimuli-organism-response approach. *Wacana Journal of Social and Humanity Studies, 23*(4), 235–247.
12. Wei, Y., & Lin, C. Y. (2025). The effects of perceived corporate social responsibility and Confucian ethics on psychological meaningfulness and employee engagement: A moderated-mediation study. *Corporate Social Responsibility and Environmental Management, 32*(1), 806–819. https://doi.org/10.1002/csr.2965
13. Duarte, A. P., & Mouro, C. (2022). The role of person–organization fit in the relationship between environmental corporate social responsibility and pro-environmental behaviors at work. *Sustainability, 14*(6), 3421. https://doi.org/10.3390/su14063421
14. Zhao, H., Li, H., & Zhang, Q. (2021). Employee perceptions of corporate social responsibility and person-organization fit: Evidence from Chinese firms. *Journal of Business Ethics, 171*(1), 21–36. https://doi.org/10.1007/s10551-019-04463-y
15. Homer, S. T., Yee, K. V., & Khor, K. S. (2022). Developing a measurement instrument for perceived Corporate citizenship using multi-stakeholder, multi-industry and cross-country validations. *Quality and Quantity*, 1–24
16. Cheema, S., Afsar, B., & Javed, F. (2020). Employees' corporate social responsibility perceptions and organizational citizenship behaviors for the environment: The mediating roles of organizational identification and environmental orientation fit. *Corporate Social Responsibility and Environmental Management, 27*(1), 9–21.

17. Zhang, L., & Gowan, M. A. (2012). Corporate social responsibility, applicants' individual traits, and organizational attraction: A person-organization fit perspective. *Journal of Business and Psychology, 27*, 345–362.
18. Anaza, N. A. (2015). Relations of fit and organizational identification to employee-customer identification. *Journal of Managerial Psychology, 30*, 925–939.
19. De Roeck, K., El Akremi, A., & Swaen, V. (2016). Consistency matters! How and when does corporate social responsibility affect employees' organizational identification? *Journal of Management Studies, 53*(7), 1141–1168. https://doi.org/10.1111/joms.12216
20. Van Dick, R., van Knippenberg, D., Kerschreiter, R., Hertel, G., & Wieseke, J. (2008). Interactive effects of work group and organizational identification on job satisfaction and extra-role behavior. *Journal of Vocational Behavior, 72*, 388–399.
21. Hu, B., Liu, J., & Zhang, X. (2020). The impact of employees' perceived CSR on customer orientation: An integrated perspective of generalized exchange and social identity theory. *International Journal of Contemporary Hospitality Management, 32*(7), 2345–2364.
22. Jufrizen, J., Khair, H., Siregar, A. P. W., & Hawariyuni, W. (2023). Person-organization fit and employee performance: Mediation role job satisfaction and organizational commitment. *Organization Management Journal, 19*(2), 360–378.
23. Lary, A. I., & Omar, R. (2021). A conceptual framework for describing the role of employer branding, person-organization fit and employee engagement in shaping organizational citizenship behavior. *Business Management Strategy, 12*(2), 172–181.
24. Farooquia, M. S., & Nagendra, A. (2014). The impact of person-organization fit on job satisfaction and performance of the employees. *Procedia Economics and Finance, 11*, 122–129. https://doi.org/10.1016/S2212-5671(14)00182-8
25. Kowalczyk, R., & Kucharska, W. (2020). Corporate social responsibility practices incomes and outcomes: Stakeholders' pressure, culture, employee commitment, corporate reputation, and brand performance. A Polish–German cross-country study. *Corporate Social Responsibility and Environmental Management, 27*(2), 595–615
26. Soelton, M., Noermijati, N., Rohman, F., Mugiono, M., Aulia, I. N., & Siregar, R. E. (2020). Reawakening perceived person-organization fit and perceived person-job fit: Removing obstacles organizational commitment. *Management Science Letters, 10*(13), 2993–3002. https://doi.org/10.5267/j.msl.2020.5.026
27. Asante Boadi, E., He, Z., Bosompem, J., Opata, C. N., & Boadi, E. K. (2020). Employees' perception of corporate social responsibility (CSR) and its effects on internal outcomes. *The Service Industries Journal, 40*(9–10), 611–632.
28. Li, Y. B., Zhang, G. Q., Wu, T. J., & Peng, C. L. (2020). Employee's corporate social responsibility perception and sustained innovative behavior: Based on the psychological identity of employees. *Sustainability, 12*(20), 8604.
29. Sarfo, C. A., Zhang, J. A., O'Kane, P., Podgorodnichenko, N., & Osei-Fosu, K. K. (2022). Perceived corporate social responsibility and employee ethical behaviour: Do employee commitment and co-worker ethicality matter? *Journal of Management and Organization, 28*(1), 184–201. https://doi.org/10.1017/jmo.2021
30. Dung, L. T. (2021). The mediating role of employee intrapreneurial behavior in nexus between constructs of internal corporate social responsibility practices and organizational outcomes. *Journal of Asia-Pacific Business, 22*(2), 134–154.
31. Adu-Gyamfi, M., He, Z., Nyame, G., Boahen, S., & Frempong, M. F. (2021). Effects of internal CSR activities on social performance: The employee perspective. *Sustainability, 13*(11), 6235.
32. Shah, M. S., Wu, C., & Ullah, Z. (2021). The inter-relationship between CSR, inclusive leadership and employee creativity: A case of the banking sector. *Sustainability, 13*(16), 9158.
33. Carlini, J., & Grace, D. (2021). The corporate social responsibility (CSR) internal branding model: Aligning employees' CSR awareness, knowledge, and experience to deliver positive employee performance outcomes. *Journal of Marketing Management, 37*(7–8), 732–760.
34. Herman, K.D., Erlina, R., & Mardiana, N. (2024). The influence of corporate social responsibility on employee performance with organizational commitment as mediation: Study of

Bank Lampung employees. *International Journal of Environmental, Sustainability, and Social Science*, 5(3). https://doi.org/10.38142/ijesss.v5i3.1086
35. Siyal, S., Xin, C., Peng, X., Siyal, A. W., & Ahmed, W. (2020). Why do high-performance human resource practices matter for employee outcomes in public sector universities? The mediating role of person–organization fit mechanism. *SAGE Open, 10*(3), 2158244020947424.
36. Jin, H., Li, X., & Li, G. (2024). Impact of corporate social responsibility on employee loyalty: Mediating role of person-organization fit and employee trust. *PLoS ONE, 19*(3), Article e0300933.
37. Thuy, N. T. T., & Phinaitrup, B. A. (2023). The effect of public service motivation on job performance of public servants in Vietnam: The role of mediation of job satisfaction and person-organization fit. *International Journal of Public Administration, 46*(5), 326–343.
38. Duarte, A. P., & Mouro, C. (2022). Environmental corporate social responsibility and workplace pro-environmental behaviors: Person-organization fit and organizational Identification's sequential mediation. *International Journal of Environmental Research and Public Health, 19*(16), 10355.
39. Creswell, J. W., & Creswell, J. D. (2017). *Research design: Qualitative, quantitative, and mixed methods approaches*. Sage Publications.
40. Ha, M. T. (2022). *Data collection using online questionnaires in marketing*. SAGE Publications.
41. Wong, A.K.F., Kim, S. (Sam), & Hwang, Y. (2021). Effects of perceived corporate social responsibility (CSR) performance on hotel employees' behavior. *International Journal of Hospitality & Tourism Administration, 23*(6), 1145–1173. https://doi.org/10.1080/15256480.2021.1935390
42. Hair, J. F., Hult, G. T. M., Ringle, C. M., & Sarstedt, M. (2019). *A primer on partial least squares structural equation modeling (PLS-SEM)* (2nd ed.). Sage Publications.
43. Etikan, I., Musa, S. A., & Alkassim, R. S. (2016). Comparison of convenience sampling and purposive sampling. *American Journal of Theoretical and Applied Statistics, 5*(1), 1–4.
44. Andrade, C. (2021). The inconvenient truth about convenience and purposive samples. *Indian Journal of Psychological Medicine, 43*(1), 86–88.
45. Turker, D. (2009). Measuring corporate social responsibility: A scale development study. *Journal of Business Ethics, 85*(4), 411–427.
46. Lee, Y. K., Kim, Y. S., Lee, K. H., & Li, D. X. (2012). The impact of CSR on relationship quality and relationship outcomes: A perspective of service employees. *International Journal of Hospitality Management, 31*(3), 745–756.
47. Oh, K. S., Han, J. R., & Park, S. R. (2021). The influence of hotel employees' perception of CSR on organizational commitment: The moderating role of job level. *Sustainability, 13*(22), 12625. https://doi.org/10.3390/su132212625
48. Afsar, B., & Badir, Y. F. (2016). Person–organization fit, perceived organizational support, and organizational citizenship behavior: The role of job embeddedness. *Journal of Human Resources in Hospitality & Tourism, 15*(3), 252–278. https://doi.org/10.1080/15332845.2016.1147936
49. Jin, H., Li, X., & Li, G. (2024). Impact of corporate social responsibility on employee loyalty: Mediating role of person-organization fit and employee trust. *PLoS ONE, 19*(3), Article e0300933. https://doi.org/10.1371/journal.pone.0300933
50. Podsakoff, P., Todor, W., & Skov, R. (1982). Effects of leader contingent and noncontingent reward and punishment behaviors on subordinate performance and satisfaction. *Academy of Management Journal, 25*(4), 810–821.
51. Silva, P., Moreira, A. C., & Mota, J. (2023). Employees' perception of corporate social responsibility and performance: The mediating roles of job satisfaction, organizational commitment and organizational trust. *Journal of Strategy and Management, 16*(1), 92–111. https://doi.org/10.1108/JSMA-10-2021-0213
52. Churchill, G. A., & Iacobucci, D. (2006). *Marketing research: Methodological foundations*. Dryden Press.

53. Podsakoff, P. M., MacKenzie, S. B., Lee, J. Y., & Podsakoff, N. P. (2003). Common method biases in behavioral research: A critical review of the literature and recommended remedies. *Journal of Applied Psychology, 88*(5), 879–903. https://doi.org/10.1037/0021-9010.88.5.879
54. Fornell, C., & Larcker, D. F. (1981). Evaluating structural equation models with unobservable variables and measurement error. *Journal of Marketing Research, 18*(1), 39–50. https://doi.org/10.2307/3151312
55. Chin, W. W. (1998). The partial least squares approach to structural equation modeling. In G. A. Marcoulides (Ed.), *Modern Methods for Business Research* (pp. 295–336). Lawrence Erlbaum Associates.
56. Falk, R. F., & Miller, N. B. (1992). *A primer for soft modeling*. University of Akron Press.
57. Cohen, J. (1988). *Statistical power analysis for the behavioural sciences*. Lawrence Erlbaum Associates.
58. Zhao, X., Lynch, J. G., Jr., & Chen, Q. (2010). Reconsidering Baron and Kenny: Myths and truths about mediation analysis. *The Journal of Consumer Research, 37*(2), 197–206.
59. Carlini, J., & Grace, D. (2021). The corporate social responsibility (CSR) internal branding model: Aligning employees' CSR awareness, knowledge, and experience to deliver positive employee performance outcomes. *Journal of Marketing Management*. https://doi.org/10.1080/0267257X.2020.1860113

Strategic Integration of Corporate Social Responsibility: Linking Theory and Practice for Sustainable Development

Kuchi. Srinivasa Krishna, V. S. K. Verma, Bhupathi Raju Vishnu Verma, N. Visalakshi, and B. Sreekanth

Abstract

This research work critically evaluates the alignment of Corporate Social Responsibility (CSR) with strategic business applications to foster inclusive and sustainable development. The purposes of this study are to (1) examine the development of CSR in relation to SDGs and ESG frameworks, (2) assess CSR implementation across various sectors, (3) evaluate its impacts on stakeholders and sustainability, and (4) bridge the gap between the theory and application. This research employs a qualitative, multiple-case study method to examine cases of six corporate houses: Unilever, Patagonia, ITC Limited, Dabur, Reliance Foundation, and a comparative comparison of SBI and Axis Bank. The study utilises thematic content analysis of CSR reports, academic literature, and ESG disclosures. Significant findings reveal that Unilever minimised CO_2 emissions by 65%, Dabur recycled more than 12,400 tons of plastic scraps, and ITC positively impacted over 6 million people via rural projects. SBI and Axis Bank together spent more over ₹1,000 crore in CSR from 2017 to 2024. These findings validate CSR's potential to enhance brand equity, operational resilience, and social legitimacy. However, the projects face several challenges concerning greenwashing, regulatory compliance, and ineffective impact assessment. The study calls for standardised ESG reporting, fiscal incentives for validated

Kuchi. S. Krishna (✉) · B. Sreekanth
School of Management, MITS- Deemed to be University, Madanapalle, Andhra Pradesh, India
e-mail: drkrishna1975@gmail.com

V. S. K. Verma · B. R. V. Verma
Department of Management Studies, MR P.G. College, Vizianagaram, Andhra Pradesh, India

N. Visalakshi
School of Business, Aditya University, Surampalem, India

S. Patnaik et al. (eds.), *People, Planet and Profit: Advancing Sustainable Business Practices*, Springer Proceedings in Geography and Sustainability,
https://doi.org/10.1007/978-3-032-11296-5_9

social impact, and stakeholder-oriented governance. Future study should be conducted on longitudinal impact studies, digital CSR instruments, and behavioural reactions to corporate sustainability.

Keywords

Corporate social responsibility • Sustainable development goals • Environmental • Social • And governance • Stakeholder engagement • Strategic sustainability

1 Introduction

CSR has significantly evolved in the last 2 decades, transitioning from philanthropic model to strategic business model. This transition has been driven by increasing global demands to align with firms' operations with sustainable development goals, ethical governance, and equitable growth. The implementation of frameworks such as the United Nations Sustainable Development Goals (SDGs), the Paris Agreement, and national regulations like India's Companies Act, 2013, has further institutionalised CSR as a legislative and reputational imperative. Nevertheless, the planning, execution, and impacts of the CSR project often are impeded by greenwashing, lack of standardised impact measures, and incoherent stakeholder engagement.

Academic literature highlights this development while simultaneously underscoring significant limitations in empirical validation, theoretical coherence, and practical implementation. Researchers such Siddique et al. [1], Elembilassery [2], and Paul and Parra [3] highlight the need for context-specific, outcome-oriented, and institutionally embedded multi-dimensional CSR models [1–3]. This study addresses this need by analysing the implementation of CSR projects in various organisational setups and investigates six cases of global and Indian corporate houses to understand how the firms operationalise CSR projects, the resultant impact, and the insights that can be derived for policy implication, practice, and future research.

2 Literature Review

The existing literature on CSR, Environmental, Social, and Governance (ESG) performance, and sustainable development provides a comprehensive investigation of themes across many industries and regions. The systematic literature study entitled "Corporate Social Responsibility as the Gateway to Sustainable Banking" provides a comprehensive theoretical analysis of 30 papers over one decade and explores 28 CSR dimensions and 14 foundational theories. These dimensions encompass environmental sustainability, employee welfare, consumer accountability, community engagement, ethical finance, and governance, among others. The authors offer a conceptual model linking these dimensions to examine how CSR improved ESG performance and sustainable development in the banking industry. However, this model is theoretical in nature and necessitates empirical investigation, specifically across diverse regulatory and institutional setups, highlighting the need for further research in distinct regional banking ecosystems [1].

Elembilassery [2] focuses on the disciplinary advancement of CSR, integrating it with strategic management, marketing, and organisational behaviour. The study states that CSR research will progressively rely on political, legal, and institutional contexts, particularly in non-Western environments [2]. Paul and Parra [3] use text mining on 1,188 research studies, and find a concern on low engagement with CSR subjects in business literature. The study reveals corruption and embeddedness as importance themes; however, it does not provide a model to apply these findings into implementable research goals [3].

Saxena et al. [4] offer a classification model, namely ECG framework (Environment, Community, Growth) after an analysis of 301 research studies. The research renders a systematic approach to navigate CSR literature, although it offers inconclusive findings about the impact of CSR on corporate performance. The paucity of causal data challenges the argument for CSR's direct impact on corporate results [4]. On the contrary, industry reports such as the Textile Exchange [5] and the Vogue Business Index [6] investigate sustainability through the lens of fashion industry. Textile Exchange critiques the unregulated growth of fashion firms and endorses for a transition to circular economic practices [5]. The Vogue analysis assesses ESG performance of 60 luxury brands, highlighting advancements by a few but lacking intentions to attain sustainable goals. Though these assessments offer detailed industry-specific information; yet, often neglect studying the socio-economic conditions of manufacturing hubs in emerging countries [6].

The Financial Times [7] renders a critical analysis of the ineffectiveness of the present ESG strategy. It recommends "competitive sustainability" as an alternative to realign market incentives for sustainable resilience [7]. However, the report does not provide practical insights on the policy or institutional mechanism, necessary to enable this transformation. Time Magazine offers this broad perspective with a comprehensive analysis, focusing on the importance of data, collaboration, and employee engagement in driving corporate climate initiatives [8].

Reuters [9] reveals the financial constraints for accomplishing SDGs and proposes a new financing framework based on quantifiable outcome instead of operational inputs. This outcome-oriented approach links well with impact investment and advocates transparency and accountability in corporate sustainability contributions [9]. Ultimately, The Times studies the intricate issues of "greenhushing," in which the firms intentionally refrain from publicly disclosing ESG actions in apprehension of criticism [10]. This approach is often neglected in academic discourses. This practice highlights the evolving tensions between strategic discretion and ethical transparency in sustainability.

The review of the literature together suggests a paradigm change in CSR and ESG from voluntary and reputational instruments to strategic, performance-driven imperatives. A prevailing trend across industries is the evolution from CSR as charity to CSR as embedded governance, whereby sustainability is incorporated into risk management, innovation, and value generation. This change is inconsistently noticed in developed regions and industries such as banking and fashion at the forefront, while emerging economies and informal sectors lag behind due to institutional limitations and resource constraints.

3 Rationale for the Research

The increasing emphasis on sustainable development, particularly after the Paris Agreement and the implementation of the SDGs, has intensified the significance of CSR as a strategic instrument in reshaping traditional business models. Despite growing scholarly interest, a fragmented understanding persists on the direct impact of CSR practices on sustainable banking and corporate performance across different sectors. Though several research offer the conceptual models and theoretical frameworks, yet, they often lack empirical validation and sector-specific relevance, especially in developing economies. Besides, gaps remain in establishing causal relationships between CSR programs and quantifiable outcomes such as financial performance, stakeholder confidence, and environmental sustainability. The literature also highlights underrepresentation of CSR discourse in mainstream business studies and the uncertainty surrounding its incorporation with ESG frameworks. The problems of greenwashing and greenhushing intensify the challenges on transparency and stakeholder engagement. In this context, there is a pressing need for a thorough, context-specific study that not only utilises current CSR methodologies but also assesses their execution, challenges, and impacts on sustainable outcomes. This research also seeks to bridge this gap by offering an analytical view on the role of CSR in supporting sustainable development and guiding prospective corporate and policy-level strategies.

4 Objectives of the Research

The present research work has the following research objectives:

- To rigorously critically examine the development of CSR practices concerning sustainability, stakeholder engagement, and environmental governance.
- To examine the CSR strategies of leading Indian and international firms (e.g., Unilever, Patagonia, ITC, Reliance, Dabur, SBI, and Axis Bank) on implementation, funding, and impact.
- To evaluate the effect of CSR programs of the leading corporate houses with SDGs and ESG frameworks.
- To identify critical gaps between CSR conceptual models suggested in the literature and their practical implementation by corporate houses, emphasising transparency, impact metrics, and stakeholder inclusion.

5 Methodology

This research utilises a qualitative, multiple-case study approach to study the linkage of CSR into sustainable business practices across several industries. The study uses secondary data and thematic content analysis for facilitating a comprehensive understanding of the application of theoretical CSR frameworks in real-world business context. The case selection comprises six leading organisations—Unilever, Patagonia, ITC Limited, Dabur, Reliance Foundation, and a comparative analysis of SBI and Axis Bank—and these firms are selected through purposive sampling based on their CSR transparency, sectoral diversity, and geographical representation in both international and Indian contexts. Data sources include publicly accessible corporate documents like CSR and sustainability reports, annual reports for FY 2022–23 and 2023–24, websites, and CSR disclosures mandated by the Companies Act of 2013 and these were supplemented by scholarly literature [1–3] and industry analyses from Financial Times, Reuters, and Vogue Business Index [6, 7, 9]. The present research also analyses CSR expenditures of the firms, strategic alignment with SDGs and ESG frameworks, stakeholder engagement, and impact measures. Cross-case analyses were performed to identify pattern, variations, and sector-specific findings. Data is triangulated from many sources which guarantees the validity and reliability of the findings. This methodological framework enables a comprehensive assessment of CSR's role as both a compliance mechanism as well as a catalyst for strategic sustainable change.

6 Findings and Analysis

6.1 Case Study 1: Unilever—Sustainable Living Initiative

Unilever's Sustainable Living Plan (USLP), initiated in 2010, serves as a holistic strategy to incorporate sustainability within its core business operation. Its objective is to decouple environmental effect from company's growth while improving results in its social bottom line. The USLP set up ambitious targets in three domain areas: improving health and well-being, lessening environmental impact, and sustainable livelihoods. By 2020, Unilever engaged over 1.3 billion people via its hygiene and health activities and had sustainably sourced 67% of its agricultural raw materials. The company has waned CO_2 emissions from energy use by 65% per ton of manufacturing since 2008, minimised water usage and trash output [11].

Together with its environmental accomplishments, Unilever imparted training on sustainable agriculture practices to over 70,000 small farmers and promised to ensure that all plastic packaging is reusable, recyclable, or compostable by 2030. It also manufactured water-conserving personal care products and biodegradable detergents to impact customer behaviour favourably. These initiatives improved brand value, stakeholder trust, and strengthened investor trust. The incorporation of sustainability into product manufacturing and supply chains illustrates a strategic CSR framework in which profitability is congruent with environmental and social value. Unilever's case exemplifies how multinational firms can use sustainability to gain competitive advantage and achieve global impact, establishing an industry benchmark [11].

6.2 Case Study 2: Patagonia—Ecological Advocacy

Patagonia, a leading apparel company, is internationally recognised for its profound commitment to environmental sustainability and ethical business practices. The company's CSR strategy is built on the principle that the firm must safeguard and conserve the natural environment. Patagonia allocates 1% of its annual profit to actions concerning environmental protection through "1% for the Planet" campaign, while its "Worn Wear" program supports for product repair and reuse to minimise textile waste [12].

In a groundbreaking decision in 2022, the company assigned ownership to a trust and a not-for-profit organisation, ensuring that all future profits, projected at around $100 million per annum, are directed towards combating climate change and protecting biodiversity. This ownership exemplifies Patagonia's conviction that the company may operate beyond profit motive. These bold moves have improved consumer loyalty and shaped global corporate responsibility discourses.

Patagonia's actions are consistent with stakeholder theory and ESG principles which illustrates that sustainable business can be aligned with environmental and social goals. Their initiatives illustrate that authentic and well-integrated CSR actions inculcate a strong brand identity and catalyse sector-wise transformation [12].

6.3 Case Study 3: ITC Limited—Triple Bottom Line Approach

ITC Limited, a highly diversified conglomerate in India, uses a Triple Bottom Line strategy that incorporates economic, social, and environmental dimensions into its core operations. The company's CSR strategy emphasises stakeholder engagement, rural development, and sustainable resource management. ITC's flagship project, "e-Choupal," has empowered more than 4 million rural farmers as it provides access to real-time market information, quality inputs, and digital platforms. This not only increases farmers' income but also strengthens ITC's agricultural supply chain [13].

ITC has invested resources on afforestation, water conservation, and solid waste management. The social investment campaign namely "Mission Sunehra Kal" focuses on education, health, and vocational training, positively touching lives of nearly 6 million people. In FY 2022–23, ITC invested ₹352.76 crore to CSR initiatives, focusing on scalable and sustainable outcomes. Besides, ITC has accomplished carbon-positive, water-positive, and solid waste recycling-positive status for over 17 consecutive years.

The company advocates environmentally sustainable packaging and uses agricultural waste to manufacture eco-friendly paper goods. ITC's model corresponds with the ECG framework and illustrates how enterprises may drive social development while achieving commercial objectives. Its proactive CSR policy and practices have attracted ethical investors and improved its market reputation [13].

6.4 Case Study 4: Reliance Foundation—Community Advancement

The corporate social responsibility division of Reliance Industries Limited has established itself as a pioneer in community development through its significant and inclusive programs. The CSR strategy is centred on five pillars: rural development, health, education, disaster response, and urban redevelopment. The Reliance Foundation has touched the lives of nearly 66 million individuals through its projects like "Project Drishti," which offers free corneal transplants, and its health outreach programs aimed at vulnerable communities [14].

Reliance Foundation's rural development projects include watershed management, digital literacy, and vocational training. Employees are urged to involve in CSR activities, inculcating a culture of engagement and ethical responsibility. In FY 2023–24, Reliance spent ₹1,592 crore on CSR, ranked among the biggest spent by Indian companies. Their educational programs support digital classrooms and girl child education, while healthcare programs address challenges of malnutrition, sanitation, and maternal care.

By aligning with national development objectives and SDGs, Reliance's activities also attract ethical investors. Its ability to integrate social welfare with brand equity establishes it as a role model for other corporate houses. The Foundation's integrated strategy for stakeholder engagement and shared value creation add to its long-term sustainability and support its societal license to operate [14].

6.5 Case Study 5: Dabur India—Sustainable Packaging Solutions

Dabur India, a leading manufacturer in the FMCG products, integrated sustainability into its core business operations, particularly through innovations in environmentally friendly packaging. In response to pressing environmental demand, the company adopted wholly recyclable plastic packaging and initiated comprehensive plastic garbage collection and recycling initiatives. In 2022, Dabur recycled almost 12,400 tons of used consumer plastic waste and achieved plastic-waste neutrality ahead of national regulatory deadlines [15].

Dabur's CSR initiatives are implemented by its NGO partner, Sundesh, which implements projects on health, sanitation, and education, especially in rural India. Programs like menstrual hygiene education for teenagers, mobile health units, and toilet building have profoundly impacted community health. In fiscal year 2022–23, Dabur spent ₹35.21 crore on CSR initiatives.

The company's initiatives align with India's Swachh Bharat and Beti Bachao objectives and advocate the circular economy. Dabur regularly educates customers on sustainability and ethical consumerism by its awareness initiatives. Dabur has improved its brand value and set an example for small companies that they can also achieve significant, community-oriented sustainability outcomes [15].

6.6 Case Study 6: Comparative Analysis of CSR: SBI Versus Axis Bank

A comparative analysis of CSR involvement of State Bank of India (SBI) and Axis Bank offers pertinent insights into the approaches of public and private sector banks on their social responsibilities. The SBI, the biggest public sector bank in India, emphasises rural development, education, and sanitation. From FY 2017–18 to FY 2023–24, SBI spent around ₹576.34 crore for its CSR programs which includes programs on financial literacy, skill development for youth, and support for those with disabilities [16].

On the other hand, Axis Bank, one of the leading private sector banks in India, focused on inclusive development through programs like "Axis Dil Se," which support education for the children residing in underserved regions. It spends around ₹495.65 crore regularly designating a greater proportion of its profit after tax (PAT) to CSR (reaching 2.73% in 2018–19). SBI's CSR expenditures fluctuated due to varying profits, while Axis demonstrated stable growth. Despite the varying spending on CSR, both the banks illustrated equivalent commitment to impactful CSR.

The CSR foundations of both SBI and Axis deliver impactful outcomes. Both the banks allocated resources on education, health, and environmental sustainability, illustrating a matured banking sector that recognises CSR as a means for sustaining societal and reputational benefits [16, 17].

7 Discussion

The evolution of CSR from peripheral philanthropy to an integral strategic corporate function is extensively studied in scholarly academic literature. Analyses by Siddique et al. [1], Elembilassery [2], and Paul and Parra [3] show that while CSR is increasingly linked with sustainability goals, its implementation across many sectors and regions is uneven and often remains inconsistent [1–3]. The selected case studies of Unilever, Patagonia, ITC, Reliance Foundation, Dabur, and the comparative analysis of SBI-Axis Bank CSR provide real-world examples of these dynamisms, validating several themes in the academic body of literature while also revealing gaps in implementation and accountability (Table 1).

Table 1 Key CSR initiatives and impact metrics of selected companies

Company	Key CSR focus	Key metrics
Unilever	CO_2 reduction, sustainable sourcing	65% CO_2 reduction (since 2008)
Patagonia	Environmental activism, product reuse	100% profits redirected to climate action
ITC limited	Rural development, afforestation, livelihoods	6M+ people impacted, 8 lakh ha afforestation
Dabur	Eco-friendly packaging, plastic recycling	12,400+ tons of plastic recycled (2022)
reliance foundation	Health care, education, disaster relief	66M+ people impacted (cumulative)
SBI	Financial literacy, sanitation, rural outreach	₹576.34 Cr CSR spend (2017–24)
Axis Bank	Education, inclusive finance, skill development	₹495.65 Cr CSR spend (2017–24)

Source Authors' own compilation

Siddique et al. [1] viewed emergence of CSR as a cornerstone for sustainable banking and business development after the Paris Agreement. Their thorough research includes 14 theories and 28 CSR dimensions and results in a synthesised model that aligns CSR with SDGs [1]. This theoretical framework links with the CSR strategies of Unilever and ITC, whereby sustainability is integrated into the core business model. Unilever's Sustainable Living Plan connects environmental, social, and economic dimensions, reflecting the model's advocacy for a multi-dimensional approach. ITC's Triple Bottom Line Approach is a real-world application of the Environment-Community-Growth (ECG) model proposed by Saxena et al., producing measurable outcomes including carbon neutrality, rural empowerment, and afforestation [4].

Elembilassery [2] focuses on the external political, legal, and cultural contexts that impacts CSR operations. This is viewed in the context-specific strategies of Dabur and Reliance Foundation, and their CSR projects are need-based. They have extensively supported the communities with their projects on health care, menstrual hygiene, sanitation, and education which often align with national agendas such as Swachh Bharat and Beti Bachao. These illustrations demonstrate how the firms can manoeuvre within institutional frameworks and regional needs to create shared value, supporting Elembilassery's assertion that CSR projects must be continuously aligned to changing external environment [2].

Paul and Parra identify a critical limitation in global CSR research. The authors stated that the subject of CSR is inadequately studied in academia [3]. There is a lack of exploration on the core themes like corruption, embeddedness, and structural inequity. Patagonia provides a compelling contrast. Its deep commitment to environmental activism and ownership restructuring for climate funding demonstrate significant engagement with ethical and governance issues. However, Patagonia stands as an exception; the majority of companies including those included in this review avoid direct involvement, hence supporting Paul and Parra's assertion about the limited adoption of transformational CSR in traditional business frameworks.

The criticism of the fashion sector as evident in Textile Exchange report [5] and the ESG evaluations by the Vogue Business Index [6] further advance this discourse [5, 6]. Patagonia illustrates proactive environmental stewardship, while corporate houses like Dabur are at the nascent state to integrate circular economy (CE) ideas. These differences validate the literature's concern on the uneven implementation of sustainability. This corresponds with Financial Times report, which criticises the ineffectiveness of the ESG framework and proposes for "competitive sustainability" where market incentives reward long-term resilience [7]. The experiences of Unilever and ITC partially appreciate this by demonstrating how sustainability expenditures can improve brand equity, stakeholder engagement, and operational effectiveness.

The phenomenon of greenhushing, as reported by The Times, where firms undermine reporting CSR actions to avoid public criticism [10], is not directly illustrated in these case studies. However, it indirectly relates to businesses such as SBI that, despite substantial contribution, report their CSR programs less aggressively than their private competitors like Axis Bank. The comparison between SBI and Axis Bank further supports literature that CSR initiatives in public and private sectors differ in consistency and style, though both public and private corporate houses are progressively incorporating CSR into their governance structures [1].

Additionally, the funding issues as highlighted by Reuters [9] advocate for outcome-driven impact finance which aligns with the frameworks shown by Reliance Foundation and ITC, whereby CSR spends on sustainable livelihoods, education, and health care produce measurable social benefits. However, only a few of these firms except Patagonia and Unilever have used transparent, impact-oriented ESG metrics—a gap found in Paul and Parra's where the authors criticise lack of empirical evaluation in CSR operations [3].

The research and case studies illustrate that CSR is evolving into a more complex, multi-dimensional strategy grounded on stakeholder theory, sustainability, and ethical governance. However, significant gaps exist. There exists a difference between academic frameworks and CSR implementation, particularly in regard to transparency, theoretical foundations, and cross-sectoral evaluation. Companies such as Patagonia and Unilever illustrate deep integration of CSR, while others like Dabur and SBI illustrate regionally adopted incremental strategies with substantial grassroots effects.

This comprehensive research analysis affirms that the effectiveness of CSR depends not on spending, but also on strategic alignment, institutional integration, and transparent impact evaluation. Future CSR discourse must resolve the academic-practice gap by developing sector-specific, data-driven, and culturally contextualised models that connect corporate objectives with social development.

8 Policy Implications

The outcomes of the study have substantial policy implications for policymakers and corporate houses seeking to embed sustainability through CSR. There is a distinct prerequisite to institutionalise impact-oriented CSR assessment frameworks which can move beyond the financial input matrices to monitor long-term outcomes, mostly in relation to the SDGs. Policymakers must develop standardised reporting formats that necessitate sector-specific ESG metrics, consequently, ensuring sectoral comparability and accountability. The Companies Act, 2013, in India offers a baseline for mandating CSR which is strengthened by transparency by mandatory disclosures, independent audits, and participatory stakeholder assessments.

Furthermore, governmental bodies might provide monetary incentives or tax benefit to the firms that demonstrate high social and environmental outcomes. The companies like ITC and Dabur invest in sustainable agriculture and CE projects which could benefit from this policy initiative. Public–private partnerships (PPPs) may implement impactful CSR activities on health, education, and rural emersion, predominantly in undeveloped areas. Regulatory authorities must address issues concerning greenwashing and greenhushing through legislative measures that necessitate authenticity and transparency in CSR and ESG reporting. This would ensure that CSR serves as a catalyst for transformation rather than a mere symbolic responsibility.

9 Directions for Future Research

This study offers valuable insights as it connects CSR and sustainable business practices; however, several avenues for further research remain open. Longitudinal studies can be made to monitor CSR projects and their socio-economic and environmental outcomes over time, particularly in developing economies. This study would facilitate the establishment of causal links between CSR investment and quantifiable performance outcomes such as profitability, brand equity, stakeholder loyalty, and carbon reduction. Additionally, quantitative frameworks that include CSR, ESG ratings, and financial performance measurements could improve understanding of the strategic benefits of responsible business practices.

Future study should examine cross-country comparative research, focusing on the impact of national institutional environment and legislative frameworks on CSR strategies. This would render insights into why firms in some regions such as Patagonia in the U.S. or Reliance in India adopt distinct sustainability practices. The growing importance of digital platforms, AI, and blockchain in CSR governance offers a compelling avenue for investigation, mostly on transparency, traceability, and stakeholder engagement.

Table 2 Linking CSR theory, practice, and research agenda: A structured overview

Sections	Connected to literature	Connected to cases	Reflects discussion themes
Policy implications	Siddique et al. (multi-dimensional CSR, SDGs), Elembilassery (policy/legal influence), Paul and Parra (transparency)	ITC, Dabur, Reliance (grassroots initiatives, aligned with national goals), Unilever (impact-based CSR)	Need for regulatory reform, outcome-based CSR, sectoral alignment, addressing greenwashing
Direction for future research	Paul and Parra (data gaps), Elembilassery (national systems), Siddique et al. (future inquiry models)	Patagonia (ownership model, ethical CSR), SBI vs. Axis (comparative trends), Unilever (data-driven practices)	Longitudinal and cross-country studies, behavioural insights, digital transformation in CSR
Conclusion	Synthesises key themes across Siddique et al., Paul and Parra, Elembilassery	Draws on all six cases to frame CSR as strategic and transformative	Emphasis on strategic integration, stakeholder engagement, and bridging theory–practice gap

Source Authors' own compilation

Finally, behavioural research studying consumer and investor perceptions of CSR and sustainability communication may support firms in refining their strategies. Bridging this gap between CSR discourses and stakeholder expectations is a critical domain for future academic contributions (Table 2).

10 Conclusion

This study rigorously analysed the evolution, execution, and strategic incorporation of CSR activities across global and Indian business environments. This study exemplifies that CSR is progressively moving from compliance obligation to a fundamental catalyst for sustainable value creation as evident by the findings of comprehensive literature reviews and selected case studies of Unilever, Patagonia, ITC, Reliance Foundation, Dabur, and Indian banks. These cases illustrate a series of approaches from Patagonia's activist-led ownership model to ITC's triple bottom line strategy, and demonstrate the many ways in which CSR can serve societal concerns while promoting business interests.

However, the study highlights persistent gaps between theory and practice, particularly in terms of measurable impact, transparency, and cross-sector comparison. While literature supports integrated, transformational CSR approaches, many corporate houses continue to adopt superficial or performative actions. The study validates that stakeholder alignment, innovation, and linking ESG with SDG

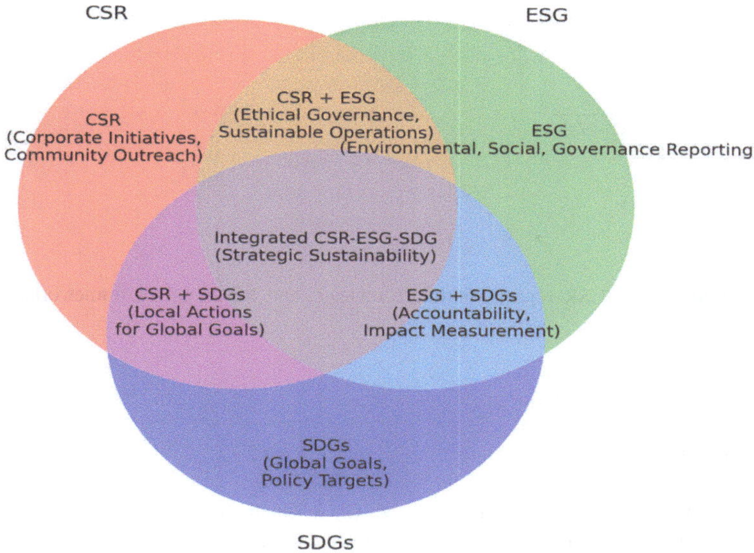

Fig. 1 Thematic integration of CSR, ESG, and SDGs. *Source* Authors' own compilation

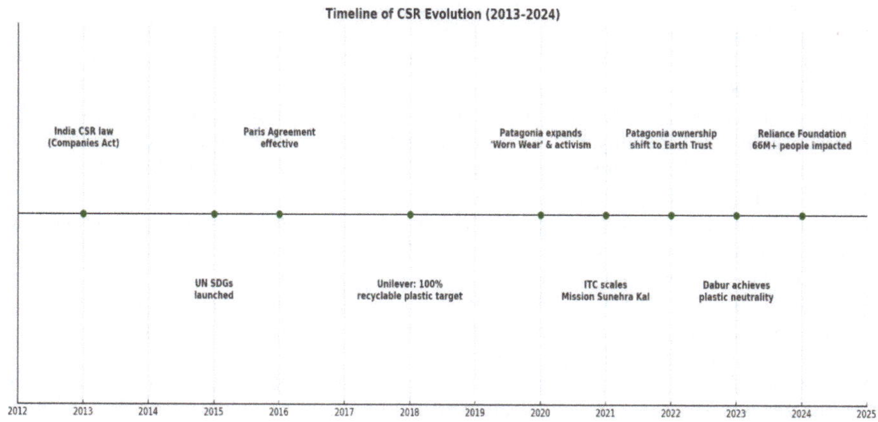

Fig. 2 Timeline of CSR evolution (2013–2024). *Source* Author's own compilation

frameworks are indispensable for realising the full potential of CSR. CSR needs to be viewed, not as a cost or obligation, but as a strategic prospect to inculcate resilience, improved brand reputation, and drive inclusive development. The future effectiveness of CSR will depend on the alignment of corporate intent with its implementation and how policies evolve to incentivise authenticity and measurable impact.

Fig. 3 Compiled from CSR reports, literature reviews, and SDGs policy frameworks. *Source* Authors' own compilation

References

1. Siddique, M.N.-E.-A., Nor, S. M., Senik, Z. C., & Omar, N. A. (2023). Corporate social responsibility as the pathway to sustainable banking: A systematic literature review. *Sustainability, 15*(3), 1807. https://doi.org/10.3390/su15031807
2. Elembilassery, V. (2023). Evolutionary influences on corporate social responsibility literature: A review and prognosis. *Society and Business Review, 18*(3). https://doi.org/10.1108/SBR-05-2022-0136
3. Paul, K., & Parra, C. M. (2021). Corporate social responsibility in international business literature: Results from text data mining of the *Journal of International Business Studies. International Journal of Corporate Social Responsibility, 6*(12). https://doi.org/10.1186/s40991-021-00066-6
4. Saxena, K., Balani, S., & Srivastava, P. (2021). The relationship among corporate social responsibility, sustainability, and organizational performance in the pharmaceutical sector: A literature review. *International Journal of Pharmaceutical and Healthcare Marketing, 15*(4), 572–597. https://doi.org/10.1108/IJPHM-12-2020-0104
5. Textile Exchange (2024). Reimagining growth: A new vision for sustainable fashion. *Textile Exchange*. https://www.voguebusiness.com/story/sustainability/textile-exchange-tackles-fashions-last-climate-taboo-growth
6. Vogue Business (2025). H1 2025 Vogue Business Index: Evaluating luxury fashion brands on ESG performance. *Vogue Business*. https://www.voguebusiness.com/story/sustainability/the-winners-of-sustainabilitys-bumpiest-year-yet
7. Financial Times (2024). ESG is dead. Long live ESG. *Financial Times*. https://www.ft.com/content/47b334a9-69eb-4cd0-af22-6ad0b92c0ef6
8. Time Magazine (2024). Six truths about climate action that all companies should know. *Time*. https://time.com/7027918/six-truths-about-corporate-climate-action/
9. Reuters (2024). How companies can help alleviate global poverty and meet their ESG targets. *Reuters*. https://www.reuters.com/sustainability/society-equity/comment-how-companies-can-help-alleviate-global-poverty-meet-their-esg-targets-2024-10-08/
10. The Times (2025). The rise of greenhushing: Embrace ESG, but don't talk about it. *The Times*. https://www.thetimes.co.uk/article/the-rise-of-greenhushing-embrace-esg-but-dont-talk-about-it-txv2lwpp2
11. Unilever (2023). *Unilever sustainable living plan: Conclusive report*. https://www.unilever.com/sustainable-living/
12. Patagonia (2023). *Environmental and social accountability*. https://www.patagonia.com/activism/

13. ITC Limited (2023). *Sustainability report 2022–2023.* https://www.itcportal.com/sustainability/itc-sustainability-report-2024/
14. Reliance Foundation (2024). *Report on CSR initiatives for 2023–24.* https://www.ril.com
15. Dabur India Ltd. (2023). *Business responsibility and sustainability report 2022–23.* https://www.dabur.com
16. State Bank of India (2023). *Fiscal report for the year 2022–2023.* https://sbi.co.in
17. Axis Bank (2023). *Annual report 2022–2023.* https://www.axisbank.com/docs/default-source/annual-reports/annual-report-2022-23.pdf
18. Elkington, J. (1997). *Cannibals with Forks: The triple bottom line of 21st century business.* Capstone Publishing.

A Study of Volatility Spillover of Sustainable Financial Markets on Indian Stock Market

Shradhanjali Panda, Alok Arun, and Prajnya Paramita Pradhan

Abstract

The great financial crisis in 2007 followed by COVID in 2019 made us realized that future lies in sustainable economic development. In this context, sustainable financial markets are increasingly getting more responses from investors, policy-makers, and researchers, as it helps in forecasting risks and returns, optimizing portfolios, and formulating economic policies. Globalization has amplified cross-border financial risks, causing ripple effects across global markets. In this context, this study analyzes the dynamic linkage among the S&P Green Bond Index (S&PGB), S&P 500 ESG Index (S&PESG), Renewable Energy Industrial Index (RENIXX), Dow Jones Sustainability World Index (DJSW), Dow Jones (DOW) and DAX Europe (DAX) and their volatility spillover impact on Indian Stock Market. Volatility for each financial market is measured using the Generalized Autoregressive Conditional Heteroskedasticity (GARCH) model, which provides a comprehensive assessment of market fluctuations. The Polynomial Distributed Lag (PDL) model is applied to investigate the relationship between the Nifty index and various global financial indices. To further validate these findings, a Vector Error Correction Model (VECM) is utilized to capture the long-term equilibrium progress among the selected markets, with the PDL model offering additional confirmation. The results reveal a significant causal linkage between the Indian stock market and most of the global markets

S. Panda
Department of Business Administration, Ravenshaw University, Cuttack, India

A. Arun (✉)
Interscience Institute of Management and Technology, Bhubaneswar, India
e-mail: alok.arun@iimt.ac.in

P. P. Pradhan
KIIT School of Management (KSOM), KIIT (Deemed to be) University, Bhubaneswar, India

S. Patnaik et al. (eds.), *People, Planet and Profit: Advancing Sustainable Business Practices*, Springer Proceedings in Geography and Sustainability,
https://doi.org/10.1007/978-3-032-11296-5_10

analyzed, underscoring the interconnected nature of financial markets and the importance of mitigating cross-border financial risks.

Keywords

Sustainable stock market • Volatility spillover • GARCH • VECM

JEL Classification

G14 • G15 • C22 • C23 • C35

1 Introduction

The world is becoming more aware about the seriousness of climate change, degrading environmental condition and importance of sustainability and social responsibility. Thus, sustainability is today's buzzing word post COVID pandemic. Financial markets are no deviation from this concern as the challenges in economic development has been shifted to sustainable economic development. The impact is already visible as in recent past, stocks related to environmental, social, and governance (ESG) have gained their momentum in lieu of rising investor demands [1]. Majority of listed companies in U.S. stock markets (categorically S&P 500) are now publishing their annual ESG reports [2]. A company's long-term market success is often tied to its strategic decisions regarding ESG performance. Rather than focusing on short-term profit-driven choices, sustainable practices allow a company to realize perpetuity. It can be done in terms of enhancing their brand reputation as well as attracting socially conscious investors who have a commitment to sustainability and social responsibility. In long run, it establishes better stakeholder relationship and easy access to capital as institutional investors increasingly integrate ESG criteria into their investment decisions. Ultimately, by prioritizing ESG considerations, companies can build resilience and achieve sustainable growth, aligning both economic and societal goals. A crucial query that arises at this juncture is how Indian stock market is affected by the spillover effect from these sustainable financial markets.

Information transfer from one stock market to another is presently a widespread phenomenon. This has connected to the globalization of the markets, providing investors with new possibilities to choose various options throughout the world. Globalization of sustainable financial markets has sparked the interest all around the world [3]. It compelled investors sought to expand their investments into several markets in search of greater returns. In recent decades, the topic of financial information spillover from one market to another has gained a lot of attention. A major impact on one financial market is likely to boost the returns and prominence of that market as well as other markets, according to the theory of spillover effects

[4]. There is significant price and volatility spillover in emerging sustainable financial markets, according to numerous researchers [5]. The same conclusion was drawn by these studies: The United States will inevitably have an impact on other emerging markets [6]. To address the above cited discussions and queries, the present research work intends to examine the influence of volatility spillover effect from the six leading sustainable financial indices to the leading indicator of Indian stock market, i.e., NSE Nifty.

At first the volatility of all of these indices is estimated by using a GARCH (1,1) model and then a Vector Error Correction Model (VECM) is fitted to the data. Findings show a significant co-movement between the Nifty and majority of the indices. Since, time series data does not always have a direct relationship with others on the basis of current data; they sometime exhibit a relationship when a lagged variable is taken into consideration. These lagged variables sometimes are not linearly related but the relationship may be non-linear as well. That is why a Polynomial Distributed Lag Model (PDL) is fitted on the data and a significant correlation is observed between the Indian stock market and other leading indicators.

2 Literature Review

Environmental, social, and governance (ESG) performance significantly impacts stock market performance, as a long-term ESC practice helps companies for a better performance as well as effective risk management, operational efficiency, and enhanced brand reputation [7]. These firms tend to attract more investment, experience reduced volatility, and enjoy a lower cost of capital, as they are perceived as lower-risk by investors. While the initial costs of ESG initiatives can be high, they often lead to long-term profitability. Sector-specific impacts vary, with environmental practices being critical in energy and utilities, governance and social issues driving performance in financial services, and data privacy and diversity influencing the tech sector [8]. Empirical evidence suggests a positive correlation between ESG performance and stock returns, though short-term underperformance may occur due to upfront costs [9]. ESG-focused companies also tend to be more resilient during financial crises. However, challenges such as data inconsistencies and green washing can hinder accurate assessment [10]. With increasing regulatory pressures and growing investor demand for sustainable investments, ESG performance is becoming an essential factor in financial models and a key driver of stock market success [11]. Since governments and regulatory bodies around the world have begun to pay more attention to ESG, ESG investment has progressively taken over as the dominant trend in the global investment community. Many studies show that the market overreacts to ESG news, which could negatively impact investor behavior and market efficiency. In light of carbon neutrality, the international community is increasingly advocating for sustainable investment. Thus, more researchers are looking into the connection between green finance, renewable energy stock indexes, and the ESG index. They think that investing in renewable energy stocks might be better done by buying stocks of companies with the highest environmental, social, and governance (ESG) ratings [6].

Another important segment among the sustainable financial market is green bond market. With its strong ties to the larger financial sector, the green bond market is a vital tool for raising money for ecologically friendly initiatives like energy efficiency, renewable energy, and climate change mitigation. By integrating environmental goals into mainstream finance, green bonds attract institutional and retail investors seeking both financial returns and alignment with environmental, social, and governance (ESG) criteria, thereby contributing to portfolio diversification. The rapid growth of the green bond market has enhanced its influence on the financial system, offering issuers a lower cost of capital due to high investor demand and increasing liquidity in the market [12]. Regulatory frameworks and standards like the Green Bond Principles (GBP) further strengthen this connection by ensuring transparency and accountability, making green bonds a vital component of sustainable finance within the global financial market [13]. Asymmetric volatility was not observed in many studies of green bonds; however, unlike equity, they were sensitive to positive return shocks in terms of volatility [14].

The renewable energy market significantly impacts the stock market by driving growth, influencing investor sentiment, and reshaping industry dynamics. As global demand for clean energy rises, companies involved in renewable technologies—such as solar, wind, and hydrogen—experience increased market valuation and attract substantial investment [15]. Government policies, tax incentives, and environmental regulations further enhance the sector's appeal, prompting traditional energy firms to diversify into renewable to remain competitive. This transition creates new market opportunities while reducing the volatility associated with fossil fuel prices. As a result, renewable energy stocks are becoming a key driver of sustainable growth in global financial markets, contributing to the rise of environmental, social, and governance (ESG) investing.

An investigation of the U.S. stock market's relationship with the other three Central European markets, both in the short and long term [16]. The study concluded that markets are unrelated over the long term. Using data with extremely high frequencies from markets in the US, UK, Germany, and a few other European countries, it was also discovered that information is shared between markets in less than an hour but not in a day or more. Before quantifying market cointegration, some examinations on the impact of the GARCH model on stock returns established interrelation of various global markets [17, 18]. On the other hand, other researchers have focused on the variables that affect volatility spillovers and in addition to stock market crises, the factors like volatility, industry nature, bilateral trade, etc., influence the spillover of volatility from one stock market to another [19, 20]. According to Reference [21], stock market co-movements are positive when trade volume is high but negative when bilateral exchange rate volatility is elevated. A small number of studies focused on the volatility of the Indian stock market, but many examined the impact of volatility on the financial market [22].

Some studies found a significant volatility spillover between the Indian and U.S. stock markets using the Granger causality test. Reference [23] looked at the relationships between three markets, including those in Singapore, Taiwan, and India, and discovered no cointegration between the stock indices of these nations. Reference [24] looked into the volatility spillover from the United States and Japan to India, Pakistan, and Sri Lanka. He found that there was a significant return spillover from the U.S. and Japan to all three countries, but an even greater volatility spillover from Japan to Pakistan and the U.S. to Sri Lanka and India [25].

3 Research Gap and Objectives of the Study

Significant number of researches has been done in the area of volatility spillover impact on different stock markets around the world [26, 27]. The basic statistical tool used for testing volatility spillover effect is GARCH model, i.e., developed by References [28, 29]. The past literatures have mainly used Granger Causality test to find out the volatility and return spillovers across the countries and the studies are limited to test the spillover effect among the stock exchanges [30, 31]. But few works have been done considering sustainable financial markets and in Indian context [32], research regarding spillover impact of sustainable financial markets on Indian stock market is almost nil. In the context of sustainable financial market, basically some selected market indices are considered for the study. The above school of thoughts identified five sustainable markets and those are used in the present study. They are S&P Green Bond Index (S&PGB), S&P 500 ESG Index (S&PESG), Renewable Energy Industrial Index (RENIXX), Dow Jones Sustainability World Index (DJSW), Dow Jones (DOW) and DAX Europe (DAX). Dow Jones is also used as a proxy to U.S. market.

Thus, present study uses a VECM and a GARCH model to establish the volatility spillovers from the six sustainable financial indices to Indian stock market. This paper's goal is to determine how the Indian stock market is interdependent with the other six sustainable financial markets.

4 Research Methodology

4.1 Data

Paris Climate Agreement in 2015 boosted renewable energy segment and thus the related stocks. Apart from it, another sustainable financial instrument is also gaining its ground day by day and, i.e., green bonds, issued for mitigating climate change projects. In this regard, the role of sustainability index cannot be avoided as it is constructed to provide sustainability related information to investors [33]. Lastly, sustainability and ESG are like two sides of a coin and goal of sustainability can be achieved only through the path of ESG. Keeping the above logic as backdrop, the objective of the paper aims at testing the volatility spillover effect of sustainable financial indices on Indian market. For that purpose, S&P Green Bond Index (S&PGB), S&P 500 ESG Index (S&PESG), Renewable Energy

Industrial Index (RENIXX), Dow Jones Sustainability World Index (DJSW), Dow Jones (DOW), and DAX Europe (DAX) are used in the present study. Some previous studies have found Green Bond Index, S&P ESG, and RENIXX are closely related to sustainability [6, 9].

The aforementioned sustainable financial markets were chosen mainly because of their commitment to and goal of sustainable practices. A market benchmark that monitors the global green bond market, the S&P Green Bond Index has multiple connections to sustainable development. Similar to this, the S&P 500 ESG Index aims to increase allocation to businesses with better sustainability traits while preserving a wide range of exposure, which has historically produced performance comparable to benchmarks. Sustainable development can be greatly aided by renewable energy sources like solar, wind, hydro, geothermal, and biomass. Because renewable energy is a crucial part of sustainable development, the Renewable Energy Industrial Index (RENIXX) is a global stock index that tracks the top 30 renewable energy companies worldwide. The Dow Jones Sustainability World Index (DJSI) is an additional sustainable financial market. This international standard, which is closely tied to sustainable development, evaluates a business's environmental, social, and governance (ESG) performance: The DJSI assists investors in locating businesses with robust ESG policies and a dedication to long-term sustainable development. The top 10% of the S&P Global BMI's 2,500 biggest companies are represented by the DJSI. DAX is the final financial market mentioned. The DAX Index's goal is to represent the DAX Index while simultaneously lowering its carbon intensity by at least 30% and optimizing its ESG score. Regarding the parent index, the anticipated tracking error is limited.

4.2 Method

As the spillover effect is tested on Indian Stock Market, thus Nifty is used as the proxy. For robustness of the result, two international indices, i.e., Dow Jones and DAX Europe are also used. All data are collected from the website of yahoofinance.com, investing.com, Bloomberg database, and Nifty indices ranging the time period from 1st January, 2022 to 1st January, 2025. First of all, the data series are matched and only the common dates of the closing prices are taken and then the return for all the series are calculated by using the formula: $r_t = \ln (C_t/C_{t-1})$, where r_t is the return of an index at time t, C_t and C_{t-1} are the closing values of an index at time t and at time t-1, respectively.

In next step, Augmented Dickey Fuller test is conducted to test data stationarity series in order. The aim of the test is to confirm the need for the change in the logarithmic price series. Followed by ADF test, Johansen's Co-integration test is conducted for lag specification results. The objective here is to test relationships among the data series. All data used here are time series data. So, this test is suitable to check the co-integrating links among them. Three information criteria—the Bayesian Information Criterion (BIC), the Akaike Information Criterion (AIC), and the Hannan-Quinn (HQ) information criterion—must be used to determine the

Table 1 Descriptive statistics of the indices

Index	No. of Obs	Mean	Std. Dev	Median	Range	Skew	Kurtosis	SE
S&PGB	2260	578.9	134.48	607.2	626.06	−0.9723601	3.436722	2.83
S&PESG	2260	9019	270.22	9322	10,122.59	−0.00709245	1.654449	56.84
RENIXX	2260	16,792	5825.35	16,290	23,004.37	0.4276434	2.015441	122.54
DJSW	2260	1874.6	683.7	1867.6	2850.12	0.346029	1.976443	14.38
DOW	2260	2828	568.94	2869	3791.19	0.8762884	4.973045	11.97
DAX	2260	8762	1618.42	8629	1.027.047	−0.1218934	3.374689	34.04
Nifty	2260	7319	2510.04	6695	9809.15	0.2806029	1.933255	52.80

*The Author

lag of each test prior to using the Johansen's cointegration test. At the end, ARCH, GARCHA, and VECM models are used to test volatility spillover of the indices. These models have the specialty in testing the wide fluctuations, popularly known as volatility in financial markets.

4.3 Result and Analysis

Looking at seven financial indices' descriptive data before evaluating volatility, the mean, skewness, kurtosis, and standard deviation (S.D.) are displayed in Table 1. Each return series has a skewness larger than zero, which means that the distribution of the series is skewed relative to the normal distribution. The distribution is more convex than the normal distribution, as indicated by the kurtosis of each return series being greater than +1. The time series data is regarded as nonstationary at the level. Consequently, the initial research involves utilizing unit root analyses to verify the stationarity of every variable, or, to put it another way, —uring out the integration order. To determine the integration of order, we employed the Augmented Dickey Fuller test [34]. The results of the ADF Test of the return series presented in Table 2 shows that it is a stationary series of order 1(0), confirming the need for the change in the logarithmic price series.

Cointegration between Nifty and all other 6 variables are measured by using a Johansen's Cointegration test. All the indices are paired with Nifty one by one and then a Johansen's Cointegration test is applied on the data.

5 Johansen's Cointegration Test

Since the AIC, BIC, and HQ criteria have the lowest values in all variables except S&PESG, Table 3 shows that lag 3 is the most preferred option for the cointegration between the pairs of indices taken into consideration in this study. Therefore,

Table 2 Result of augmented dickey-fuller unit root test

Variables	ADF statistics	p value	Decision
S&PGB	−42.2616	0.00	Stationary
S&PESG	−33.1329	0.00	Stationary
RENIXX	−34.4542	0.00	Stationary
DJSW	−34.4373	0.00	Stationary
DOW	−43.6784	0.00	Stationary
DAX	−42.4347	0.00	Stationary
Nifty	−33.6717	0.00	Stationary

** H_0 is rejected at 1% level of significance
* H_0 is rejected at 5% level of significance
* The Author

we will decide that the ideal lag for the Johansen's cointegration process is 3. Each pair of indices is subjected to Johansen's cointegration test following the specification of the models' lag order. Johansen's methodology is defined by applying the Vector Autoregression (VAR) of order p concept, which is provided by:

$$y_t = \mu + A_1 y_{t-1} + \ldots + A_p y_{t-p} + \epsilon_t \ldots \tag{1}$$

where y_t is a (n x 1) vector of variables of order one integration, commonly represented by I (1), and ε_t is a (n x 1) vector of residuals. One way to rewrite this VAR is as follows:

$$\Delta y_t = {}^- + \prod y_{t-1} \sum_{i=1}^{p-1} \left(\sum_{j=i+1}^{p} A_j \right) \Delta y_{t-i} + \epsilon_t \tag{2}$$

where $\Pi = \sum_{i=1}^{p} A_i$ –I.

If the coefficient matrix Π has reduced rank r < n, then there exist n x r matrices α and β each with rank r such that $\Pi = \alpha\beta'$ and $\beta' y_t$ is stationary. Each column of β represents a cointegrating vector, r is the number of cointegrating vectors, and the elements of α are referred to as the adjustment parameters in the vector error correction model (Bollerslev, 1990). Johansen offers the trace test and maximum eigenvalue test. The significance of these canonical correlations and, thus, the reduced rank of the Π matrix, were tested using two different likelihood ratios, as follows:

$$J_{trace} = -T \sum_{i=r+1}^{n} \ln(1 - \lambda\prime_i) \tag{3}$$

$$J_{eigen} = -T\ln\left(1 - \lambda'_{r+1}\right) \tag{4}$$

where λ' is the largest canonical correlation and T is the sample size. H0: r cointegrating vectors or rank = r against the alternative hypothesis H_1: n cointegrating

Table 3 Lag specification for cointegration with NSE Nifty

Index	Lag	AIC	BIC	HQ
S&PGB	1	7.130898	7.138504	7.133674
	2	7.125952	7.136092	7.129653
	3	**7.104503***	**7.117179***	**7.109129***
S&PESG	**1**	**9.899349***	**9.906955***	**9.902125***
	2	9.900233	9.910374	9.903934
	3	9.900760	9.913436	9.905386
RENIXX	1	11.02076	11.02836	11.02353
	2	10.99284	11.00298	10.99654
	3	**10.98698***	**10.99965***	**10.99160***
DJSW	1	6.611200	6.618805	6.613975
	2	6.581225	6.591366	6.584926
	3	**6.574486***	**6.587162***	**6.579112***
DOW	1	11.30337	11.31098	11.30615
	2	11.27921	11.28935	11.28291
	3	**11.26233***	**11.27501***	**11.26696***
DAX	1	12.46772	12.47533	12.47050
	2	12.46688	12.47702	12.47058
	3	**12.45587***	**12.46855***	**12.46050***

* The Author

vectors or rank = n is the null hypothesis for the trace test. At the same time H_0: r cointegrating vectors against the alternative hypothesis H1: r + 1 cointegrating vectors is the null hypothesis for the maximum eigen value test. Despite the fact that neither of these tests uses a chi square distribution, Reference [27] shows asymptotic critical values. Table 4 lists the findings of the cointegration tests.

Co-integration tests are useful in the context of volatility spillover because they can be used to measure financial market trends and the time it takes for markets to catch up to one another. The existence of long-term patterns in financial markets is indicated by the interchangeable terms co-integration and co-movement. GARCH models can be used to identify patterns in the volatility of asset prices. GARCH models are better at forecasting volatility than traditional models because they account for conditional volatility clustering and autocorrelation.

All of the indices are cointegrated with the Nifty, according to Table 4, indicating a long-term link between the Nifty and every other index. In order to determine the short-term relationship between each set of cointegrated indices, a VECM model is fitted to the data [35].

The data is now fitted to a Vector Error Correction Model (VECM) in order to determine the volatility spillover from all seven indexes to the Indian stock market (Nifty). Here is a definition of the VECM model: When co-integration between

Table 4 Results of cointegration with Nifty for lag 3

Index	Eigen value		Trace	
	H_0 for rank = 0	H_0 for rank = 1	H_0 for rank = 0	H_0 for rank = 1
S&P GB	55.93***	4.15	60.07***	4.15
S&P ESG	13.95*	3.79	17.75	3.79
RENIXX	16.20**	5.41	21.61**	5.41
DJSW	19.49**	5.94	25.43***	5.94
DOW	92.20***	4.03	96.23***	4.03
DAX	223***	4	227***	4

Note
* Means null hypothesis is rejected at 10% level of significance
** Means null hypothesis is rejected at 5% level of significance
*** Means null hypothesis is rejected at 10% level of significance
* The Author

two series is observed, it is concluded that the time series vectors have a long-term relationship [36]. This is when VECM becomes necessary. To test whether, the two cointegrated series are related in short run a VECM is applied on the pair of data vectors. In this paper, VECM is applied to those pairs of indices which are established as cointegrated series on the basis of a Johansen cointegration test.

$$\Delta y_t = \mu + \Pi\, y_{t-1} + \sum_{i=1}^{p-1}\left(-\sum_{j=i+1}^{p} A_j\right)\Delta y_{t-i} + \epsilon_t \tag{5}$$

$$\text{Or,}\ \ \Delta y_t = \mu + \alpha\beta\prime\, y_{t-1} + \sum_{i=1}^{p-1}\left(-\sum_{j=i+1}^{p} A_j\right)\Delta y_{t-i} + \epsilon_t \tag{6}$$

where Π is defined as in Eq. (5).

The results of VECM are reported in Table 5.

From Table 5, it can be observed that S&PGB is insignificant at all the lags, whereas S&PESG is significant in lag 1 and 3 at 10% level of significance; RENIXX is significant at lag 1 only at 5% level of significance. Others are insignificant at all lags. From this result, it can be said that for all the pair of variables has a long-run link with Nifty but except S&PESG and RENIXX other indices do not have any short run relationship with Nifty. This may lead to the conclusion that the information flow from other international financial markets to Indian stock markets is fair and no canvassing is evident.

After a short-run relationship is established by using a VECM, the non-linear dependence between Indian stock market and all other six markets is modeled by using a Polynomial Distributed Lag (PDL) model. Before applying a PDL model on the data first of all the stationarity of the first difference of all the data series is checked with the help of Augmented Dickey Fuller Test (ADF). After ensuring

Table 5 Result of VECM of indices cointegrated with Nifty

Parameters	Nifty versus S&P GB	Nifty versus S&P ESG	Nifty versus RENIXX	Nifty versus DJSW	Nifty versus DOW	Nifty versus DAX
μ	0.0000014	−0.0002**	−0.0003***	−0.0005***	0.000045	0.000012
α_1	−0.1369***	−0.1554***	−0.1309**	−0.1366**	−0.1343***	−0.1486***
α_2	0.2079	0.0650	−0.000015***	0.5268***	0.0456*	0.0552***
β_1	734.4855**	1638.2315***	1611.6720***	1736.7691***	675.9760**	552.9901*
β_2	770.8691**	979.2620**	980.6882***	1049.2177***	694.6026**	647.9242**
β_3	121.8073	282.8600*	165.2654*	155.6304	8.6537	122.9639
β_4	−34.8690	−1093.1597*	−772.0216**	−814.9956	106.9848	159.6283
β_5	−56.7508	−126.9101	−16.1377	−115.7876	35.2599	110.6652
β_6	−0.9467	−337.4107*	−128.9532	−46.6929	8.6537	−1.5461

*** H_0 is rejected at 1% level of significance
** H_0 is rejected at 5% level of significance
* H_0 is rejected at 10% level of significance
* The Author

that the data vectors are stationary, a GARCH(1,1) is fitted to the data and the volatility of each of the variables are estimated. A GARCH(p,q) model is defined as following:

$$Y_t = a_0 + a_1 X_t + \varepsilon_t,\ \varepsilon_t \sim N\left(0, \sigma_t^2\right) \tag{7}$$

$$\sigma_t^2 = \alpha_0 + \sum_{i=1}^{p} \alpha_i \varepsilon_{t-i}^2 + \sum_{j=1}^{q} \beta_j \sigma_{t-j}^2 \tag{8}$$

Tabulated data series and result of GARCH are as follows.

Using the estimates of the parameters from Table 6, GARCH volatility of all the indices has been estimated. Now, on the series of volatilities of all the variables, a PDL model is fitted. A PDL model is defined as follows:

$$Y_t = \alpha + \beta_0 X_t + \beta_1 X_{t-1} + \beta_2 X_{t-2} + \beta_3 X_{t-3} + e_t \tag{9}$$

where $\boldsymbol{\beta_i} = \mathbf{a_0} + \mathbf{a_1 i} + \mathbf{a_2 i^2}$, where I is the required lag length and if β_i is approximated by using a quadratic polynomial.

It means that $\beta_0 = \mathbf{a_0}$

$$\beta_1 = a_0 + a_1 + a_2 \tag{10}$$

$$\beta_2 = a_0 + a_1 2 + a_2 4 \tag{11}$$

$$\beta_3 = a_0 + a_1 3 + a_2 9 \tag{12}$$

Now replacing the values of β's into Eqs. (10–12), it can be interpreted as,

$$\begin{aligned} Y_t = {} & \alpha + a_0 X_t + (a_0 + a_1 + a_2) X_{t-1} \\ & + (a_0 + a_1 2 + a_2 4) X_{t-2} \\ & + (a_0 + a_1 3 + a_2 9) X_{t-3} + e_t \end{aligned} \tag{13}$$

$$\begin{aligned} Y_t = {} & \alpha + a_0 (X_t + X_{t-1} + X_{t-2} + X_{t-3}) \\ & + a_1 (X_{t-1} + 2X_{t-2} + 3X_{t-3}) \\ & + a_2 (X_{t-1} + 4X_{t-2} + 9X_{t-3}) + e_t \end{aligned} \tag{14}$$

The results of PDL model are tabulated in Table 7.

With the exception of DAX at lags 1 and 2, it is evident that S&PGB, S&PESG, RENIXX, and DJSW are significant at the 1% level of significance for present values as well as for lags 1, 2, and 3. At the 5% level of significance, DOW is significant for lag 1.

Table 6 Result of GARCH model

Parameters	Estimates of the indices						
	S&P GB	S&P ESG	RENIXX	DJSW	DOW	DAX	Nifty
μ	−0.0002484	0.0007011**	0.0009634***	0.0009312***	0.0003976	−0.000161	0.0009117***
ω	0.00001855***	0.0000047***	0.00000528***	0.000005093***	0.00001527***	0.000007434***	0.000004485***
α_1	0.003692***	0.07744***	0.1586***	0.1473***	0.02906***	0.02443***	0.1240***
β_1	0.9590***	0.9053***	0.8174***	0.8322***	0.9687***	0.9732***	0.8638***
Jarque bera test	15,536.85***	674.8499***	481.4447***	612.8166***	9057.089***	9865.799***	1993.441***

*** H_0 is rejected at 1% level of significance
** H_0 is rejected at 5% level of significance
* The Author

Table 7 Result of polynomial distributes lag model

Lags	Nifty versus independent variables					
	S&P GB	S&P ESG	RENIXX	DJSW	DOW	DAX
0	**0.280*****	**0.44000*****	**0.29100*****	0.2990***	**0.226****	**0.3320*****
1	**0.208*****	**0.06620*****	**0.03170*****	0.0403***	**0.130**	**0.1930*****
2	**0.196*****	**−0.07790*****	**−0.06770*****	−0.061***	**0.105**	**0.1130**
3	**0.244*****	**0.00793*****	**−0.00694*****	−0.004***	**0.153**	**0.0905**

*** H_0 is rejected at 1% level of significance
** H_0 is rejected at 5% level of significance
* H_0 is rejected at 10% level of significance
* The Author

6 Conclusion

Two components of risk, i.e., systematic and unsystematic play pivotal role in investment as well as financial market. Though unsystematic risk can be ignored through proper diversification yet systematic risk is unavoidable in nature and is an integral part of every financial market as it is caused by external factors. Irrespective of systematic or unsystematic risk, recent past has made the investor-race realized that global financial market is going to be affected by elements like climate change, COVID, etc. It leads to the path of sustainability and the future buzzing word in investment and finance is going to be "Sustainable Financial Market" and they are going to affect the market mood and trend of investment. With the objective of testing the volatility spill out effect of sustainable financial markets on Indian market, six different financial markets are used in the study. The sample used in the study is directly related to the concept of sustainability and are sustainable financial markets. They are S&P Green Bond Index (S&PGB), S &P 500 ESG Index (S&PESG), Renewable Energy Industrial Index (RENIXX), Dow Jones Sustainability World Index (DJSW), Dow Jones (DOW) and DAX Europe (DAX). S & P CNX Nifty is selected as the representative of Indian stock market and the spill-over impact of the above cited sustainable financial markets on it is tested in the present work. PDL, GARCH, and VECM tools are employed for this purpose and result can be interpreted that India is affected by the sustainable cross-border volatility. Moreover, co-integration test confirms it and it has been seen that all the indices taken in this study are co-integrated with Nifty which means that the co-movement of Nifty and all the indices is established. On the other hand, from the results of VECM, it can be seen that there exists a volatility spillover from all the sustainable markets to Indian stock market except S&P and DAX since all other indices have been significant in different lags. While using the Polynomial Distributed Lag (PDL) model, it is experienced that all the markets have some contribution to the movement of the Indian stock market. The result can be interpreted that Green Bond, RENIXX, and S&PESG have strong influence on Indian stock market whereas the impact of S&P and Dow have comparatively less impact. In a nutshell, it can be stated that India is influenced by volatility spillover from different sustainable financial markets from different parts of the world. It displays, with India's integration into global sustainable finance, more foreign partnerships and investments in green projects may be attracted. India may see a rise in capital inflows if sustainable markets expand abroad, particularly in ESG industries (green tech, renewable energy, etc.). However, India might have to cope with the fallout from international events over which it has no control, such as shifts in international policy or crises in other nations. Therefore, it can be said that volatility spillover is a sign of financial interconnectedness rather than something that is intrinsically good or bad. What counts is how effectively India handles risks while seizing international opportunities in the field of sustainable finance.

References

1. Aneja, R., & Makkar, A. (2017). Relationship between stock prices and financial crises: A case study of Indian commercial banks. *Asian Journal of Management, 8*(3), 809.
2. Chang, X. (2022). Sustainable finance: ESG/CSR, firm value and investment returns. *Asia-Pacific Journal of Financial Studies, 51*, 325–371.
3. Campbell, J. Y., Lo, A. W., MacKinlay, A. C., & Whitelaw, R. F. (1998). The econometrics of financial markets. *Macroeconomic Dynamics, 2*(4), 559–562.
4. Rashid Sabri, N. (2004). Stock return volatility and market crisis in emerging economies. *Review of Accounting and Finance, 3*(3), 59–83.
5. Brailsford, J.(1996). Volatility spillovers across Tasman. *Australian Journal of Management*, 21(1), 13–27.
6. Tiwari, A. K., Abakah, E. J. A., Gabauer, D., & Dwumfour, R. A. (2022). Dynamic spillover effects among green bond, renewable energy stocks and carbon markets during COVID-19 pandemic: Implications for hedging and investments strategies. *Global Finance Journal, 51*, Article 100692.
7. Chiang, T. C., Yang, S. Y., & Wang, T. S. (2000). Stock return and exchange rate risk: Evidence from Asian stock markets based on a bivariate GARCH model. *International Journal of Business, 5*(2), 97–117.
8. Enders, W. (2008). *Applied econometric time series*. John Wiley & Sons.
9. Hanif, W., Hernandez, J. A., Mensi, W., Kang, S. H., Uddin, G. S., & Yoon, S. M. (2021). Nonlinear dependence and connectedness between clean/renewable energy sector equity and European emission allowance prices. *Energy Economics, 101*, Article 105409.
10. Park, D., Park, J., & Ryu, D. (2020). Volatility spillovers between equity and green bond markets. *Sustainability, 12*(9), 3722.
11. Yousaf, I., Ali, S., & Wong, W. K. (2020). Return and volatility transmission between world-leading and Latin American stock markets: Portfolio implications. *Journal of Risk and Financial Management, 13*(7), 148.
12. Joshi, N. A., Mehta, D., Patel, N., & Patel, B. (2021). Volatility analysis and volatility spillover across equity markets between India and Europe. *SMART Journal of Business Management Studies, 17*(1), 31–41.
13. Pham, L. (2021). Frequency connectedness and cross-quantile dependence between green bond and green equity markets. *Energy Economics, 98*, Article 105257.
14. Joshi, N. A., Mehta, D., Patel, B., & Patel, N. (2021). Causality and cointegration among stock market indices: A study of developed markets with sensex. *International Journal of Accounting and Finance Review, 7*(1), 31–52.

15. Shahbaz, M., Topcu, B. A., Sarıgül, S. S., & Vo, X. V. (2021). The effect of financial development on renewable energy demand: The case of developing countries. *Renewable Energy, 178*, 1370–1380.
16. Bouteska, A., & Regaieg, B. (2018). Loss aversion, overconfidence of investors and their impact on market performance evidence from the US stock markets. *Journal of Economics Finance and Administrative Science, 25*(50), 451–478.
17. Bollerslev, T., Chou, T. Y., Kroner, K. (1992). ARCH modeling in finance: A review of the theory and empirical evidence. *Journal of Econometrics,* 52, 5–59.
18. Chou, R. Y. (1988). Volatility persistence and stock valuations: Some empirical evidence using GARCH. *Journal of Applied Econometrics*, 279–294.
19. Nelson, D. B. (1991). Conditional heteroskedasticity in asset returns: A new approach. *Econometrica: Journal of the econometric society*, 347–370.
20. Roll, R. (1989). Price volatility, international market links, and their implications for regulatory policies. *Journal of Financial Services Research, 3*, 211–246.
21. Kaushik, N. (2018). Corporate governance and Indian stock market: A study with asymmetric conditional volatility models. *Asian Journal of Management, 9*(1), 80–86.
22. Maraqa, B., & Bein, M. (2020). Dynamic interrelationship and volatility spillover among sustainability stock markets, major European conventional indices, and international crude oil. *Sustainability, 12*(9), 3908.
23. Kanas, A. (2000). Volatility spillovers between stock returns and exchange rate changes: International evidence. *Journal of Business Finance and Accounting, 27*(3–4), 447–467.
24. Kim, K. H. (2003). Dollar exchange rate and stock price: Evidence from multivariate cointegration and error correction model. *Review of Financial Economics, 12*(3), 301–313.
25. Chinhema, S. (2020). The relationship between change management and organisational effectiveness in the Zimbabwean banking industry. *The International Journal of Business and Management, 8*(4), 417–438.
26. Eun, C. S., & Shim, S. (1989). International transmission of stock market movements. *Journal of Financial and Quantitative Analysis, 24*(2), 241–256.
27. Johansen, S., & Juselius, K. (1990). Maximum likelihood estimation and inference on cointegration—with applications to the demand for money. *Oxford Bulletin of Economics and Statistics, 52*(2), 169–210.
28. Bollerslev, T. (1986). Generalized autoregressive conditional heteroskedasticity. *Journal of Econometrics, 31*, 307–327.
29. Bollerslev, T. (1990). Modeling the coherence in short-run nominal exchange rates: A multivariate generalized ARCH approach. *Review of Economics and Statistics, 72*, 498–505.
30. Baele, L. (2005). Volatility spillover effects in European equity markets. *Journal of Financial and Quantitative Analysis, 40*, 373–401.
31. Nelson, D. B., & Cao, C. Q. (1992). Inequality constraints in the univariate GARCH model. *Journal of Business and Economic Statistics, 10*(2), 229–235.
32. King, M. A., & Wadhwani, S. (1990). Transmission of volatility between stock markets. *The Review of Financial Studies, 3*(1), 5–33.
33. Mishra, A. K., Ghate, K., Renganathan, J., Kennet, J. J., & Rajderkar, N. P. (2022). Rolling, recursive evolving and asymmetric causality between crude oil and gold prices: Evidence from an emerging market. *Resources Policy, 75*, Article 102474.
34. Dickey, D. A., & Fuller, W. A. (1979). Distribution of the estimators for autoregressive time series with a unit root. *Journal of the American Statistical Association, 74*(366a), 427–431.
35. Engle, R. F., & Granger, C. W. (1987). Co-integration and error correction: representation, estimation, and testing. *Econometrica: Journal of the Econometric Society*, 251–276.
36. Engle, R. F. (1982). Autoregressive conditional heteroscedasticity with estimates of the variance of United Kingdom inflation. *Econometrica: Journal of the Econometric Society*, 987–1007.

Circular Economy and Sustainability: Strategies, Challenges, and Implications

K. Surya Narayana, B. Muthukrishnan, and Busige Nishantha

Abstract

This paper studies the principles, issue, and potential of the Circular Economy (CE) as a transformative model for sustainable development. CE principles, as outlined by the Ellen MacArthur Foundation, emphasize on designing out waste and pollution, keeping products and materials in use, and regenerating natural systems. The study reviews relevant literature from leading relevant scholarly journals to study the theoretical foundations and practical applications of CE. A mixed-methods approach is used, combining a systematic literature review with case studies from the European Union and China, to present successes and obstacles in CE implementation. The findings reveal that CE can deliver significant environmental, economic, and social gains, including reduced resource consumption, improved recycling rates, and job creation. The EU's Circular Economy Action Plan created over €147 billion in economic benefits and enhanced recycling rates, while China's eco-industrial parks decreased energy and water use by over 40% and generated thousands of green employment opportunities. This study critically examines implementation models and identifies enduring global challenges, such as regulatory fragmentation, cultural resistance, and technological limitation, thereby validating existing frameworks and establishing a citadel for more inclusive, adaptable, and effective CE practices. This originality positions the study as a significant resource for policymakers, academics, and practitioners seeking to implement circular solutions in both developed and developing countries.

K. Surya Narayana (✉) · B. Muthukrishnan
Department of MBA, MLR Institute of Technology, Hyderabad, India
e-mail: suryan568@gmail.com

B. Nishantha
Faculty of Management & Finance, University of Colombo, Colombo, Sri Lanka
e-mail: nishantha@mos.cmb.ac.lk

S. Patnaik et al. (eds.), *People, Planet and Profit: Advancing Sustainable Business Practices*, Springer Proceedings in Geography and Sustainability,
https://doi.org/10.1007/978-3-032-11296-5_11

Keywords

Circular economy • Sustainability • Resource efficiency • Waste reduction • Systems thinking

1 Introduction

The notion of Circular Economy (CE) offers a transition from conventional linear economic models, which are based on a "take, make, dispose" paradigm, to a more reformative system where the value of products, materials, and resources is preserved in the economy for as long as possible. The CE focuses on system thinking, emphasizing on closed-loop processes where waste and pollution are lessened, and materials are continuously reused. Systems thinking, closely associated with Circular Economy (CE), entails a comprehensive examination of the interdependencies and feedback loops of players, processes, and results across value chains and sectors [8]. Systems thinking is used here both as a theoretical framework and as a methodological instrument to examine the functioning of Circular Economy policies and practices within intricate socio-economic and ecological environments, especially in the EU and China.

As stated by the Ellen MacArthur Foundation (2015), the core principles of CE are designed to decrease dependency on limited resources and lessen environmental impacts while promoting economic growth [2]. Nonetheless, while the theoretical allure of CE is well recognized, its practical implementation poses enormous challenges. Regulatory disparity across regions, technological limitation in recycling intricate materials, and paucity of supportive infrastructure, especially in emerging economies, impede steady advancement. Furthermore, cultural and consumer resistance to refurbished or recycled items continue to be a significant impediment, exacerbated by inadequate awareness and education.

The objective of this paper is to review the theoretical grounding of CE, analyze empirical cases and its application, and recognize issues and opportunities in scaling its practices at even global level. By synthesizing already existing literature and analyzing real-time cases, the study aims to present insights into the potential of CE in adding to United Nations Sustainable Development Goal (SDG) 12, which also prioritizes on responsible consumption and production [2].

2 Literature Review

The CE is reinforced by principles of systems thinking, where economic actions are designed to copycat natural cycles of rejuvenation. As described by the Ellen MacArthur Foundation (2015), CE is based on the citadel of the following three key principles:

(a) Designing out waste and pollution,
(b) Keeping products and materials in use, and
(c) Regenerating natural systems.

This approach aligns closely with the United Nations Sustainable Development Goals (SDGs), particularly SDG 12 (responsible consumption and production).

The notion of the **Circular Economy (CE)** has gained importance as a strategic approach to sustainable development. The goal is to decouple economic growth at the cost of resource consumption and environmental degradation. Rooted in principles of closing material loops, designing out waste, and extending product lifecycles, CE contrasts sharply with the conventional linear "take-make-dispose" economic model [3, 6].

In this study, we adopt Kirchherr et al. [6] comprehensive definition, studying CE as a system-level solution that incorporates reducing, reusing, and recycling activities supported by technological, regulatory, and behavioral shifts [6].

The Circular Economy (CE) has emerged as a revolutionary paradigm. It is intended to decouple economic progress from environmental degradation by designing out waste, maintaining the use of goods and resources, and regenerating natural systems. The expanding corpus of scholarly literature offers a comprehensive perspective on CE, its development, definitions, frameworks, issues, and implementation tactics.

Sustainability, in this paper, is treated as an integrative framework which depicts environmental integrity, economic resilience, and social equity [4]. The interplay between CE and sustainability is critically examined examine to understand how CE practices align with and add to sustainability objectives by assessing not only the environmental outcomes like reductions in waste and emissions, but also the social benefits (e.g., job creation, health improvements) and economic impacts (e.g., increased efficiency, industrial innovation).

Geissdoerfer et al. [3] provide a fundamental perspective CE by distinguishing it from conventional linear models and other sustainability paradigms, including cradle-to-cradle, industrial ecology, and the performance economy. They propose a conceptual framework that explains the essential concepts of CE—explicitly its focus on closed-loop systems, prolonged product life cycles, and regenerative resource flows—as catalysts for economic resilience and ecological sustainability. The effort is to establish CE not just as an environmental instrument but also as a potentially transformational model for industrial and economic systems [3].

Ghisellini et al. [4] study the development of CE across academic, industrial, and policy domains, stressing its growing importance in global sustainability discourses. The authors examine the potential of the CE to minimize environmental impact while improving resource efficiency and economic output, using real-world case studies from the European Union and China. They also focus on systemic challenges to implementation, including fragmented policy frameworks, lack of incentives for innovation, and institutional inertia. The study closes by advocating for more inter-sectoral collaboration and the incorporation of CE into national sustainability policies [4].

Kirchherr et al. [6] add to the body of conceptual discourse by examining 114 distinct definitions of CE and reveal substantial inconsistencies and ambiguities. Their critical analysis contends that the lack of universally accepted definitions affects both policy formulation and practical implementation. The research emphasizes on the need for standardizing CE principles around key objectives such as waste reduction, resource circularity, and systems thinking and to increase cooperations and collaborations across industries and nations. Their findings underlie the importance of a shared language in promoting global consensus and coordinated action [6].

Korhonen et al. [7] offer a critical perspective, characterizing CE as a "contested concept" marked by contradictions between theory and practice. They contend that although CE extensively borrows from the domains of industrial ecology and systems theory, its application often overlooks deeper socio-economic and institutional aspects. The authors advocate for a more comprehensive analytical framework that considers power dynamics, market inefficiencies, and governance issues. This perspective focuses on the need of strong institutional provision to overcome systemic barriers and promote and facilitate equitable and inclusive transformations [7].

Stahel [11] brings attention to the economic aspect of the CE by focusing on innovation in business models, namely product-service systems. He advocates a shift from product ownership to service-oriented consumption to prolong product lifespans and lessen material use. The study shows how CE concepts may increase competitiveness, create employment opportunities, and promote sustainable economic growth in automobile sectors, while simultaneously reducing environmental hazards [11].

Bocken et al. [1] study the intersection of product design and business strategy in promoting CE. They advocate for pragmatic design principles—such as modularity, durability, and reparability—that may increase product lifespan and ease circular material flows. The writers also study business concepts such as circular supply chains, product leasing, and resource recovery. Their research emphasizes that design-driven innovation may offer sustainable value for firms while linking corporate objectives with environmental responsibility [1].

The Ellen MacArthur Foundation [2] presents an important paper, advocating the adoption of the Circular Economy, explaining a persuasive economic justification for departing from the linear "take-make-dispose" paradigm. The paper quantifies economic advantages via case studies of firms such as Renault and H&M, citing examples on how resource efficiency and waste reduction may help in cost savings, getting competitive edge, and resilience. It cites enablers, including digital technology, reverse logistics, and collaborative ecosystems, as necessary for expediting the worldwide deployment of CE practices [2].

Mathews and Tan [9] conduct an empirical study on China's eco-industrial parks. They illustrate industrial symbiosis as a concrete implementation of the CE. The authors emphasize on the reduction of energy use, water usage, and emissions. This can be achieved by coordinated resource-sharing among adjacent enterprises. Their findings strengthen the significance of governmental policies, infrastructure development, and stakeholder collaboration in advancing CE initiatives in poor nations, where industrial transition is often constrained by inadequate institutional capacity [9].

Murray et al. [10] adopt a comprehensive, multidisciplinary perspective, contending that CE must include social and ethical dimensions apart from adding environmental and economic concerns. They propose that CE concepts may go beyond industry uses to embrace community systems, education, and social innovation. Their research highlights the capacity of CE to tackle global challenges such as social inequality, environmental justice, and resource accessibility, therefore advocating a more inclusive and equitable model for sustainable transitions [10].

Lieder and Rashid [8] study the application of the CE in the industrial sector, identifying hinderances such as inefficiency in recycling technology and customer aversion to reconditioned items. The authors endorse technology interventions such as additive manufacturing, Internet of Things (IoT), and digital twins, to better product traceability and resource efficiency. Moreover, they emphasize on the importance of consumer awareness and behavioral modification in facilitating the acceptance and extensive implementation of CE practices [8].

Collectively, these research works offer a strong academic foundation that both conceptualizes and analyzes CE from several perspectives. They emphasize on the transformational potential of the CE in altering production and consumption patterns while also cautioning against oversimplification and uniform solutions. Innovation, policy alignment, cooperation from stakeholders and systemic change are important elements for attaining sustainable, scalable, and inclusive circular transitions.

These reviews not only offer a blueprint for transitioning from linear to circular systems but also state the significance of innovation, collaboration, and systemic change in accomplishing sustainability.

3 Research Gaps

Despite the growing body of literature on CE, significant gaps remain unexplored which limit its effective implementation by several nations. Existing studies predominantly emphasize on developed regions like the European Union and China, leaving a paucity of research on CE practices in emerging economies, where infrastructural, regulatory, and socio-economic conditions vary widely. Social and community dimensions which include inclusiveness, equity, and the role of marginalized sections of society are often ignored. Additionally, while the ambiguity on CE definitions is acknowledged, there is a dearth of efforts to develop standardized, universally applicable frameworks. Furthermore, consumer behavior, cultural resistance, and psychological barriers to adopting CE practices have not been deeply investigated. Despite extensive theoretical discourses, empirical research on CE's practical applications remains limited. Scholars like Geissdoerfer et al. [3] argue that more interdisciplinary studies are required to bridge the gap between theoretical models with real-world practices [3]. This research bridges this gap by examining case studies of successful CE implementations.

4 Methodology

This study uses a secondary data collection approach. First, a systematic literature review was made using scholarly databases, focusing on studies between 2010 and 2024. Second, relevant documents were selected and analyzed to offer practical insights. The cases of the EU and China are carefully chosen as representative case studies, based on their global leadership and contrasting contexts in CE implementation. Furthermore, the justification for the mixed-methods framework is strengthened by articulating on how qualitative insights from policy analysis and secondary data were complemented by quantitative indicators (such as employment data, recycling rates, and material efficiency outcomes), thereby confirming a more integrated and robust analysis.

5 Findings

5.1 Case Study 1: Economic Implications: The EU Circular Economy Action Plan

The European Union's Circular Economy Action Plan (CEAP), first introduced in 2015 and revised in 2020 under the European Green Deal, illustrates a very ambitious and systemic regional strategy for shift from a linear to a CE model. It put emphasis on the EU's dedication to dissociating economic growth from over use of available resources and environmental degradation, while simultaneously promoting economic innovation and social inclusion.

CEAP fundamentally consists of legal, regulatory, and financial mechanisms that advocate sustainable production and consumption. Thrust areas of focus include sustainable product design, extended producer responsibility (EPR), improved recycling infrastructure, and the removal of non-recyclable materials. The plan additionally advocates eco-innovation and endorses circular business models, predominantly in high-impact sectors such as electronics, textiles, packaging, construction, and food.

The European Commission's 2020 progress report reveals that CEAP has facilitated an increase in EU-wide recycling rates, achieving above 50% recycling for several waste streams, mostly municipal and packaging trash.

a. A decrease in material dependency, characterized by significant reductions in resource imports, particularly in necessary raw materials.
b. The production of approximately €147 billion in economic benefits in 2019, resulting from resource efficiency, waste minimization, and the setting up of circular employment across member states.

Additionally, CEAP has facilitated a change in market by creating an increased demand for refurbished, remanufactured, and recycled items, thereby broadening the CE's impact across several sectors. The economic change is noticed in the expansion of the secondary materials market and the emergence of new job possibilities connected with sustainable practices.

Philips illustrates the use of a product-as-a-service (PaaS) model within its lighting sector. In this model, Philips leases lighting systems to clients rather than selling lighting products, keeps ownership of the materials, and oversees maintenance and end-of-life recycling. This methodology has helped achieving a 30% decrease in material consumption, reduced lifetime expenses for clients, and generated new income avenues for the organization, while aligning with corporate objectives with environmental goals [11].

Additionally, the implementation of CEAP has inculcated a conducive environment for startups and SMEs focusing on circular technologies, supported by platforms like as the European Circular Economy Stakeholder Platform (ECESP) and financing initiatives like Horizon 2020 and the LIFE Program. These initiatives emphasize how strong legislative frameworks can align economic incentives with environmental sustainability, positioning CE as a central pillar of Europe's long-term competitiveness and resilience.

This case study illustrates how the EU's holistic approach to circularity has facilitated systemic transformation and shown that economic prosperity and environmental stewardship can coexist, given strategic alignment among policy, industry, and society.

5.2 Case Study 2: Environmental Benefits: China's Circular Economy Initiatives [10]

China has emerged as a leader in implementing the ideas of the CE, motivated by increasing environmental concerns, resource constraints, and its goal to transition toward sustainable development. A fundamental element of China's CE plan is setting up of eco-industrial parks (EIPs) and extensive CE pilot projects, which are crucial in promoting industrial symbiosis—the reuse of waste outputs from one firm as inputs for another.

A notable illustration is the Suzhou Industrial Park (SIP), which integrates CE notion via resource-sharing systems, closed-loop manufacturing methods, and sophisticated waste-treatment infrastructure. Collaborative initiatives among several firms in the park have resulted in decreased energy and water consumption by over 40%, while industrial by-products have been successfully repurposed, markedly reducing emissions and waste production [9]. These achievements emanate from systematic planning, inter-company collaboration, and pro-governmental laws that foster cleaner manufacturing.

Besides industrial sector, China's CE programs have included urban waste management and electronic trash (e-waste) processing. A notable success story in this context is the transformation of Guiyu, once notorious as one of the world's biggest informal e-waste processing centers. Historically, basic recycling techniques resulted in significant environmental pollution and public health emergencies owing to the unregulated discharge of heavy metals and perilous materials. In response, the Chinese government instituted regulatory measures and set up a centralized, controlled e-waste recycling facility in Guiyu, including advanced technology and environmental protections.

The Guiyu CE Industrial Park now assists licensed recyclers according to stringent environmental regulations. This change has led to a decrease of over 70% in perilous material leakage into the environment, concurrently improving workers' health and safety conditions [12]. This study reveals how the formalization of an informal recycling sector via legislation, infrastructure, and capacity building—can better resource efficiency and minimize environmental damage.

These pilot projects show how China has used the CE not just as an environmental strategy but also as an instrument for industry improvement, pollution mitigation, and sustainable urban development. The CE model used in these pilot studies links with national policy objectives set up in China's CE Promotion Law (2009) and the 13th and 14th Five-Year Plans, offering a cohesive strategy for addressing issues concerning climate change, minimizing waste, and fostering sustainable development.

Ultimately, China's CE pilots highlight the importance of policy-supported, large-scale systemic reforms in achieving environmental sustainability. By cultivating linkages between environmental conservation and industrial competitiveness, these efforts offer significant insights for other emerging nations attempting to implement circular transformations within their own socio-economic frameworks.

5.3 Case Study 3: Social Impacts: Job Creation in the Recycling Industry [6]

The shift toward CE is an environmental need and as well as the source of enormous social benefits, particularly for job creation. By reassessing the linear "take-make-dispose" paradigm, the CE encourages for the reuse, refurbishment, recycling, and redesign of materials and goods, therefore, creating new job prospects across different stages of the value chain.

The International Labor Organization offers an illustration of this potential, estimating that the global shift to CE practices may create over 6 million new employment opportunities globally by 2030 in the fields of materials recovery, product refurbishing, maintenance services, sustainable design, and reverse logistics, so nurturing more inclusive and rational economic growth [5].

5.4 Case Study 3: The Swedish Paradigm: A Blueprint for Sustainable Employment

A notable real world case study is Sweden's "Recycling Revolution," a national initiative aimed at significantly reducing landfill trash by supporting repair and refurbishment of domestic appliances and electronics. This government-endorsed initiative has led to the development of proficient and high skilled workforce in recycling industry, creating over 10,000 employment opportunities across diverse areas of the recycling industry. These jobs include logistic professionals for reverse supply chains, technicians for electronics repair personnel, and quality inspectors to make sure that refurbished products adhere to the accepted standard.

Sweden's strategy highlights how the CE may simplify sustainable development, predominantly in areas shifting from resource-heavy, polluting sectors to low-carbon, green economies. The endeavor has reduced the negative impacts of waste accumulation and fostered local economic revitalization, particularly in cities once reliant on declining industrial sectors.

Broader Consequences for Just Economic Advancement

By emphasizing on local processing, repair, and reuse, CE techniques might minimize economic disparities. Job creation in the CE is sometimes more labor-intensive than conventional trash management or manufacturing, making it particularly beneficial for developing countries and the countries where underemployment prevails. Besides, these strategies correspond with international objectives for decent work and economic growth (SDG 8) as well as responsible consumption and production (SDG 12).

6 Challenges to Achieving the Complete Potential of Circular Economy: A Global Perspective

Notwithstanding the optimistic perspective, the implementation of CE practices is inconsistent worldwide, hindered by many substantial barriers:

Regulatory Fragmentation

The absence of standardized legislation and policies is one of the significant barriers that facilitate CE. Although region like European Union possess developed frameworks for waste management, eco-design, and extended producer responsibility (EPR), several underdeveloped nations are lacking institutional capacity, enforcement mechanisms, and financial resources necessary for supporting CE projects. Therefore, the adoption of CE is often found uneven and inconsistent throughout different regions.

Korhonen et al. [7] identify this legal mismatch as a significant impediment to the global scalability of the CE, contending that in the absence of a unified worldwide regulatory framework, firms are less motivated to engage in CE technologies that may not generate profits in diverse markets [7].

Technological Obstacles
A further problem relates to the technical problems pertaining to recycling and material recovery. Composite plastics, multi-material packaging, and electronic trash often render recycling challenges owing to their intricate compositions. The equipment requirement for the effective disassembly, sorting, and recycling of such products is costly and not readily available, particularly in low-income regions.

These technical limitations impede advancement toward realizing the "closed-loop" objectives of CE, whereby resources are perpetually reintegrated into manufacturing.

Cultural and Consumer Resistance
Cultural opposition and consumer attitudes render a critical barrier. In several nations, refurbished or recycled goods are seen as lower in quality. As a result of which, people hesitate to use them. This not only affects demand side of the goods but also discourages firms from investing in such items.

Addressing This Obstacle Requires a Dual Strategy
There is a need to undertake initiatives to increase public awareness and education for changing consumer attitudes toward sustainability and value preservation. There is a need to implement rigorous quality standards and certification procedures to imbibe customer confidence on the safety, durability, and performance of reused or recycled products.

Besides, opposition to novel CE business models, such as product-as-a-service (PaaS) where users lease instead of owning products constitutes a cultural barrier to impede the adoption of Circular Economy practices globally.

The instance of job creation within the recycling sector exemplifies the transformational societal potential of the CE, as exhibited by Sweden's effective model. To fully realize these benefits globally, rigorous legislative initiatives, technical advancements, and cultural transformations are indispensable. A comprehensive and inclusive CE framework offers the potential for environmental sustainability, economic resilience, and social equality, predominantly at a time of rising unemployment and ecological decline.

7 Discussion

The findings effectively address several critical facets of the identified research gap. Although primary focus is on developed regions such as the EU and China, thereby only partially addressing geographic limitation, they provide adaptable frameworks and policy frameworks that may benefit developing countries. The Swedish case successfully examines social and community challenges, focusing on job creation, inclusive development, and linking with the Sustainable Development Goals (SDGs). Although there is a paucity of uniform global CE framework, the EU's CE Action Plan exhibits important regional cooperation. Consumer behavior and cultural resistance are recognized, with suggested solutions including awareness campaigns and certification criteria. The research offers significant sector-specific insights, especially in electronics and textiles, and encompasses relevant data about environmental and economic impact. However, the function of digital technologies (AI, IoT, blockchain) and agriculture-specific CE applications is inadequately examined, requiring more quantitative and longitudinal assessments and research grounded in varied socio-economic contexts.

8 Direction for Further Research

Future study needs to focus on broadening Circular Economy (CE) studies in developing economies by analyzing region-specific challenges such as infrastructure limitations, informal sector integration, and regulatory preparedness. There is a pressing need to create validate universally applicable, but locally adaptable CE frameworks which include inclusivity, social equality, and the engagement of underprivileged sections of society. Further investigation is necessary to study the influence of digital technologies explicitly AI, IoT, and blockchain on the facilitation of CE implementation, particularly in domains such as waste tracking, product lifecycle management, and reverse logistics. Consumer behavior, cultural resistance, and psychological barriers have to be examined using multidisciplinary research that link behavioral science with sustainability research. Longitudinal and quantitative research is indispensable to assess the long-term impacts of CE activities. Additionally, future research should concentrate on formulating sector-specific CE models in inadequately examined research areas such as agriculture and electronics, guaranteeing that circular practices are scalable, quantifiable, and congruent with both SDGs and local socio-economic realities.

9 Conclusion

This study offers a novel contribution by linking evidence from various global contexts particularly the cases of European Union, China, and Sweden. The cases show how effective policy frameworks, industrial symbiosis, and socially inclusive strategies can enable successful Circular Economy (CE) implementation. Unlike the previous research which often remains more conceptual or narrowly focused, this study bridges the gap between policy, practice and socio-economic outcomes, illustrating real world cases with measurable social, environmental, and economic impacts. It vividly highlights the revolutionary potential of CE beyond environmental sustainability, including economic innovation and social equity. This study critically examines implementation models and identifies enduring global challenges, such as regulatory fragmentation, cultural resistance, and technological limitation, thereby validating existing frameworks and establishing a citadel for more inclusive, adaptable, and effective CE practices. This originality positions the study as a significant resource for policymakers, academics, and practitioners seeking to implement circular solutions in both developed and developing countries.

References

1. Bocken, N. M., De Pauw, I., Bakker, C., & Van Der Grinten, B. (2016). Product design and business model strategies for a circular economy. *Journal of Industrial and Production Engineering, 33*(5), 308–320. https://doi.org/10.1080/21681015.2016.1172124
2. Ellen MacArthur Foundation. (2013). Towards the circular economy: Economic and business rationale for an accelerated transition. Ellen MacArthur Foundation. https://ellenmacarthurfoundation.org
3. Geissdoerfer, M., Savaget, P., Bocken, N. M., & Hultink, E. J. (2017). The circular economy–A new sustainability paradigm? *Journal of Cleaner Production, 143*, 757–768. https://doi.org/10.1016/j.jclepro.2016.12.048
4. Ghisellini, P., Cialani, C., & Ulgiati, S. (2016). A review on circular economy: The expected transition to a balanced interplay of environmental and economic systems. *Journal of Cleaner Production, 114*, 11–32. https://doi.org/10.1016/j.jclepro.2015.09.007
5. International Labour Organization. (2018). World employment and social outlook 2018: Greening with jobs. Geneva: ILO. https://www.ilo.org/global/publications/books/WCMS_628654/lang--en/index.htm
6. Kirchherr, J., Reike, D., & Hekkert, M. (2017). Conceptualizing the circular economy: An analysis of 114 definitions. *Resources, Conservation and Recycling, 127*, 221–232. https://doi.org/10.1016/j.resconrec.2017.09.005
7. Korhonen, J., Nuur, C., Feldmann, A., & Birkie, S. E. (2018). Circular economy as an essentially contested concept. *Journal of Cleaner Production, 175*, 544–552. https://doi.org/10.1016/j.jclepro.2017.12.111
8. Lieder, M., & Rashid, A. (2016). Towards circular economy implementation: A comprehensive review in context of manufacturing. *Journal of Cleaner Production, 115*, 36–51. https://doi.org/10.1016/j.jclepro.2015.12.042
9. Mathews, J. A., & Tan, H. (2016). Circular economy: Lessons from China. *Nature, 531*(7595), 440–442. https://doi.org/10.1038/531440a
10. Murray, A., Skene, K., & Haynes, K. (2017). The circular economy: An interdisciplinary exploration of the concept and application in a global context. *Journal of Business Ethics, 140*(3), 369–380. https://doi.org/10.1007/s10551-015-2693-2
11. Stahel, W. R. (2016). The circular economy. *Nature, 531*(7595), 435–438. https://doi.org/10.1038/531435a
12. Zeng, Z., Cheng, L., Shi, L., & Luetkenhorst, W. (2021). China's green transformation through eco-industrial parks. *World Development, 140*, Article 105249. https://doi.org/10.1016/j.worlddev.2020.105249

Advancing ESG Practices: Trends, Challenges, and Future Directions in Sustainable Finance

Srinivas Karri, Alle Yashoda, Deepak Kumar Sahoo, and Harika Repakula

Abstract

The objective of this paper is to critically examine the changing landscape of Environmental, Social, and Governance (ESG) investment by integrating theoretical principles with industry applications. This research is grounded in Freeman's stakeholder theory and Elkington's Triple Bottom Line (TBL) and examines the effect of ESG practices on sustainable financial outcomes, while addressing key challenges such as reporting inconsistencies, ethical dilemmas in AI integration, and infrastructural deficiencies in developing economies. The research uses a qualitative method, integrating research insights from existing literature and using case studies of Unilever and Microsoft to illustrate the implementation of ESG policies in practical business environments. The findings reveal that while ESG integration provides long-term financial and reputational gains, its implementation is complicated by greenwashing issues, the absence of standardize metrices, and divergent stakeholder expectations. This paper highlights the requirement of ethical AI implementation, sector-specific criteria, and legislative consistency to improve ESG effectiveness. This research is original in its integrated methodology, amalgamating theoretical frameworks,

S. Karri (✉)
Loyola Academy (Degree & PG College), Hyderabad, Telangana, India
e-mail: talktoksv@gmail.com

A. Yashoda
Anwarul-Uloom College of Business Management, New Mallepally, Hyderabad, India

D. K. Sahoo
College of IT and Management Education, Bhubaneswar, Odisha, India
e-mail: principal@cime.ac.in

H. Repakula
School of Business and Management, IBS University, Boroko, Papua New Guinea
e-mail: palakurthi.harika@ibs.ac.pg

S. Patnaik et al. (eds.), *People, Planet and Profit: Advancing Sustainable Business Practices*, Springer Proceedings in Geography and Sustainability,
https://doi.org/10.1007/978-3-032-11296-5_12

technical advancements like AI, and behavioral insights to offer a holistic perspective on ESG investing. It combines comprehensive case studies of Unilever and Microsoft, integrating theoretical ESG concepts with real-world business applications. The study identifies substantial deficiencies in ESG standards and offers new insights for future academic and policy-oriented investigations in sustainable finance.

Keywords

Sustainable finance • ESG practices • Green bonds • AI ethics

1 Introduction

Environmental, Social, and Governance (ESG) practices have become vital components of sustainable finance, profoundly shifting the approach of the firms and the investors toward social and environmental challenges. As global awareness of sustainability increases, ESG frameworks have become indispensable in linking firm's strategy with overarching ESG goals. Incorporating ESG into financial decision-making has offered significant benefits, particularly in bettering risk management and achieving long-term profits. The journey toward successful ESG implementation is affected by several barriers such as discrepancies in reporting standards, ethical dilemmas related with developing technologies like artificial intelligence, and the inherent conflict between short-term profitability and long-term sustainability objectives.

These difficulties are exacerbated in emerging countries where political instability, regulatory inadequacy and limited institutional capacity obstruct successful ESG integration. This paper organizes a systematic assessment of pertinent studies in ESG practices, focusing on advancements, ongoing challenges, and new prospects in specific domain.

To strengthen the rationale for ESG adoption, recent evidence illustrates that ESG-aligned practices can generate positive financial returns across various industries. The renewable energy industry has appreciated increased investment and profitability due to ESG-aligned policies [31], while technology leader like Microsoft have experienced increased market value and stakeholder confidence by focusing on carbon neutrality and social inclusion [17]. In financial services sector, the firms that align ESG measures into risk assessment models have illustrated better resilience even amid economic volatility. Research shows that ESG-oriented portfolios outperformed traditional ones during the COVID-19 market downturn [18, 12].

Collectively, these cases show the growing consensus that ESG is not only a moral obligation but a strategic avenue for attaining financial gain, stakeholder confidence, and sustained competitiveness.

2 Review of Literature

The review of literature on ESG practices shows a dynamic and developing domain impacted by diverse perspectives and methodologies. Many researches have been conducted to study integration of ESG factors into corporate governance, investment strategies, and reporting frameworks, putting emphasis on both prospects and obstacles. Freeman's [7, 9] show the ethical and sustainability aspects of business practices [7, 9], while empirical studies disclose the financial paybacks of ESG integration, including better risk management and resilience in market recession. Nonetheless, limitation in standardized reporting, ethical dilemmas connected with AI-driven ESG technologies, and the intricacies of executing ESG initiatives in developing countries shows the need for more study. This review collects insights from seminal works and recent advancement and offers a thorough analysis of current and future challenges, potential solutions and direction in ESG practice. Freeman [9] stakeholder theory challenged Friedman [10] theory of profit maximization [9]. When developing business strategies, Freeman argued, corporations should take into account the needs of all "those who are directly or indirectly impacted by business operations." They are called stakeholders: communities, suppliers, consumers, employees, investors, and the environment. Apart from economic responsibilities, business must discharge its social and ethical responsibilities, naming it as Corporate Social Responsibilities (CSR), balanced relationships, and sustainability in the long run. Incorporating ESG elements in decision-making, the theory has expanded its influence on corporate governance and research, pushing for a more equitable, environmentally conscious, and socially conscious model of boardroom management.

Research by Cohen et al. [5] explored a positive relation between environmental performance and profitability [5]. Firms that give importance to environmental responsibility often have better financial outcomes. This finding has significant ramifications, particularly in emerging economies, where firms are progressively assessed on their environmental impact along with their financial performance. The research studied on the possibility of "win–win" scenarios, when enterprises might accomplish both desired financial results and positive environmental outcomes. This study added to the growing recognition of the importance of sustainable business actions and integration of ESG elements in business strategies.

Elkington [7] offered TBL framework which revolutionized the way the businesses evaluate their performance by linking three indispensable dimensions: People—the social bottom line; Planet—the ecological or the environmental bottom line; and Profit—the ecological bottom line [7]. The TBL framework proposed that the firms should put focus on not just profit (the financial bottom line) but also their impact on people (the social bottom line) and the planet (the environmental bottom line). This comprehensive perspective encouraged businesses to make more sustainable and socially responsible business actions, expanding the parameters of firms' assessment beyond conventional financial measurements. Elkington's paradigm has become vital to corporate social responsibility (CSR) and sustainability, supporting organizations in reconciling economic accomplishment along with constructive contribution to society and the environment at large.

The United Nations Principles for Responsible Investment (UNPRI-2006) offered a framework of responsible investment principles designed to aid institutional investors to include ESG elements into their investment decisions [30]. The UNPRI put emphasis on the need of linking financial returns with wider societal and ethical considerations. Though these principles accomplished substantial global recognition, nevertheless, the implementation of responsible investment practices showed notable regional differences. Some nations and institutions adopted these principles with great enthusiasm, but others encountered problems in integrating ESG elements into their investment frameworks due to varying degrees of awareness, regulating support and market circumstances.

Friede et al. [10] conducted a meta-analysis of ESG investment by reviewing many research works to explore the kind of correlation between ESG parameters and financial success. Their findings showed that firms with robust ESG practices often outperform their competitors in financial returns. They found significant gap in assessment of ESG performance. The study revealed paucity of standardized frameworks for evaluating ESG performance. This emphasized the need for better metrices and reporting standards to simplify more reliable and transparent ESG investment practices [11].

Clark et al. [4] highlighted the importance of ESG elements in attaining long-term financial gains. The study endorsed the integration of ESG elements into corporate strategies and manage risk efficiently, better operational efficiency, and foster sustainable profitability. They argued that ESG elements are not mere ethical considerations, but also are important determinants of financial outcomes. Future study might investigate dynamic interplay among several ESG dimensions, analyzing how changes in one dimension may affect others and impact financial gains over time [4].

The Corporate Finance Institute [6] highlighted significant impediments to greater adoption of ESG elements in portfolio management. The absence of standardized ESG metrics, inconsistent data quality, and complicacies in assessing the financial implications of ESG elements are the major concerns as identified in the study. The research urged for the indispensable innovations to address these obstacles, such as better ESG reporting standards, more data transparency, and the development of tools for more effective evaluation of ESG risks and opportunities. These insights are envisioned to support investors and asset managers in integration of ESG elements into their investing strategies for improved long-term results [6].

Schoenmaker et al. [24] offered the notion of aligning financial systems with the United Nations Sustainable Development Goals (UNSDGs), focusing on the incorporation of sustainability into financial decision-making. He contended that financial institutions and markets must assume a significant role in easing the attainment of the SDGs by channeling resources into sustainable projects and investments. The study revealed many regulatory hurdles and proposed for revised and updated frameworks, standardized reporting, and policies to promote long-term sustainable investment while addressing possible risks associated with environmental and social aspects [24].

Riedl et al. [23] studied the behavioral dimensions of socially responsible investments (SRIs) and focused on the effect of psychological variables on investors' options to prioritize ESG criteria. The investors' socially responsible investing often arises from personal ideals, ethical principles, and societal factors, rather than just from financial success. Riedl and Smeets urged for psychological models to explain these behaviors and link them into financial decision-making processes while supporting investors and financial advisers in developing strategies that harmonize financial goals with social responsibility [23].

Bebbington et al. [1] studied relation between accounting practices and the UNSDGs. The study highlighted that the firms should synchronize their financial reporting with sustainable goals. The study highlighted substantial deficiencies in existing quantified reporting systems, which often do not adequately have the comprehensive impacts of business actions on the SDGs. The authors advocated for the creation of standardized, quantifiable frameworks to better evaluation and reporting of the firms' contributions to the SDGs, thereby enabling more informed decision-making for investors, regulators, and other stakeholders dedicated to sustainability [1].

The Morningstar Portfolio Sustainability Report Disclosure Statement [18] analyzed developments in ESG indexes and highlighted their increasing significance in the formation of institutional investment portfolios. The study stated that ESG indices have become a crucial instrument for individuals looking for aligning their investments with ethical and ecologically responsible business. These indices offer a systematic approach to integrate ESG aspects with investment strategies, easing decision-making, risk management, and the generation of enduring social and environmental advantages [18].

KPMG [15] studied trends in ESG reporting and focused on the indispensable function of standardization in enhancing the comparability of ESG disclosures across organizations and sectors. The research stated that varied reporting processes have impeded the right evaluation and comparison of ESG performance for investors and stakeholders. KPMG advocated for the practice of standardized reporting standards and metrics to better openness, dependability, and consistency, thereby augmenting improved decision-making and cultivating more confidence in ESG data among investors, regulators, and the public [15].

Singh [27] studied the role of technology, especially AI, in the incorporation of ESG elements into finance. The study focused on the transformative impact of AI on sustainable finance via automation of ESG data collection, the simplification of more precise ESG performance projections, and the enhancement of decision-making and risk management processes. AI techniques assisted in forecasting trends in ESG data that may not be instantly noticeable, thereby endorsing innovation and better integration of ESG elements into financial strategies, eventually advancing more sustainable and responsible investing practices [27].

Bloomberg Intelligence [2] predicted that the ESG market will exceed $50 trillion by 2025, witness substantial growth potential of sustainable finance. Nevertheless, they also recognized other limitations in the infrastructure prerequisites for this expansion, such as the scarcity of defined ESG measurements, uneven reporting methodologies, and insufficient tools for evaluating the long-term effects of ESG investments. The report highlighted the need for improved legal frameworks, more data accuracy, and technical innovations to endorse efficient and sustainable growth of the ESG industry while addressing the increasing demand from investors and stakeholders [2].

Zhu et al. [33] linked corporate green innovation with competitive advantage. They stated that ESG compliance plays a pivotal role in fostering this innovation. Firms that support green innovation support environmental sustainability and improve their market position as it offers competitive advantages of the growing customer demand due to ethical business practices. The research highlighted that compliance with ESG standards enables businesses minimize risks, improved brand image, and secure a competitive advantage, so setting them for continuing success in markets gradually oriented toward sustainability [33].

Whelan et al. [32] examined the performance of ESG index funds across different industries and found notable inconsistent in returns and risk profiles based on the industry. The study advocated the development of sector-specific standards to more precisely evaluate ESG fund performance, assumed that several sectors have distinct sustainability problems and possibilities. The study highlighted that sector-specific ESG criteria and attributes need to be identified to better investment strategies and make sure that funds link with the varied sustainability objectives of each industry [32].

The Responsible AI Institute [22] studied the impact of AI on ESG investing and focus on integrating AI into ESG investment strategies. They proposed frameworks for the ethical incorporation of AI and emphasized the necessity for AI systems to conform to ESG principles and to avoid reinforcing prejudices or inequities. The authors put emphasis AI's capacity to improve ESG data analysis, refine decision-making, and optimize risk management, while stressing the need for openness, accountability, and ethical standards in using AI for sustainable finance [22].

Shefrin [25] studied the behavioral effects of investor perspectives on ESG fund flows and justified the significance of these beliefs on investment choices. The author advocated good communication methods in fostering favorable investor perceptions of ESG funds. The study advocated a need for clear, honest, and persuasive communication regarding ESG goals, performance, and social effect are the need of the hour and these may substantially boost investor trust and involvement. The authors advised that fund managers should refine ESG-related disclosures and communication to match investor perspectives with the funds' sustainability objectives, thereby increasing fund inflows and developing long-term support [25].

Shekarian et al. [26] studied the increasing importance of ESG in supply chain finance and its role in risk evaluation and decision-making processes. The authors recommended that the firm should use ESG criteria in supply chain finance to assess risk. It may most possibly enhance resilience and sustainability. The authors proposed to develop enhanced ESG-oriented risk assessment frameworks to accurately analyze suppliers' environmental and social activities. These frameworks will also support bankers to link supply chain finance with sustainability objectives, promise economic viability and compliance with responsible business practices [26].

The Global Sustainable Investment Review [12] identified the evolving trends in ESG integration and offered extensive global standards for sustainable investments. The research puts focus on the rising implementation of ESG principles worldwide, driven by legislative changes, investor demand, and vivid understanding of sustainability's financial and social results. GSIA found notable discrepancies in global ESG actions and highlighted requirement for standardized frameworks to make sure uniformity and comparability. The standardized frameworks offer a vital resource for investors, politicians, and businesses looking to link their strategies with evolving ESG demands and support sustainable development [12].

Gunung Capital [13] analysis found considerable limitations in current sustainable investment frameworks, notably with standardization and consistency. The study found that contradictory definitions, assessment and reporting processes impede successful ESG integration and comparability across assets. The authors highlighted the need for legislative improvements to standardize sustainable investment actions globally. Their proposals urged for the preparation of standardized ESG reporting rules, improved transparency mandates, and the implementation of monitoring systems to ensure compliance, thereby creating confidence and efficacy in sustainable investing methods [13].

The OECD [19] studied the complications connected with ESG actions for developing market economies, highlighting both possibilities and problems. The paper found that the implementation of ESG in these areas might contribute to sustainable development, captivate foreign investment, and better resilience to global economic downturns. Nonetheless, it also identified obstacles which included limited access to financing, a lack of standardized ESG standards, and inadequate institutional capability. The OECD recommended the adoption of necessary policies, such as capacity-building efforts, regulatory harmonization, and incentives for sustainable practices, to enable developing nations to fully capitalize on ESG's potential for economic and environmental advancement [19].

Organization for Economic Co-operation and Development [20] studied the essential relationships between biodiversity and sustainable finance. The study recommended that the firms should incorporate environmental aspects into financial decision-making. The research stated that the deterioration in biodiversity poses serious threats to global economies and financial stability, necessitating novel methods to link nature-positive tactics into investments. OECD report urged cooperations across all the stakeholders which include sectors, governments, financial institutions, and corporations and they should set up frameworks that integrate biodiversity protection with economic objectives, therefore fostering sustainable development. The report proposed that biodiversity-related financial instruments and the augmentation of ESG frameworks need to incorporate biodiversity measures, and partnerships to facilitate systemic change across sectors [20].

The United Nations Environment Program Finance Initiative [29] proposed a comprehensive framework to link ESG standards with climate finance goals, urging the need for cohesive solutions, addressing environmental and financial concerns. These models focused on integration of climate risk evaluations within ESG frameworks, simplifying capital allocation for low-carbon projects, and adding green financial instruments such as green bonds and carbon markets. UNEP-FI advocated for increasing more and more public–private partnerships to bridge financial gap, ensuring that ESG investments significantly meet global climate objectives. The program proposed to integrate sustainable finance with climatic hazard reduction, adaption, and resilience enhancement activities [29].

FII Institute [8] made an analysis on AI-driven advancements in ESG, putting focus on their revolutionary potential and the issues. The research examined and analyzed the use of AI in ESG data analysis, risk assessment, and decision-making efficacy which offers more accurate and impactful sustainable finance strategies. It also proposed indispensable accountability and fairness factors as well as reduction of biases in AI systems and the assurance of translucent decision-making procedures. The network recommended ethical AI frameworks designed for ESG applications, focusing inclusion, justice, and responsibility while using technology to effectively tackle sustainability concerns [8].

3 Literature Gap

Notwithstanding the increasing body of literature validating the strategic significance of ESG investment, many gaps exist. Most of the studies focus on the relationship between ESG performance and financial results; however, little importance is given on the incorporation of new technologies such as AI in improving ESG assessment frameworks, particularly in real-time and sector-specific scenarios. Furthermore, while theoretical frameworks like stakeholder theory and the triple bottom line are often used as reference, their implementation into adaptive investment strategies across many sectors is yet hardly examined. There is a paucity of literature available on the behavioral aspects of ESG decision-making, mostly regarding the effect of cognitive biases, cultural values, and investor psychology on ESG adoption. The unavailability of global standards in ESG indicators hinders comparative analyses and complicates the establishment of globally recognized best practices, highlighting an important domain for future academic research.

4 Objectives

1. To critically examine the incorporation of ESG principles in investment strategies using stakeholder theory and the triple bottom line, and assess their practical relevance across various industrial sectors.
2. To explore the impact of new technologies, particularly Artificial Intelligence, on improving the effectiveness, transparency and standardization ESG assessment and decision-making in sustainable finance.

5 Methodology

The present study uses a **qualitative research design** and adopts an exploratory approach to examine the integration and effect of ESG investing. Secondary data source is used to collect data from scholarly, peer-reviewed journals, institutional reports, companies' disclosures, and ESG databases. The case studies of Unilever and Microsoft were selected to analyze the real-world applications of ESG principles. **Content analysis** is used to identify patterns, challenges, and best practices in ESG implementation. Additionally, key theoretical frameworks such as Freeman's stakeholder theory and Elkington's Triple Bottom Line approach guided the interpretation of findings. This methodological approach allows for a holistic and comprehensive understanding of ESG trends, behaviors, and strategic alignment.

6 Findings and Analysis

6.1 Case Study 1—Unilever and the Integration of ESG—An Analysis from the Lens of Stakeholder Theory and the Triple Bottom Line

Unilever, founded in 1930, operates in over 190 nations, impacting more than 2.5 billion people every day through its consumer products. The company has set a benchmark for incorporating ESG values into core business practices in accordance with the Freeman's Stakeholder Theory and Elkington's Triple Bottom Line (TBL) approach. In 2010, Unilever implemented its Sustainable Living Plan (USLP)—not just as a CSR strategy but as a catalyst for company's best performance. The strategy highlighted stakeholder engagement and sustainable value creation by linking three fundamental objectives [31]:

(a) Health and Well-Being

Lifebuoy's handwashing initiatives have significantly impacted public health globally, exclusively in low- and middle-income communities. With a global reach of over a billion of population since Unilever lunched its flagship campaign in

2010, it has reduced diseases related to respiratory infections by 15%, eye infections among children under the age of 5 by 46%, and diarrheal ailments by 25%. By integrating hygiene educational program into 21,500 schools, the 2020 "H for Handwashing" campaign brought about significant improvement in children's awareness level and reached over 250 million people. By writing successful stories in Vietnam and India during COVID-19, it could reach out to 170 million views in its YouTube channel through its "Roti Reminder" and a TikTok impactful campaign. Lifebuoy embedded innovation through its "Hippo" the virtual AI instructor to improve learning. These initiatives are widely acknowledged worldwide which could brough laurels and public health accreditations to the company.

(b) Environmental Impact Mitigation

Lifebuoy's handwashing initiatives are implemented in alignment with Unilever's sustainability strategy which showcase the company's commitment to environmental and health responsibility. Between 2008 and 2020, Unilever reduced CO_2 emissions from its global manufacturing by 65% through implementing sustainable actions, using renewable energy efficiency [21]. Moreover, it promoted circular economy by eliminating landfill waste from more than 600 regions across 70 nations by 2016 [31]. The company uses locally sourced materials and environmentally friendly packaging to attain these goals. Collectively, these initiatives exemplify how Unilever amalgamates social equity, environmental protection, and public health through its products like Lifebuoy to drive large-scale green environment and social change.

(c) Inclusive Economic Growth

The company's commitment to inclusive growth and sustainability is evident from its sourcing of 95% raw materials including tea and palm oil sustainably by 2020. Its compliance to Sustainable Agriculture Code and collaboration with the organization like the Rainforest Alliance and Roundtable on Sustainable Palm Oil (RSPO) supported to achieve inclusive growth. These practices are glaring examples of Unilever fair labor practices, environmental preservation, and the empowerment of small farmers. Unilever reinforced its position as a global pioneer in ethical and inclusive business practices by integrating sustainable sourcing into product lines such as Lifebuoy, ensuring that its supply chain endorses social justice and environmental stewardship.

These outcomes unequivocally reflect Unilever's adherence to the Triple Bottom Line (TBL) framework which highlights the interconnected TBL goals: profit, planet, and people.

People: Unilever has made great strides in promoting cleanliness and public hygiene with products like Lifebuoy. PwC [21] report revealed that hygiene awareness campaigns reached out to more than 47 million people in India alone. The main goal of these campaigns is to impact public favorably through public events, school programs, and the media. These programs are often targeted at improved hygienic practices by vulnerable groups. A decline in the incidences of preventable illnesses, and an improvement in general well-being, specifically for children and low-income families have been some of the key outcomes of social bottom line.

Planet: Unilever has drastically reduced its resource usage as a step toward alignment to its environmental goals. The firm has reduced 50% of water usage per ton of manufacturing. It implemented recycling programs, used water-efficient technology, and sustainable manufacturing techniques across all its international operations. Such actions not only conserve significant natural resources, but also reduce negative ecological effect.

Profit: Unilever sustainable business model has been financially rewarding. By adopting sustainable business model, Unilever 75% of its total growth in accordance with the report by Byrns & McKinzie [3]. The report also revealed that the brands that adopted sustainability measures grew 69% faster than the brands that did not embrace it [3]. This strengthens the fact that the business by adopting sustainability measures become ethically and ecologically responsible.

Unilever linked its stakeholder expectations with sustainability goals and set an example of how ESG integration can yield tangible financial and social return, thereby validating the stakeholder-centric and multi-capital logic inherent in core ESG theory.

6.2 Case Study 2—Microsoft: AI-Driven ESG Transformation and Responsible Innovation

Microsoft, Redmond-based IT MNC from USA, is leveraging its technological expertise to position itself as a global leader in ESG innovation by incorporating AI-driven technologies into ESG frameworks. This reflects the significance of understanding stakeholder accountability and Microsoft pursuit of environmental equity.

Microsoft's ESG Approach Is based on Three Transformational Objectives

- Attain carbon neutrality by 2030
- Remove all past emissions by 2050
- Pioneer innovation in sustainable finance with AI integration.

Microsoft developed the Cloud for Sustainability to operationalize this. The company developed the AI-driven Emissions Impact Dashboard which combines real-time ESG data in global operations. The platform by using machine learning and natural language processing, it was able to:

- Recognize inefficiencies in energy, water, and carbon management
- Measure adherence to ESG disclosures and evolving regulatory standards

- Enable transparency and comparability within supply chain.

Empirical results indicate:

- 17% decrease in Scopes 1 and 2 greenhouse gas emissions in FY2022 in comparison with the 2020 baseline.
- Supported over 3,500 corporate clients in tracking and mitigating carbon emissions.
- Allocated a corpus of $1 billion Climate Innovation Fund for AI-driven ESG firms.
- Over 700 grants awarded under the "AI for Earth" initiative, supporting more than 100 nations in climate modeling, preservation of biodiversity and water conservation [17, 28].

By incorporating AI ethics and ESG, the company could address the challenges relating to data transparency, ESG automation scoring, and the function of predictive analytics. It called for industry-specific ESG frameworks, whereby technology-intensive businesses may excel by embedding ESG into their digital workspace.

This is an example of Microsoft successfully integrating AI with ESG goals and governed by ethical standards [22], transforms sustainability reporting from a compliance requirement to a strategic advantage, linking technical innovation and stakeholder governance (Table 1).

7 Discussion

The case studies of Unilever and Microsoft validate the practical application of Freeman's Stakeholder Theory and Elkington's Triple Bottom Line (TBL) framework in the real-world business environments. Unilever's Sustainable Living Plan (USLP) exemplifies a stakeholder-oriented ESG framework, linking social impact (health and hygiene), environmental stewardship (carbon reduction, water usage efficiency), and inclusive economic growth (sustainable sourcing) into its business strategy. The firm's consistent financial performance, as evidenced by the growth of its sustainable brands by 69% faster than the rest of the portfolio confirms research aligning ESG with long-term financial sustainability [4, 11]. These results exhibit Unilever TBL focus on people, planet, and profit upholding that sustainability is not about compliance, rather a value enhancer (Fig. 1).

The case of Microsoft adds to the sustainability discourse by implementing ESG through sophisticated AI-powered platforms. Its cloud for Sustainability, embedded with predictive analytics and natural language processing, addresses the essential issues of data transparency and real-time data monitoring which is one of the most discussed challenges in ESG implementation. Microsoft sets an example of how technology can simplify ESG process and democratize access to sustainability tools across the industries, illustrated by a 17% decline in Scopes

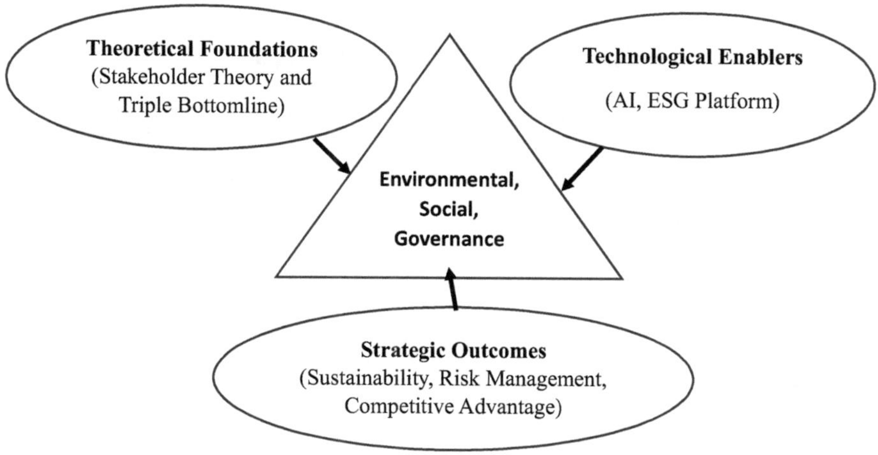

Fig. 1 Integrated ESG implementation model

1 and 2 emissions by FY2022 and support to over 3,500 businesses in tracking emissions.

These two case studies illustrate that incorporating ESG into core business operations improves strategic distinctions while technological innovation is vital for upholding ESG responsibilities, mostly in data-intensive industries.

Stakeholder engagement and right disclosures are significant for gaining investor confidence and ensuring regulatory compliance. These results strengthen the literature's support for sector-specific frameworks, ethical AI governance, and uniform reporting, providing a realistic plan for ESG implementation in the industrial and technology sectors (Fig. 2).

The timeline shows the development of ESG investing from ethics-based approaches (such as Elkington's Triple Bottom Line and Freeman's stakeholder theory) to technologically enabled, strategic, data-driven approaches. This is also reflected in reports like the UNPRI (2006), Responsible AI Institute, 2024, and the growing scholarly attention to the linkage between ESG and financial performance (Friede et al., 2015). This timeline endorses the claim that ESG is now essential to sustainable finance and business, not just as a peripheral issue.

8 Recommendation for Actionable Frameworks

To tackle the complex challenges connected with ESG integration, several actionable frameworks can be proposed. A standardized ESG Reporting Framework is vital to solve the discrepancy in disclosures. This can be accomplished by developing global standards like IFRS S1 and S2 or the GRI principles, implementing third-party assurance, and setting up regional ESG hubs to align with regional and global reporting practices. Secondly, to tackle ethical challenges

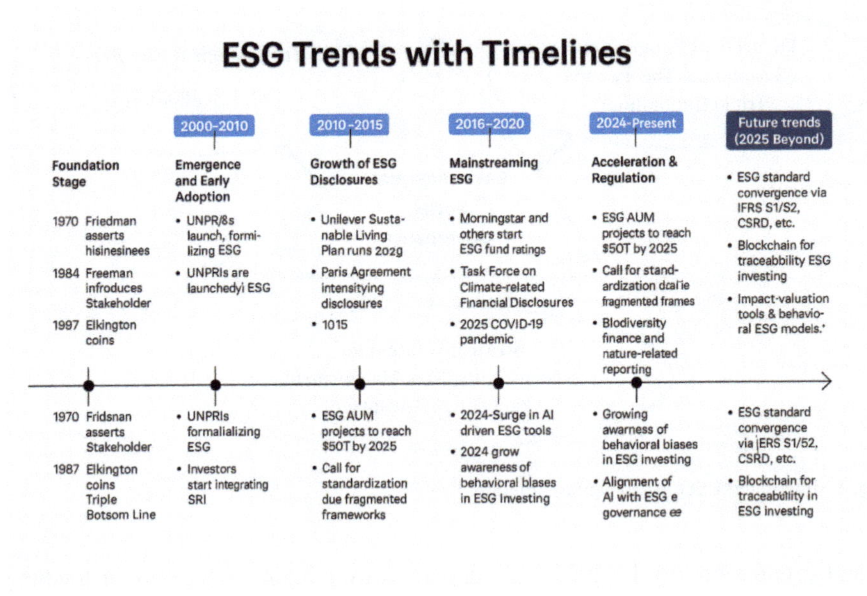

Fig. 2 ESG trends with timelines

in AI-driven ESG solutions, a Responsible AI for ESG Framework is requisite, highlighting explicable AI (XAI), compliance with OECD AI standards, and stakeholders' inclusive design processes. To combat greenwashing, a Greenwashing Mitigation Framework must be enforced through compulsory third-party audits, ISO-compliant green claim verifications, penalties for false disclosures, and effective consumer complaint mechanism. Behavioral inconsistencies in ESG investing may be minimized by an ESG Behavioral Investment Framework that fosters investor awareness, focuses on long-term performance, and implements norm-based nudges. Finally, in emerging countries, an ESG Infrastructure Capacity Building Framework should focus on public–private partnerships, the setting up of ESG innovation hubs, the formulation of localized ESG metrics and the execution of skill development programs to bridge institutional and technical disparities. These frameworks together offer a thorough approach for ensuring adoption of ethical, successful, and inclusive ESG (Fig. 3).

9 Directions for Future Research

Future research on ESG investment should be made on addressing many major gaps vital for the advancement of both theory and practice. A critical aspect is the need to examine the causal link between ESG performance and long-term financial returns across various industries and regions, particularly in emerging nations where ESG practices and enforcement mechanisms vary markedly. The integration

Fig. 3 Future ESG action framework

of AI and Big Data in ESG assessment requires further examination, particularly relating to ethical issues like algorithmic bias and data transparency. The importance of industry-specific ESG metrics need more attention, as illustrated in the case studies of Unilever and Microsoft, where standardized ESG measures inadequately address intricate sectoral complexities. Additionally, further study may explore on investor psychology and behavioral determinants such as the influence of values, biases, and risk perceptions on ESG decision-making—an area that remains unexplored. A further interesting avenue is the comparative analysis of international ESG regulation and reporting standards, including the GRI, SFDR, and TCFD, to evaluate their impact on corporate behavior and investor confidence. An interdisciplinary and context-specific research approach is essential to bridge theoretical gaps and offer practical insights that improve the effectiveness, inclusivity, and alignment of ESG investment with global sustainability goals.

Table 1 Comparative case study matrix: Unilever versus Microsoft ESG practices

Category	Unilever (FMCG)	Microsoft (Tech)
ESG strategy start year	2010 (Sustainable Living Plan)	2020 (AI for Sustainability)
Key framework used	Triple Bottom Line	Responsible AI and Stakeholder Integration
CO_2 reduction	65% (2008–2020)	17% (FY22 versus 2020 baseline)
Technological enabler	Virtual AI Instructor "Hippo"	Cloud for Sustainability, Emissions Dashboard
Social impact reach	1B + people through Lifebuoy campaigns	3,500 + clients supported for ESG data tracking
Green investment	95% sustainable raw material sourcing	$1B Climate Innovation Fund

10 Conclusion

This present research highlights the growing importance of ESG investing as a multi-dimensional strategy that integrates financial objectives with environmental sustainability, social responsibility, and effective governance. Rooted on theoretical constructs like as Freeman's stakeholder theory and the Elkington's triple bottom line, ESG has evolved from a niche ethical issue to a strategic imperative. The case studies of Unilever and Microsoft illustrate how industry leaders are adopting innovative ESG practices, endorsed by data-driven strategies and AI technologies, to create long-term business value in long run while tackling pressing global challenges such as climate change and resource depletion. Nevertheless, ESG landscape remains fragmented due to inconsistent standards, data deficiencies and insufficient legal compliance. The contextualization of ESG strategies based on industries and region-specific dynamics is the need of the hour. These deficiencies hinder effective ESG integration and highlight the need for more standardized frameworks and reliable measures. Therefore, a meticulously organized and evidence-based ESG strategy strengthened by comprehensive reporting, stakeholder involvement, and technological innovation is indispensable for sustainable investing. As ESG evolves, it has the capacity to develop business practices and the broader economic and social framework by fostering resilience, inclusivity, and accountability within the investing landscape.

References

1. Bebbington, J., & Unerman, J. (2018). Achieving the United Nations sustainable development goals: An enabling role for accounting research. *Accounting, Auditing & Accountability Journal, 31*(1), 2–24. https://doi.org/10.1108/AAAJ-05-2017-2929

2. Bloomberg Intelligence. (2021). *ESG market outlook: $50 trillion and growing*. Bloomberg Intelligence Report. https://www.bloomberg.com/company/press/esg-assets-rising-to-50-trillion-will-reshape-140-5-trillion-of-global-aum-by-2025-finds-bloomberg-intelligence/
3. Byrns, D., & McKinzie, B. (2025, February 3). *Making progress toward sustainability: A Unilever case study*. American Century Investments. https://www.americancentury.com/institutional-investors/insights/making-progress-toward-sustainability-a-unilever-case-study/?nocache=true
4. Clark, G. L., Feiner, A., & Viehs, M. (2015). *From the stockholder to the stakeholder: How sustainability can drive financial outperformance*. University of Oxford Report.
5. Cohen, M. A., Fenn, S. A., & Naimon, J. S. (1995). *Environmental and financial performance: Are they related?* Investor Responsibility Research Center.
6. Corporate Finance Institute. (2015). *Environmental, social, and governance (ESG) issues in investing: A guide for investment professionals*. CFA Institute.
7. Elkington, J. (1997). *Cannibals with forks: The triple bottom line of 21st-century business*. Capstone.
8. FII Institute. (2024). *How AI can innovate and improve ESG and sustainable finance performance (AI-for-ESG-2024-Impact-Report)*. https://fii-institute.org/wp-content/uploads/2024/05/AI-for-ESG-2024-Impact-Report_FII_IR2401_XS-1-1.pdf
9. Freeman, R. E. (1984). *Strategic management: A stakeholder approach*. Pitman.
10. Friedman, M. (1970). The social responsibility of business is to increase its profits. *The New York Times Magazine*, 122–126.
11. Friede, G., Busch, T., & Bassen, A. (2015). ESG and financial performance: Aggregated evidence from more than 2000 empirical studies. *Journal of Sustainable Finance & Investment, 5*(4), 210–233. https://doi.org/10.1080/20430795.2015.1118917
12. Global Sustainable Investment Review. (2023). *Global ESG trends and benchmarks*. GSIA Annual Report. https://www.gsi-alliance.org/wp-content/uploads/2023/12/GSIA-Report-2022.pdf
13. Gunung Capital. (2023). *Navigating regulatory and policy challenges in sustainable investing*. LinkedIn. https://www.linkedin.com/pulse/navigating-regulatory-policy-challenges-sustainable-investing/
14. IFC-Amundi. (2024). *Market green bonds*. https://www.ifc.org/content/dam/ifc/doc/2024/emerging-market-green-bonds-2023.pdf
15. KPMG. (2022). *ESG reporting trends: A call for standardization*. KPMG Insights Report. https://assets.kpmg.com/content/dam/kpmg/se/pdf/komm/2022/Global-Survey-of-Sustainability-Reporting-2022.pdf
16. Microsoft. (2022). *Sustainable workloads*.
17. Microsoft. (2024). *Sustainability at Microsoft*. https://www.microsoft.com/en-us/sustainability/
18. Morningstar Portfolio Sustainability Report Disclosure Statement. (2016). https://advisor.morningstar.com/Enterprise/VTC/MorningstarSustainabilityDisclosure10-31-2019.pdf
19. OECD. (2020). OECD business and finance outlook 2020: Sustainable and resilient finance. *OECD Publishing*. https://doi.org/10.1787/eb61fd29-en
20. OECD. (2023). *Biodiversity and development finance 2015–2021: Progress towards target 19 of the Kunming-Montreal Global Biodiversity Framework*. OECD Publishing. https://www.oecd.org/content/dam/oecd/en/publications/reports/2023/12/biodiversity-and-development-finance-2015-2021_e32571d4/6d43b54d-en.pdf
21. PwC. (2024). *How Unilever builds and delivers its social performance strategy*. https://www.pwc.nl/nl/assets/documents/pwc-bcti-case-studies-unilever.pdf
22. Responsible AI Institute. (2024). *Unveiling the guiding framework: Aligning AI and ESG for a sustainable future*. https://www.responsible.ai/unveiling-the-guiding-framework-aligning-ai-and-esg-for-a-sustainable-future-2/
23. Riedl, A., & Smeets, P. (2017). Why do investors hold socially responsible mutual funds? *Journal of Finance, 72*(6), 2505–2550. https://doi.org/10.1111/jofi.12547

24. Schoenmaker, D. (2017). *Investing for the common good: A sustainable finance framework.* Rotterdam School of Management Working Paper.
25. Shefrin, H. (2024). ESG and behavioral finance: Why ESG-investing is primarily a psychological phenomenon. https://doi.org/10.2139/ssrn.4786610
26. Shekarian, E., Ijadi, B., Zare, A., & Majava, J. (2022). Sustainable supply chain management: A comprehensive systematic review of industrial practices. *Sustainability, 14*(13), 7892. https://doi.org/10.3390/su14137892
27. Singh, G. (2024). *AI-driven ESG integration: Enhancing sustainability in investment decisions.* LinkedIn. https://www.linkedin.com/pulse/ai-driven-esg-integration-enhancing-sustainability-investment-singh-b9g0f/
28. Spencer, G. (2018). *AI for Earth: Helping save the planet with data science.* Microsoft News. https://news.microsoft.com/apac/features/ai-for-earth-helping-save-the-planet-with-data-science/
29. United Nations Environment Programme Finance Initiative (UNEP-FI). (2024). *2024 in review: UNEP FI and its members continue to drive change in challenging times.* https://www.unepfi.org/industries/banking/2024-in-review/
30. United Nations Principles for Responsible Investment (UNPRI). (2006). *Principles for responsible investment.* https://www.unpri.org
31. Unilever. (2021). *Unilever Sustainable Living Plan: 10 years of progress (2010–2020).* https://www.unilever.com/files/92ui5egz/production/16cb778e4d31b81509dc5937001559f1f5c863ab.pdf
32. Whelan, T., Atz, U., Van Holt, T., & Clark, C. (2021). *ESG and financial performance: Uncovering the relationship by aggregating evidence from 1,000 plus studies published between 2015–2020.* https://www.stern.nyu.edu/sites/default/files/assets/documents/NYU-RAM_ESG-Paper_2021%20Rev_0.pdf
33. Zhu, Y., Zhang, H., Siddik, A. B., Zheng, Y., & Sobhani, F. A. (2023). Understanding corporate green competitive advantage through green technology adoption and green dynamic capabilities: Does green product innovation matter? *Systems, 11*(9), 461. https://doi.org/10.3390/systems11090461

Predicting Cryptocurrency Trends: Approaches to Diversification and Optimization

K. V. Geetha Devi and K. Supritha

Abstract

This study delves into the intricate domain of cryptocurrency prediction, focusing on diversification and optimization strategies to enhance investment portfolios. Cryptocurrencies, characterized by their volatility and speculative nature, present unique challenges and opportunities for investors. The research employs advanced machine learning algorithms and statistical models to predict cryptocurrency prices and market trends. By integrating a diversified portfolio approach, the study aims to mitigate risks associated with individual cryptocurrency investments. Optimization techniques, including modern portfolio theory and risk management frameworks, are applied to identify optimal asset allocations that maximize returns while minimizing risk. Furthermore, the study highlights the potential of cryptocurrencies to promote financial inclusion, support environmentally conscious investments, and align with ESG goals, thus contributing to sustainable financial management. As the world seeks innovative ways to balance profitability with long-term stability, understanding cryptocurrency trends and optimizing investment strategies become crucial to fostering a resilient and inclusive financial ecosystem. The findings highlight the efficacy of diversified investment strategies and advanced predictive models in navigating the volatile cryptocurrency market, providing valuable insights for both individual and institutional investors.

Keywords

Cryptocurrency • Prediction • Diversification • Optimization • Machine learning • Portfolio management • Risk management • Volatility

K. V. Geetha Devi (✉) · K. Supritha
School of Management Studies, Madanapalle Institute of Technology & Science (MITS) Deemed to be University), Madanapalle, India
e-mail: drkotha.geetha@gmail.com

S. Patnaik et al. (eds.), *People, Planet and Profit: Advancing Sustainable Business Practices*, Springer Proceedings in Geography and Sustainability,
https://doi.org/10.1007/978-3-032-11296-5_13

1 Introduction

Cryptocurrency markets have witnessed exponential growth and widespread adoption in recent years, attracting both individual and institutional investors seeking to capitalize on the potential profits offered by this emerging asset class. As the crypto market continues to evolve and mature, the need for effective management strategies, diversification techniques, and optimization methods has become increasingly crucial to navigate the inherent volatility and uncertainties associated with digital assets. Cryptocurrencies play a role in providing financial services to the unbanked and underbanked populations, contributing to sustainable economic development. Cryptocurrencies are defined as a medium of exchange that is not reliant on central banks or national borders, in contrast to traditional currency [6]. Cryptocurrency markets have experienced rapid growth and evolution, attracting a diverse range of investors keen on harnessing the potential of digital assets. In this dynamic landscape, effective management strategies, diversification practices, and optimization techniques are essential for navigating the inherent volatility and uncertainties associated with cryptocurrencies.

The realm of cryptocurrency investment presents a dynamic landscape ripe with opportunities and challenges. As digital assets continue to reshape the financial markets, the effective management of cryptocurrency portfolios has become paramount for investors aiming to harness the potential returns while mitigating risks associated with this volatile asset class.

For effective risk mitigation in the quickly changing but erratic Indian cryptocurrency market, diversification and optimization are crucial pillars. For Indian investors, diversification entails more than just a broad selection of foreign cryptocurrencies; it also entails a deep understanding of the local market, potentially involving carefully scrutinized Indian exchange tokens and relevant domestic blockchain projects. This includes researching different crypto asset classes like DeFi and NFTs in addition to strategically employing a range of investment strategies like rupee-cost averaging through SIPs, selective DeFi participation, and staking (where compliant). Optimization in India necessitates focusing on rupee-denominated trading pairs, closely monitoring the dynamic regulatory landscape, seeking advice from regional cryptocurrency experts, and setting risk-adjusted return expectations tailored to the unique conditions of the Indian market. This study delves into the realm of cryptocurrency management, aiming to predict trends and optimize strategies for enhanced risk-adjusted returns over the long term. By leveraging historical market data, trend analysis, and sophisticated modeling tools, we seek to develop a robust framework that empowers investors to make well-informed decisions when managing their cryptocurrency portfolios.

2 Review of Literature

2.1 Recent Advances in Cryptocurrency Portfolio Management and Price Prediction

Recent research has focused on enhancing cryptocurrency portfolio management and price prediction through sophisticated methodologies. Mba and Mwambi (2020) introduced a Markov-switching continuous-time GARCH (COGARCH) model that effectively identifies shifts in market volatility, demonstrating improved portfolio selection and risk management due to its dynamic adaptation capabilities [1]. In the realm of price prediction, Pinky and Akula (2024) assessed various machine learning techniques, including linear models, tree-based algorithms, and deep learning, over a decade of cryptocurrency data. Their work underscores the importance of industrial engineering principles in optimizing these prediction models to achieve greater accuracy and better risk management [2]. Similarly, Ramesh and Karthic (2024) proposed a hybrid approach combining robust stochastic clustering with a gravitational search method to enhance Bitcoin price prediction, successfully addressing market volatility and complex data patterns to provide superior decision-making tools for investors [3]. Expanding on portfolio strategies, Tenkam et al. (2022) investigated the impact of incorporating cryptocurrencies into investment portfolios. Their analysis, using the Markowitz optimization framework alongside GARCH-Copula methodologies, revealed that portfolios integrating conventional assets with cryptocurrencies can achieve higher Sharpe ratios and improved stability in returns [4]. These studies collectively highlight the ongoing advancements in quantitative methods for navigating the complexities of the cryptocurrency market. According to Wang et al., (2020) incorporating multiple cryptocurrencies enhances portfolio outcomes.

3 Research Objectives

The present research work has the following objectives:

(a) To study sophisticated machine learning models and statistical methodologies used for forecasting cryptocurrency price fluctuations and market trends, therefore aiding investors in making educated choices.
(b) To examine and use diversification and optimization methodologies, including contemporary portfolio theory and risk management frameworks, to enhance returns while mitigating risks linked to Bitcoin investments.

4 Research Objectives

The present research work has the following objectives:

(a) To study sophisticated machine learning models and statistical methodologies used for forecasting cryptocurrency price fluctuations and market trends, therefore aiding investors in making educated choices.
(b) To examine and use diversification and optimization methodologies, including contemporary portfolio theory and risk management frameworks, to enhance returns while mitigating risks linked to Bitcoin investments.

5 Methodology

Forecasting cryptocurrency trends in India's rapidly evolving financial landscape is made possible by the combination of powerful statistical methods and cutting-edge machine learning models. Indian investors, both individuals and institutions, can benefit greatly from this combination of state-of-the-art analytics to help them navigate the volatile cryptocurrency market. By precisely forecasting price movements and identifying emerging trends, these techniques can assist in improving investment choices, managing portfolios more effectively, and ultimately establishing a more advanced and stable cryptocurrency ecosystem in the Indian context.

The requested research on the prediction, diversification, and optimization of cryptocurrency management will utilize the use of secondary data.

The utilization of data visualization tools such as correlation matrices and volatility analysis in the research provides Indian financial analysts and investors with a substantial advantage in understanding the complexities of cryptocurrency markets. With the help of these visual tools, complicated datasets can be transformed into easily comprehensible formats, providing a clearer picture of asset relationships and risk levels. When navigating the often unstable world of digital assets, Indian stakeholders can now more effectively communicate findings, make better decisions, and ultimately feel more confident thanks to this improved interpretability. The study will primarily rely on secondary data sources to conduct the necessary analyses and develop the predictive models, diversification frameworks, and portfolio optimization techniques. The secondary data will include historical price data, trading volumes, and other relevant metrics for major cryptocurrencies, as well as data on traditional asset classes and macroeconomic indicators. Reputable data providers, such as cryptocurrency exchanges, financial data vendors, and academic databases, will be the primary sources of the secondary data. The data will be carefully curated and cleaned to ensure its reliability and consistency, as the quality of the secondary data will be crucial for the validity and robustness of the study's findings.

To simplify the Cryptocurrency management for the purpose of analysis the following techniques tools are used.

- ARIMA Model using python code.
- Correlation using python code.

6 Data Analysis

Volatility using python code. The chart displays a Bitcoin price forecast from January to December. Here's a more detailed interpretation:

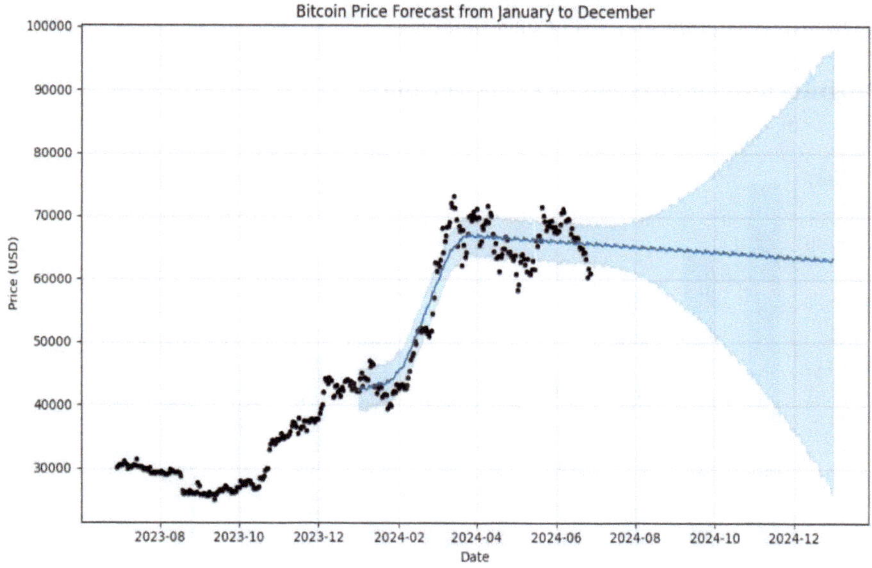

Axes: **X-axis (Date)**: This axis shows the time period from June 2023 to December 2024.

Y-axis (Price in USD): This represents the price of Bitcoin in U.S. dollars, ranging from$20,000 to $100,000.

Data Points
Black Dots: These represent the actual historical prices of Bitcoin from mid-2023 up to around June 2024.Forecast: **Blue Line:** This solid line shows the median forecasted price of Bitcoin. **Shaded Blue Area:** The shaded region represents the confidence interval, or the range within which the prices are expected to fluctuate with a certain level of confidence. The shading indicates that there is higher

certainty closer to the median forecast, with gradually decreasing certainty as the interval widens.

- Bitcoin's price shows a significant increase starting around the beginning of 2024, peaking just before mid-2024.
- Following the peak, the price appears to level off or possibly decline slightly.
- The forecast shows that the price is expected to stabilize somewhere around

$60,000, but there is considerable uncertainty, especially towards the latter months of 2024.

- Price predictions widen significantly as the forecast extends further into the future, indicating increasing uncertainty.

Program for Optimization by Using Volatility

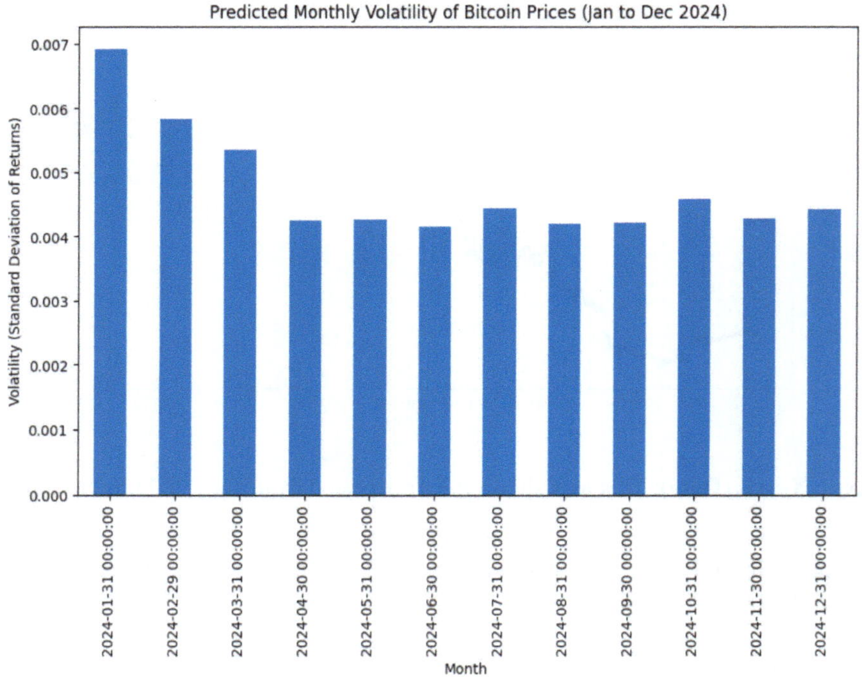

The bar chart depicts the predicted monthly volatility of Bitcoin prices from January to December 2024. Volatility is measured as the standard deviation of returns.

Key observations from the chart:

- January 2024 has the highest predicted volatility, with a standard deviation of returns around 0.007.
- Between February and October 2024, the predicted volatility stabilizes and remains relatively consistent, fluctuating between approximately 0.004 and 0.005.
- November 2024 sees a rise in volatility, although not as high as January, with a standard deviation of around 0.005.
- In December 2024, the volatility decreases slightly, reaching just under 0.005.

Overall, the chart suggests that while there is significant volatility in Bitcoin prices predicted for January 2024, the remaining months show moderate and relatively stable volatility, with a slight increase in November and a minor decrease in December.

Program for Diversification by Using Correlation

Correlation for Bitcoin and Ethereum
Output

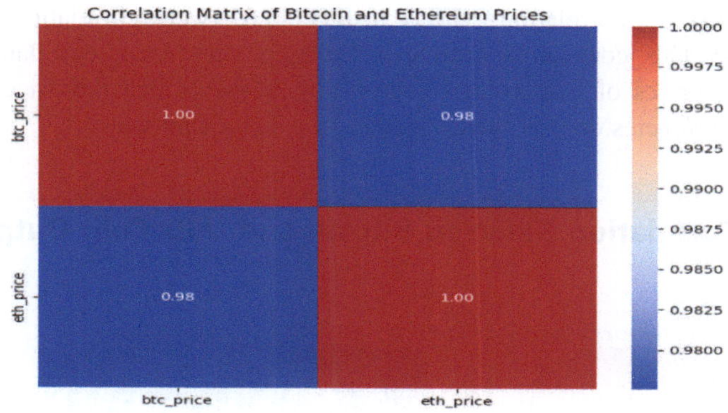

Axes: X-axis and Y-axis: Both represent two assets: "btc_price" (Bitcoin price) and "eth_price" (Ethereum price).

Correlation Values: BTC to BTC: 1.00 (perfect correlation, as expected with itself). ETH to ETH: 1.00 (perfect correlation, as expected with itself). BTC to ETH: 0.98 (very high correlation). ETH to BTC: 0.98 (very high correlation).

This correlation matrix indicates a very high correlation (0.98) between the prices of Bitcoin and Ethereum. This suggests that the prices of these two cryptocurrencies tend to move together in the market.

Correlation for Bitcoin and Tether Output

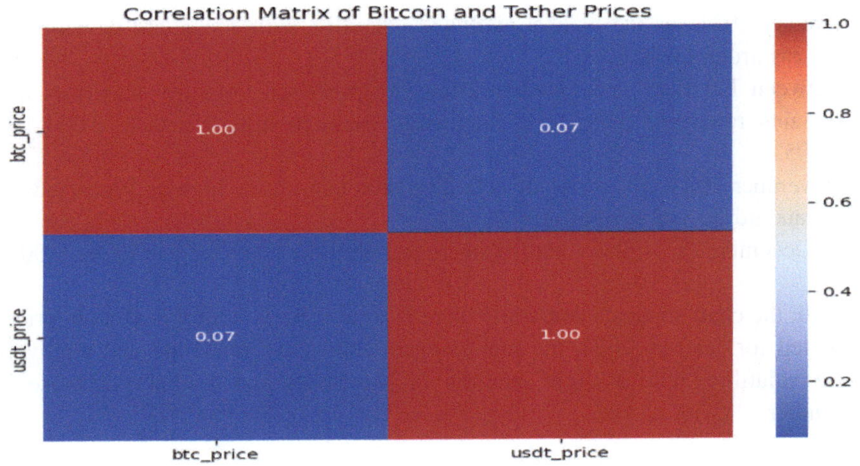

Axes: X-axis and Y-axis: Represent two assets: btc_price (Bitcoin price) and usdt_price (Tether price).

Correlation Values: BTC to BTC: 1.00 (perfect correlation, as expected with itself). USDT to USDT: 1.00 (perfect correlation, as expected with itself). BTC to USDT: 0.07 (weak correlation). USDT to BTC: 0.07 (weak correlation).

Analysis: This correlation matrix indicates a very weak correlation (0.07) between the prices of Bitcoin and Tether. This suggests that the prices of these two cryptocurrencies do not move together in a significant way.

7 Correlation Between Bitcoin and USD Coin Output

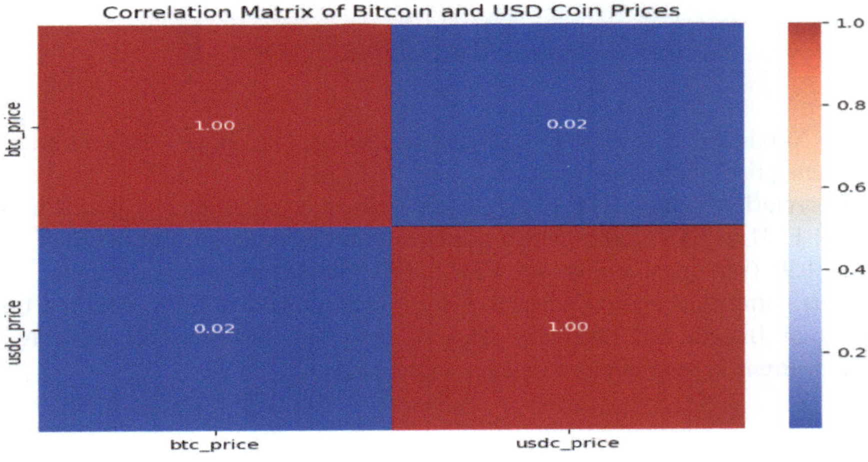

Axes: X-axis and Y-axis: Represent two assets: btc_price (Bitcoin price) and usdc_ price (USD Coin price).

Correlation Values: BTC to BTC: 1.00 (perfect correlation, as expected with itself). USDC to USDC: 1.00 (perfect correlation, as expected with itself). BTC to USDC: 0.02 (very weak correlation). USDC to BTC: 0.02 (very weak correlation).

This correlation matrix indicates a very weak correlation (0.02) between the prices of Bitcoin and USD Coin. This suggests that the prices of these two cryptocurrencies do not move significantly in tandem.

Correlation Between Bitcoin and Binance Coin Output

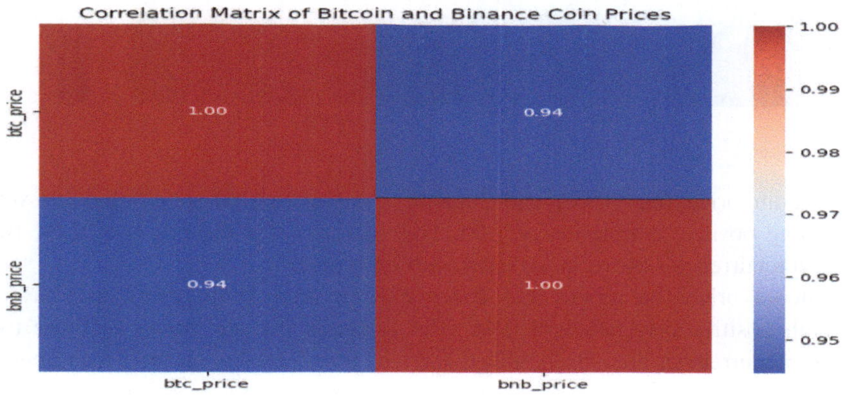

This indicates that the prices of these two cryptocurrencies tend to move in the same direction and magnitude.

2. Bitcoin price (btc_price) and Binance Coin price (bnb_price) have a moderately positive correlation of 0.94. This suggests that the prices of these two cryptocurrencies are closely linked and tend to exhibit similar price movements, though not as strongly as the correlation between the two prices.
3. The image provides a visual representation of the correlation matrix, with the values ranging from 0.95 to 1.00, indicated by the color gradient. The darker red and blue colors represent the stronger positive correlations between the cryptocurrency prices.

In summary, the correlation matrix reveals a high degree of positive correlation between the prices of Bitcoin and Binance Coin, indicating that these two cryptocurrencies tend to move in tandem with each other in the market.

Correlation Between Bitcoin and Binance USD Output

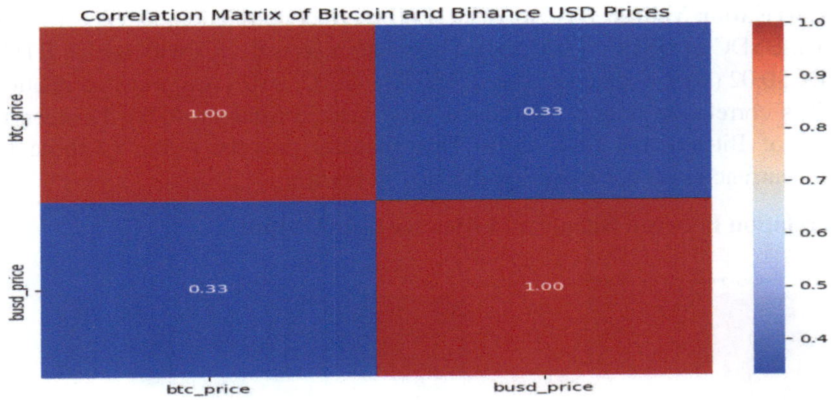

1. Bitcoin price (btc_price) and Binance Coin price (bnb_price) have a very strong positive correlation of 1.00. This indicates that the prices of these two cryptocurrencies move in lockstep with each other.
2. Bitcoin price (btc_price) and Binance Coin price (bnb_price) also have a high positive correlation of 0.94. This suggests that the prices of these two cryptocurrencies are closely linked and tend to exhibit similar price movements.
3. The visualization uses a color gradient to represent the correlation values, with darker shades of red indicating stronger positive correlations.

In summary, the correlation matrix demonstrates that Bitcoin and Binance Coin prices are highly correlated, with the prices of the two cryptocurrencies moving in a very similar fashion. This strong positive correlation implies that they are likely influenced by similar market factors and investor sentiments.

Correlation Between Bitcoin and XRP Coin Output

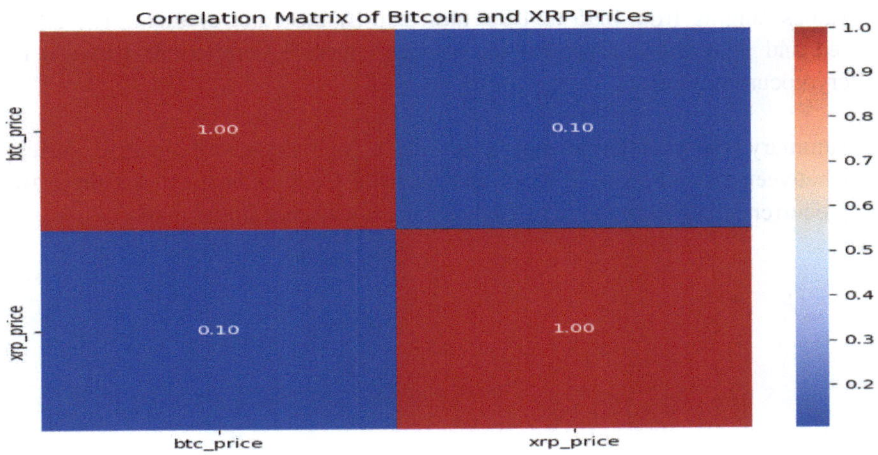

The correlation between Bitcoin price (btc_price) and XRP price (xrp_price) is very high, at1.00. This indicates a strong positive relationship between the two cryptocurrency prices. The correlation between Bitcoin price (btc_price) and XRP price (xrp_price) is also high, at 0.10. This suggests a moderately positive relationship between the two prices.

Correlation Between Bitcoin and Cardano Output

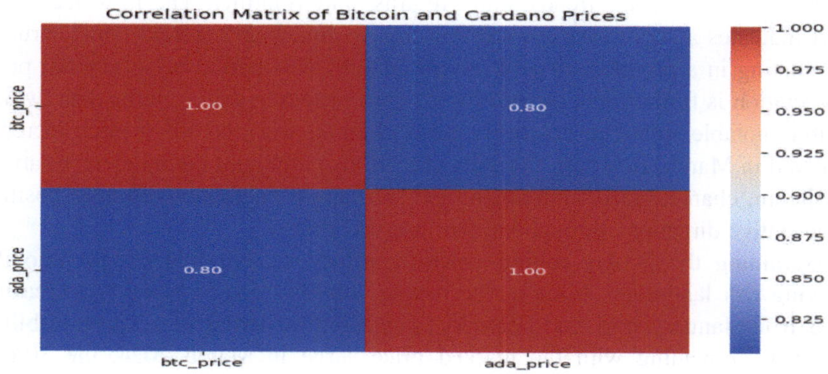

The correlation between Bitcoin price (btc_price) and Cardano price (ada_price) is very high, at 1.00. This indicates a strong positive relationship between the two cryptocurrency prices. The correlation between Bitcoin price (btc_price) and Cardano price (ada_price) is also high, at 0.80. This suggests a strong positive relationship between the two prices.

Correlation Between Bitcoin and Solana Output

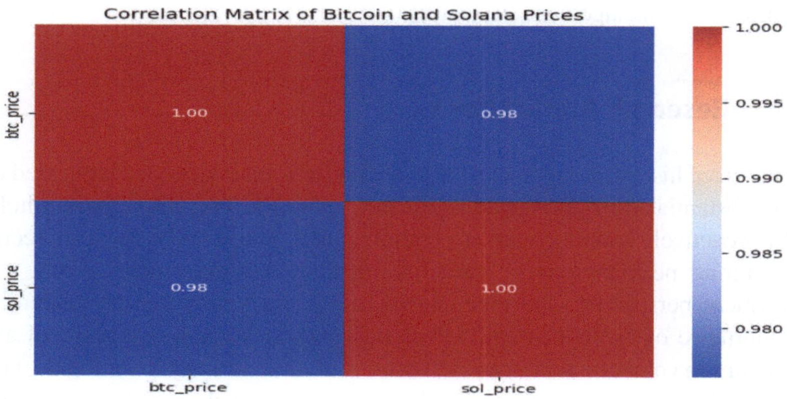

Correlation Matrix of Bitcoin and Solana Prices

The correlation between Bitcoin price (btc_price) and Solana price (sol_price) is very high, at1.00. This indicates a strong positive relationship between the two cryptocurrency prices. The correlation between Bitcoin price (btc_price) and

Solana price (sol_price) is also high, at 0.98. This suggests a strong positive relationship between the two prices.

8 Findings of the Study

Analysis of Bitcoin's price action from January to June 2024 reveals a dynamic interplay between price trends, daily returns, and volatility. The historical price chart indicates a significant upward trajectory commencing around mid-February, culminating in a pronounced peak in late March. This period of substantial price appreciation is further underscored by the corresponding daily returns data, which exhibits notable spikes in positive returns aligning with the sharp price increase observed in March. However, the daily returns also highlight the inherent volatility of Bitcoin, characterized by frequent and substantial fluctuations in both positive and negative directions throughout the observed period.

Examining the 30-day rolling volatility provides additional insight into the evolving risk landscape. Initially, the rolling volatility demonstrates a decreasing trend from January into mid-February, suggesting a period of relative stability. However, coinciding with the marked price surge in March 2024, the 30-day volatility experiences a sharp increase, reaching its apex proximate to the period of highest price. This observation strongly suggests a direct correlation between significant price movements and heightened market uncertainty, as reflected in the amplified volatility. The subsequent stabilization or potential decrease in volatility after the price peak would warrant further investigation to understand the market's response to the preceding rapid appreciation. These observations underscore the critical need for investors to not only recognize potential periods of high returns but also to be acutely aware of the accompanying elevated volatility and implement appropriate risk management strategies.

9 Research Implications

Drawing from historical cryptocurrency market data, investors have witnessed periods of substantial gains alongside significant volatility. Therefore, current holders should proactively monitor market conditions and strategically consider securing profits during peak periods to capitalize on favorable movements. From a risk management perspective, the inherent volatility of cryptocurrencies underscores the importance of diversification. Allocating investments across a range of assets or different cryptocurrencies can serve as a crucial strategy to mitigate potential losses. Furthermore, exploring hedging strategies, such as utilizing options or futures contracts, can provide tools to actively manage downside risk during anticipated periods of high market volatility. The research places a strong emphasis on incorporating contemporary portfolio theory with Environmental, Social, and Governance (ESG) principles in the context of Indian investments. This strategy provides a way for Indian investors to take part in the vibrant cryptocurrency market while also strategically coordinating their investments with risk management frameworks and more general sustainability goals. Indian individuals and financial institutions can promote a more responsible and resilient approach to cryptocurrency investments by taking into account aspects other than just returns, like ethical and environmental effects. This will support both financial growth and a more sustainable future for the country.

10 Future Recommendations

The research may not adequately capture the distinct nuances and influences influencing cryptocurrency trends in India due to the lack of investigation into real-time market dynamics and the actions of Indian investors firsthand. Significant drivers that cannot be fully explored with secondary data alone include things like local regulatory developments, the unique adoption patterns of digital assets by Indian demographics, and the influence of sentiment on social media and domestic news on trading activities.

To gain a deeper understanding of the factors influencing cryptocurrency markets, it's essential to investigate their correlations with traditional financial markets and other cryptocurrencies. This analysis can reveal broader market influences and potential contagion effects. Furthermore, incorporating sentiment analysis, which involves scrutinizing data from social media, news outlets, and other sources, can provide valuable insights into potential price movements and anticipated volatility spikes driven by market perception. To enhance predictive capabilities, the development of robust predictive models utilizing advanced statistical methods and machine learning algorithms is crucial. These sophisticated models can assist investors and portfolio managers in making more informed decisions and navigating the inherent uncertainty of the cryptocurrency market by forecasting price fluctuations and volatility.

In the realm of portfolio construction, it's vital to optimize cryptocurrency portfolio diversification. Conducting in-depth analysis on the diversification benefits of including cryptocurrencies within traditional investment portfolios can help determine the optimal allocation for risk-adjusted returns. Moreover, designing comprehensive cryptocurrency portfolio optimization frameworks is necessary. These frameworks should integrate optimization techniques that explicitly consider the unique characteristics of the cryptocurrency market, such as transaction costs, liquidity constraints, and inherent market complexities. By incorporating established financial theories like modern portfolio theory, factor investing, and risk parity, investors can construct more resilient and efficient cryptocurrency portfolios tailored to their specific risk tolerance and investment objectives.

11 Conclusion

The findings of this study underline the critical role of cryptocurrency prediction, diversification, and optimization in driving sustainable financial management and responsible investment practices. By mitigating market volatility, promoting portfolio diversification, and prioritizing environmentally conscious cryptocurrencies, this research demonstrates how cryptocurrencies can align with ESG principles and contribute to long-term financial stability. Additionally, the study emphasizes the potential of cryptocurrencies in democratizing access to finance, thereby addressing economic disparities and fostering global financial inclusion. As the financial landscape continues to evolve, integrating sustainable investment strategies with innovative technologies like cryptocurrency will be essential for building a resilient and equitable future.

References

1. Mba, J. C., & Mwambi, S. (2020). A Markov-switching COGARCH approach to cryptocurrency portfolio selection and optimization. *Financial Markets & Portfolio Management, 34*(2), 199–214. https://doi.org/10.1007/s11408-020-00346-4
2. Pinky, J. N., & Akula, R. (2024). Enhancing cryptocurrency market forecasting: Advanced machine learning techniques and industrial engineering contributions. https://arxiv.org/abs/2410.14475
3. Ramesh, R., & Karthic, M. J. (2024). Optimizing cryptocurrency price prediction: A hybrid approach with resilient stochastic clustering and gravitational search algorithm. *International Journal of Intelligent Systems & Applications in Engineering, 12*(17s), 239–248. https://www.ijisae.org/index.php/IJISAE/article/view/4869
4. Tenkam, H. M., Mba, J. C., & Mwambi, S. M. (2022). Optimization and diversification of cryptocurrency portfolios: A composite copula-based approach. *Applied Sciences, 12*(6408). https://doi.org/10.3390/app12136408
5. Maese, V. A., Avery, A. W., Naftalis, B. A., Wink, S. P., & Valdez, Y. D. (2016). Cryptocurrency: A primer. *Banking Lj, 133*, 468.
6. Ma, Y., Ahmad, F., Liu, M., & Wang, Z. (2020). Portfolio optimization in the era of digital financialization using cryptocurrencies. *Technological Forecasting & Social change, 161*, Article 120265.

The Effectiveness of Social Media Networking as a Marketing Tool for Promoting Consumer Durable Goods in Tourism Industry

A. K. Neeraja Rani, P. Pratheesha, M. Yugandharudu, J. N. J. Ravi Sharin, and P. Shabanabi

Abstract

Social media interacting has occurred as a powerful promotion tool, influencing consumer behavior and purchase intentions across various industries, including tourism. This research measures the effect of key attributes of social media networking such as Consumer Perception, Advertising, Social Media Platform, Content Quality, Interactivity, and Credibility on the buying intentions of consumers for durable things in the tourism industry. Using a structured questionnaire, data was collected from a sample of 328 respondents. The analysis was conducted using SPSS, employing EFA to identify underlying factors and Multiple Linear Regression to assess the connection among these attributes and buying intents. The findings reveal that Content Quality, Credibility, and Interactivity significantly influence consumers' purchasing decisions, while the part of advertising and social media platforms is moderated by consumer perception. The study offers esteemed perceptions for sellers and trades, emphasizing the

A. K. Neeraja Rani (✉)
Department of MBA, Santhiram Engineering College (Autonomous), Nandyal, Andhra Pradesh, India
e-mail: akneeru@gmail.com

P. Pratheesha
Department of MBA, Ananthalakshmi Institute of Technology and Sciences (Autonomous), Anantapur, Andhra Pradesh, India

M. Yugandharudu
Department of Commerce, Rayalaseema University, Kurnool, Andra Pradesh, India

J. N. J. Ravi Sharin
Department of Business Management, PG Centre Gadwal, Palamuru University, Yenkampeta, Telangana, India

P. Shabanabi
Department of MBA, Sai Rajeswari Institute of Technology, Proddatur, India

S. Patnaik et al. (eds.), *People, Planet and Profit: Advancing Sustainable Business Practices*, Springer Proceedings in Geography and Sustainability,
https://doi.org/10.1007/978-3-032-11296-5_14

need for strategic social media campaigns that enhance engagement, trust, and quality content to drive consumer purchase behavior in the tourism industry.

Keywords

Consumer perception • Advertising • Social media platform • Content quality • Interactivity • Credibility

1 Introduction

Rapid technology improvements, better infrastructure, and rising disposable incomes have all contributed significantly to the tourism industry's recent growth [1]. Consumer durable goods, such cameras, Smartphone's, and travel accessories, have grown more and more essential to improving the trip experience as travel becomes more accessible [2]. At the same time, social media's pervasiveness has drastically changed how customers interact with travel-related goods and services. These days, social networking sites like Facebook, Instagram, and Twitter are crucial for sharing experiences, influencing travel choices, and finding new items [3, 4].

Social media has changed from being a simple medium for communication to a potent marketing tool in the digital age. It gives businesses previously unheard-of chances to interact with customers, share content, and advertise consumer durable goods [4]. Sales, website traffic, and brand exposure can all be greatly impacted by social media marketing [5]. Social media marketing has emerged as a major force behind consumer behavior, especially in the travel and tourism industry, where customers are increasingly depending on peer recommendations, online reviews, and influencer material when making travel plans [6]. Comprehensive studies on the efficiency of social media marketing in promoting consumer durable products in the context of travel and tourism are few, despite the platform's increasing relevance. By investigating how social media affects consumer views and decisions about durable items in the travel industry, this study seeks to close this gap.

2 Literature Review

Social media influencers (SMIs) turn as independent third-party backers who use stands like blogs and tweets to shape audience opinions. While the traits of successful spokespersons are well-documented in public relations literature, audience perceptions of SMIs remain less understood. Using a q-sort technique, the study identified the core perceived attributes of SMIs, providing insights that can help organizations leverage SMI influence effectively [7]. Similarly, Fodor and note that businesses face challenges in adapting to the changing consumer engagement landscape driven by social media platforms like Facebook. Companies such as

Burger King, Dell, and Target are exploring strategies to effectively engage with customers on these platforms [8].

Social media influences consumer decision-making. Through a quantitative survey, the study explored how social media usage shapes consumer experiences and opinions about goods and services, particularly in developed areas, which subsequently impact decisions in less connected regions. Since advertisers cannot control the content, timing, or frequency of online consumer interactions, social media has converted an influential instrument for influencing buyers. The findings suggest that social media usage enhances customer satisfaction, particularly during the information search and alternative evaluation phases, and continues to positively influence satisfaction during the decision-making and post-purchase assessment stages [9].

Business leaders are becoming more interested in using social media sites like the website Wikipedia, Facebook, YouTube, Second Reality, and Twitter for their company's benefit, according to [3]. The idea of social media is still not well understood, despite this desire. The meaning of social media is explained in the article, along with how it differs from similar terminology like web 2.0, and content created by users. Virtual gaming universes, websites, multimedia groups, social networks, collaborations, and virtual communities are the six categories into which it divides social media. Additionally, the authors offer 10 tactical suggestions for companies aiming to use social media efficiently.

Consumer Behavior (CB) in the travel and tourism sector, emphasizing the dearth of thorough analyses in spite of the field's importance. They identified nine essential CB concepts—decision-making, principles, motives, identity, character, goals, beliefs, opinions, satisfaction, trust, and loyalty—by analyzing 191 papers from three major travel magazines between 2000 and 2012. The study also looks at three outside variables that affect how travelers behave: ethical purchasing, Generation Y, and technology. Five crucial research contexts are also noted for further investigation: consumer misbehavior, emotions, cross-cultural problems in emerging markets, understudied market groups, and collaborative decision-making. The authors stress that hedonic and emotional elements of tourism must be included in future CB research [10].

Facebook and Twitter's marketing efficacy in the hotel industry. In order to evaluate the effectiveness of social media marketing, the study suggested a theoretical model that combined the attitude-toward-the-ad (Aad) model with the attitude-toward-social-media-page notion. The model's suitability for elucidating the influence of social media on hotel marketing was confirmed by the results of an online poll. The results showed that hotel guests' social media experiences affect how they see social media sites, which in turn affects how they feel about hotel brands. Booking intentions and electronic word-of-mouth are influenced by these brand views. Additionally, the study discovered that Facebook and Twitter both exhibit comparable degrees of marketing efficacy, indicating that hotel management may use the same tactics on both platforms [11].

Motivations behind summer vacationers' social media contributions and content-sharing behaviors based on a survey conducted in Mallorca, a popular tourist destination. The study highlights the role of social capital and individual and community benefits in influencing user-generated content sharing. It found that visual content dominates social media sharing and that motivations vary among different types of content creators, with altruistic and community-related factors playing a significant role. While social media sharing reflects friendliness and emotional support, it is less influential as an information source for vacation planning. The study also reveals how new and traditional technologies complement and overlap in tourism practices [12].

A company's social customer relations management (CRM) competence may be developed through the use of social media technologies and customer-centric management systems. The conceptualization and assessment of social CRM capacity, which draws on literature on advertising, technology, and strategic management, is the study's first significant contribution. The second contribution examines how social CRM capabilities are improved by the interplay between social media technology and customer-focused management systems. The study, which used structural equation modeling on data from 308 organizations, discovered that customer relationship performance is positively impacted by a strong social CRM capacity [13].

Social media's visibility in search engine outcomes for information on travel. By combining nine well-known tourist attractions in the United States with particular keywords, the study mimicked the search behavior of a traveler. According to the data, a large percentage of search results are from social networking platforms, suggesting that search engines frequently send visitors to these websites. The results demonstrate social media's increasing significance as a major information source in the field of online travel [14].

The present study utilizes the Theory of Planned Behavior (TPB) [18] and Uses and Gratifications Theory (UGT) [19] to strengthen its conceptual framework and hypothesis. These frameworks offer comprehensive insights on how social media attributes impact consumer intentions, especially when it relates to travel and consumer durable items. According to the Theory of Planned Behavior, attitudes toward behavior, subjective norms, and perceived behavioral control are impacted by behavioral intentions. In this research:

- The attitudes that consumers develop due to their exposure to social media marketing are reflected in their perceptions and content quality.
- Customers' confidence in their ability to make well-informed decisions is also impacted by perceived behavioral control, which is reflective of credibility and interaction.
- Similar to subjective norms, advertising and platform attributes act as social signals that affect behavior via influencer endorsements and peer validation.

Conversely, the Uses and Gratifications Theory focuses on how consumers actively select media to address certain needs such as entertainment, information, social interactions, and decision-making support.

- Content quality satisfies such information need.
- Social integration and satisfaction from entertainment are addressed through interaction and platform.
- When making purchase decisions, credibility gratifies the desire for certainty and trust.

The study ensures robust theoretical foundation for examining the effectiveness of social media networking as a marketing tool in the context of sustainable tourism by connecting these components with TPB and UGT.

3 Research Gap

Despite the growing use of social media in marketing, its effectiveness in promoting consumer durable goods in the travel and tourism sector remains under-researched. Most existing studies focus on social media's role in enhancing brand visibility and customer engagement, but few examine its direct impact on consumer behavior toward durable goods in a tourism context. Current literature primarily addresses service-based tourism marketing, overlooking how social media influences travelers' purchasing decisions for durable goods. There is also limited empirical evidence on the most effective platforms, content strategies, and engagement tactics for increasing consumer interest and sales. Closing this gap is essential for businesses to optimize their digital marketing efforts in this niche market.

4 Research Problem

Social media's explosive development has revolutionized marketing, especially in the tourism and travel sector, where consumer interaction has a big impact on decisions to buy. Although consumer durable products like electronics, baggage, and travel necessities are necessary for travel, little is known about how to properly market them on social media. Although user-generated content, influencer marketing, and tailored advertisements are made possible by channels such Facebook, YouTube, and Instagram, it is still unknown how well these tactics work to increase interest and sales. Marketing results may be greatly impacted by elements including viewership, trust in internet reviews, social media algorithms, and the relevancy of the material. In order to assist businesses in improving their social media strategy, this research attempts to assess how well social media promotes consumer durable products in the travel industry, with a particular emphasis on how it affects brand recognition, customer perception, and purchase behavior.

5 Research Objectives

1. To identify the attributes of social media networking in promoting consumer durable goods in tourism industry.
2. To assess the effect of attributes of social media networking on the purchase intentions of consumers for durable goods in tourism industry.

6 Statistical Tools

- Reliability Test
- Factor Analysis
- Regression

7 Research Methodology

Sampling Procedure

In this study on "The Effectiveness of Social Media Networking as a Marketing Tool for Promoting Consumer Durable Goods in Tourism," a convenience sampling method will be employed to select respondents who are easily accessible and willing to participate. Given the diverse and widespread nature of social media users, a sample size of 328 respondents will be considered appropriate to ensure meaningful insights while maintaining feasibility. The sample will include active social media users, travelers, and potential consumers who engage with tourism-related advertisements and promotions. This approach allows for a practical and cost-effective method of data collection while capturing relevant opinions on how social media influences consumer behavior in the tourism industry (Tables 1 and 2).

Table 1 Case processing summary

		N	%
Cases	Valid	328	100.0
	Excluded[a]	0	0.0
	Total	328	100.0

Table 2 Reliability statistics

Cronbach's alpha	No. of items
0.866	22

8 Results

Reliability and Validity Test

The scale's 22 questions carry a Cronbach's Alpha score of 0.886 based on dependability statistics. This conclusion suggests that the research's instrument or questionnaire is quite reliable, as Cronbach's Alpha scores over 0.70 are generally considered good and values above 0.80 indicate strong internal consistency. The high dependability indicates that the responses are constant and dependable, and that the components are consistently evaluating the intended construct. Consequently, the instrument can be considered suitable for further analysis and interpretation (Tables 3 and 4).

Table 3 Demographic profile of respondents

Demographic variable	Category	Frequency (n)	Percentage (%)
Gender	Male	176	53.7
	Female	152	46.3
Age group	18–25 years	98	29.9
	26–35 years	134	40.9
	36–45 years	64	19.5
	46 years and above	32	9.8
Education level	Undergraduate	86	26.2
	Graduate	158	48.2
	Postgraduate and above	84	25.6
Occupation	Student	82	25.0
	Private sector employee	114	34.8
	Government employee	36	11.0
	Self-employed/business	52	15.9
	Others	44	13.4
Monthly income	Below ₹25,000	102	31.1
	₹25,000–₹50,000	120	36.6
	₹50,001–₹75,000	66	20.1
	Above ₹75,000	40	12.2
Social media usage	Less than 1 h/day	38	11.6
	1–3 h/day	126	38.4
	3–5 h/day	104	31.7
	More than 5 h/day	60	18.3
Frequency of travel	Rarely (Once a year or less)	72	22.0
	Occasionally (2–3 times/year)	156	47.6
	Frequently (4 + times/year)	100	30.5

Table 4 KMO and Bartlett's test

KMO of sampling adequacy		0.815
Bartlett's test of sphericity	Approx. chi-square	3753.831
	Df.	231
	Sig.	0.000

Descriptive Statistics
The respondents' demographic profile offers important background information for analyzing the study's conclusions. The 328 individuals in the sample had a fairly balanced gender distribution (46.3% female and 53.7% male), suggesting a range of viewpoints. The younger age groups of 18 to 35 years old accounted for a sizable share (70.8%), indicating that the study well represented the opinions of the main social media platform users. The bulk of respondents (73.8%) had graduate degrees or higher, indicating a well-educated sample with possibly knowledgeable views on tourism and marketing. The two largest occupational groups were students (25%) and private sector workers (34.8%), both of whom are generally more engaged on social media and more likely to interact with online marketing content. According to the income distribution, the majority of respondents (67.7%) made less than ₹50,000 a month, indicating a middle-class demographic that is both frugal and tech-savvy. Interestingly, 88.4% of respondents reported using social media at least an hour every day, demonstrating the platforms' applicability as a marketing tool. The fact that 78.1% of the respondents said they traveled at least two or three times a year further validates their exposure to content about tourism and qualifies them to assess the efficacy of social media marketing in the travel industry.

Exploratory Factor Analysis
A KMO value more than 0.80 shows strong correlation between variables, making them suitable for factor extraction; results of the Bartlett Test of Sphericity indicate an important finding (The chi-square = 3753.831, the df = 231, the significance level is = 0.000), suggesting that the matrix of correlation is not a matrix of identity, verifying that there are enough relationships between variables to proceed with factor analysis; and the Kaiser-Meyer-Olkin (KMO) of adequate sampling (0.815), which is significantly higher than the suggested threshold of 0.60 (Table 5).

Extraction Method: Principal component analysis.

Varimax Rotation and Kaiser Normalization were used to identify six components. Each factor is composed of all the variables with factor loadings larger than 0.5. Twenty-two variables were integrated into six factors. These six factors accounted for 68.156 percent of the variability in the tourism industry (Table 6).

Regressions
Multiple regression analysis was used to determine how social media networking characteristics affected consumers' intentions to make purchases (Table 7).

Table 5 Total variance explained

Component	Initial eigenvalues			Extraction sums of squared loadings			Rotation sums of squared loadings		
	Total	% of variance	Cumulative %	Total	% of variance	Cumulative %	Total	% of variance	Cumulative %
1	7.197	32.716	32.716	7.197	32.716	32.716	3.281	14.913	14.913
2	2.151	9.776	42.491	2.151	9.776	42.491	2.982	13.553	28.466
3	1.777	8.078	50.570	1.777	8.078	50.570	2.782	12.647	41.113
4	1.548	7.036	57.606	1.548	7.036	57.606	2.168	9.856	50.969
5	1.213	5.514	63.120	1.213	5.514	63.120	1.989	9.042	60.012
6	1.108	5.036	68.156	1.108	5.036	68.156	1.792	8.144	68.156
7	0.860	3.911	72.067						
8	0.761	3.460	75.527						
9	0.647	2.941	78.467						
10	0.550	2.498	80.965						
11	0.499	2.270	83.235						
12	0.490	2.226	85.461						
13	0.460	2.090	87.551						
14	0.448	2.036	89.587						
15	0.413	1.875	91.463						
16	0.375	1.706	93.168						
17	0.358	1.625	94.794						
18	0.337	1.534	96.328						
19	0.299	1.361	97.689						
20	0.269	1.223	98.912						

(continued)

Table 5 (continued)

Component	Initial eigenvalues			Extraction sums of squared loadings			Rotation sums of squared loadings		
	Total	% of variance	Cumulative %	Total	% of variance	Cumulative %	Total	% of variance	Cumulative %
21	0.202	0.917	99.828						
22	0.038	0.172	100.000						

Table 6 Rotated component matrix

Statement	Component					
	1	2	3	4	5	6
The content shared on social media about consumer durable goods in tourism is informative and useful	0.800					
The visual appeal and presentation of products in social media marketing influence my interest	0.783					
The content provided by brands on social media is engaging and keeps me interested	0.755					
The quality of information on social media helps me make better purchasing decisions	0.709					
I can actively interact with businesses and their content thanks to social media marketing		0.899				
I find it helpful to interact with other consumers through comments and reviews on social media		0.845				
The ability to ask questions and get responses from brands enhances my trust in their products		0.688				
Social media features (likes, shares, comments) influence my perception of a product's popularity		0.644				
I trust product recommendations from social media influencers and brand pages			0.838			
Reviews and ratings on social media impact my decision to purchase consumer durable goods			0.792			
I believe that social media marketing provides accurate and reliable information			0.777			
User-generated content (e.g., testimonials, travel experiences) makes the marketing more trustworthy			0.591			
Social media advertisements for consumer durable goods in tourism capture my attention				0.732		
The advertisements I see on social media are relevant to my interests and travel preferences				0.705		
Sponsored content and paid promotions influence my purchasing decisions				0.644		

(continued)

Table 6 (continued)

Statement	Component					
	1	2	3	4	5	6
Social media ads provide enough details to make informed decisions about durable goods				0.559		
I like to use certain social media sites (such Facebook, YouTube, and Instagram) to get information on products relevant to travel					0.742	
Different social media platforms offer varying levels of credibility and engagement for marketing					0.718	
I find video-based content (e.g., reels, YouTube ads) more influential than text-based posts					0.550	
Social media marketing influences my perception of brand quality and reliability						0.798
I believe that social media enhances my awareness of new consumer durable goods in tourism						0.661
Social media promotions and discounts impact my decision to buy tourism-related products						0.563

Table 7 Model summary

Model	R	R Square	Adjusted R square	Std. error of the estimate	Durbin-Watson
1	0.728[1]	0.530	0.521	0.656	1.172

The characteristics of social media networking and purchasing inclinations are strongly positively correlated, as indicated by the R value (0.728). A decent model fit is indicated by the R Square value (0.530), which indicates that the independent variables (Consumer Perception, Advertising, Social Media Platform, Content Quality, Interactivity, and Credibility) account for 53% of the variance in purchase intentions. Although significantly lower, the Adjusted R^2(0.521), which takes into consideration the quantity of social media networking traits and purchase intentions, nevertheless shows strong explanatory power. The accuracy of social media networking traits and buy intentions is indicated by the Standard Error of the Estimate (0.656), which is the average difference of observed values from projected values. Lastly, autocorrelation in residuals is evaluated using the Durbin-Watson statistic (1.172). Given that the optimal value is near 2, a score of 1.172 indicates some positive autocorrelation, which may call for additional diagnostic testing. Though improvements could increase its accuracy, the model seems to be a good predictor of customer purchase intentions overall.

The overall significance of the regression model used to forecast consumers' intents to buy durable products based on consumer perception, advertising, social media platform, content quality, interaction, and credibility is evaluated using the Analysis of Variance (ANOVA) results. A substantial amount of the alteration in purchase intentions may be described by the independent factors taken together, as indicated by the significantly high F-statistic (60.244). The p-value (Sig. = 0.000), which shows that the probability that the observed correlation occurred by chance is very low, suggests that the model used for regression is significant at the level of confidence of 95%. According to the Total of Squares values, which show that the variables that are independent explain 155.675 of all variances (293.924) and the remaining variance is 138.249, the model explains a sizable portion of the variance in purchase intentions. The model's explanatory power is further supported by the regression's mean square (25.946), which is higher than the residuals' mean square (0.431). Overall, these findings support the model's dependability by indicating that the predictors significantly influence customers' feelings to buy durable products (Table 9).

Standardized regression coefficients that show the magnitude and direction of the influence are shown in Table 8. Additionally, it contains t and significant values to support the formulation of the measured hypothesis. Substantial influence of social media networking's many dimensions on customers' intentions to buy durable products.

- Table 8's beta value of 0.147 demonstrates that customers' intentions to purchase durable products are positively impacted by the quality of the content. Given that the t value is 2.913 and the sig value is 0.004, both of which are smaller than 0.05, content attributes significantly influence buyers' inclinations to acquire durable items. Table 8's beta value of 0.083 demonstrates that interaction has a favorable effect on customers' inclinations to buy durable products. Given that the t value is 6.611 and the sig value is 0.001, both of which are less than 0.05, interactivity significantly influences buyers' inclinations to buy durable items.
- Table 8 displays a beta value of 0.296, indicating a favourable influence of trust worthiness on customers' inclinations to purchase durable goods. Customers' trust in durable items is greatly impacted by content quality since the t value is 5.500 and the sig value is 0.000, both of which are less than 0.05.
- Table 8 displays a beta value of 0.213, indicating that advertising has a favorable effect on consumers' intentions to purchase durable goods. Given that

Table 8 Analysis of variance

Model		Sum of squares	Df	Mean square	F	Sig.
1	Regression	155.675	6	25.946	60.244	0.000^2
	Residual	138.249	321	0.431		
	Total	293.924	327			

Table 9 Coefficients

Model		Unstandardized coefficients		Standardized coefficients	T	Sig.
		B	Std. error	Beta		
1	(Constant)	0.421	0.175		2.408	0.017
	Content quality	0.121	0.042	0.147	2.913	0.004
	Interactivity	0.069	0.043	0.083	6.611	0.001
	Credibility	0.258	0.047	0.296	5.500	0.000
	Advertising	0.197	0.038	0.213	5.147	0.000
	Social media platform	0.089	0.040	0.104	2.241	0.026
	Consumer perception	0.163	0.043	0.167	3.783	0.000

[1]Dependent Variable: Purchase intentions of consumers for durable goods

the t-value is 5.147 and the sig-value is 0.000, both of which are less than 0.05, advertising significantly influences customers' intentions to buy durable products.

- Table 8's beta value of 0.104 demonstrates that social media platforms have a favorable influence on customers' intentions to buy durable goods. Given that the t value is 2.241 and the sig value is 0.026, both of which are smaller than 0.05, social media platforms significantly influence customers' inclinations to buy durable goods.
- Table 8 displays a beta value of 0.167, indicating a favorable influence of customer perception on consumers' intentions to purchase durable goods. Given that the t value is 3.783 and the sig value is 0.000, both of which are smaller than 0.05, customer perception significantly influences consumers' intentions to buy durable items.

Histogram and Normal P-P Plot

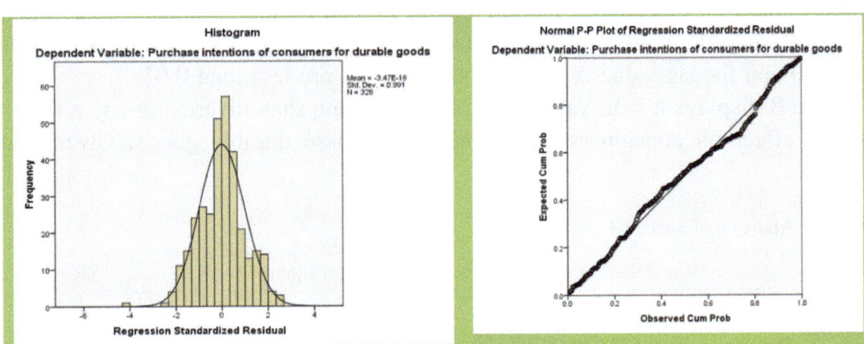

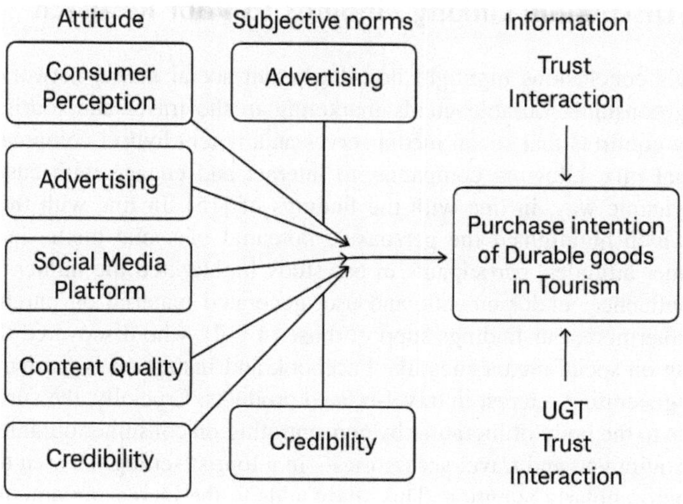

Fig. 1 Theoretical model integrating TPB and UGT for social media marketing in tourism

Figure 1 shows a normalized histogram of the distribution of the residual. A distribution is considered normal when the plotted points line up with the diagonal line on a normal P-P plot.

Recommendations

- In the tourist industry, customer perceptions toward durable goods are influenced by perceived value, trust, awareness, and involvement. Good experiences increase self-esteem, which affects brand loyalty and buying decisions.
- Consumer attraction is increased by frequency, visual appeal, personalization, and targeted advertisements. Purchase intentions for durable products connected to tourism are increased when curiosity is piqued by effective communications and emotional connections.
- Consumer engagement is impacted by platform popularity, user demographics, accessibility, and interactive features. Preferred platforms increase brand awareness and have an impact on tourists' decisions to buy durable products.
- Trust is increased by interesting, visually appealing, educational, and pertinent material. When buying durable items for travel adventures, consumers are influenced by high-quality postings, reviews, and professional comments.
- User interaction, community conversations, real-time reactions, and two-way communication all donate to the growth of brand affinity. In the tourist industry, interactive elements like surveys, live sessions, and Q&A affect customers' inclinations to make purchases.
- Trust is increased by genuine reviews, influencer recommendations, openness, and brand reputation. Consumer trust in buying durable items for travel is greatly impacted by verified information, professional advice, and ethical marketing techniques.

9 Discussion: Linking Findings to Prior Research

The study's conclusions highlight how important social media networking is to improving consumer durable goods marketing in the travel and tourism sector. This study confirms that social media serves as a potent hybrid component of the promotional mix, allowing companies to interact and engage with customers in a more dynamic way, in line with the findings of [15]. In line with the findings of [16], which highlighted the persuasive potential of social media in influencing customer attitudes, participants in our study highlighted the influence of peer reviews, influencer endorsements, and user-generated material on purchase decisions. Furthermore, our findings support those of [17], who discovered that visual storytelling on social media sites like Facebook and Instagram is particularly successful at generating interest in travel-related products. Crucially, the current study contributes to the body of literature by concentrating on consumer durable goods—like tech equipment and travel accessories—in a tourist setting, an area that hasn't gotten much scholarly attention. This study adds to the increasing amount of data indicating that social media marketing tactics may greatly increase brand visibility and impact consumer behavior in the travel industry when they are customized to travelers; lifestyle goals.

10 Limitations and Future Research Directions

This study suffers from certain limitations even though it provides valuable insights on how social media attributes impact consumers' intensions to buy durable items in the tourism sector. First, this work uses convenience sampling technique which could possibly limit representatives of the results. The findings may not be entirely applicable to the larger population using social media or tourists since respondents were chosen based on convenience rather than random selection.

Second, the sample was geographically drawn from the state of Andra Pradesh, India, although having a broad demographic profile in terms of age and employment. This might have an effect on the external validity of the conclusion in other cultural or regional contexts in India or throughout the world.

Third, the research's cross-sectional design records consumer buying intentions at a certain moment in time, disregarding seasonal or temporal variations in social media trends or tourism behavior. Future research could address these limitations by using probability sampling techniques and organizing longitudinal studies to broaden the scope to include comparisons global tourism. Besides, use of qualitative methods like focus group discussions or interviews possibly improve underlying insights on tourists' motives and attitudes.

11 Conclusion

The study on the attributes of social media networking such as Consumer Perception, Advertising, Social Media Platform, Content Quality, Interactivity, and Credibility reveals their significant influence on consumers' purchase intentions for durable goods in the tourism industry. Consumer perception plays a vital role in shaping attitudes and trust toward brands, while effective advertising strategies enhance product visibility and desirability. The choice of social media platform determines reach and engagement, with content quality acting as a key driver in capturing consumer interest and fostering informed decision-making. Additionally, interactivity between consumers and brands strengthens relationships, enhances customer experiences, and boosts purchase confidence. Credibility remains a crucial factor, as consumers rely on authentic and trustworthy content before making purchasing decisions. Overall, the integration of these attributes in social media networking effectively enhances consumer engagement, builds brand trust, and positively influences their purchase intentions for durable goods in the tourism industry.

References

1. World Tourism Organization. (2020). World Tourism Barometer.
2. Kim, J., & Lee, C. (2015). The role of consumer durable goods in tourism experience. *Journal of Travel Research, 54*(3), 349–361.
3. Kaplan, A. M., & Haenlein, M. (2010). Users of the world, unite! The challenges and opportunities of social media. *Business Horizons, 53*(1), 59–68.
4. Trainor, K. J., Andzulis, J., Rapp, A., & Agnihotri, R. (2014). Social media technology usage and customer relationship performance: A capabilities-based perspective. *Journal of Business Research, 67*(6), 724–731.
5. Huang, C. C., & Chen, C. C. (2016). Exploring the impact of social media on consumer behavior. *Journal of Business Research, 69*(11), 4848–4855.
6. Xiang, Z., & Gretzel, U. (2010). Role of social media in online travel information search. *Tourism Management, 31*(2), 179–188.
7. Freberg, K., Graham, K., McGaughey, K., & Freberg, L. A. (2011). Who are the social media influencers? A study of public perceptions of personality. *Public relations review, 37*(1), 90–92.
8. Hoffman, D. L., & Fodor, M. (2010). Can you measure the ROI of your social media marketing? *MIT Sloan Management Review, 52*(1), 41–49.
9. Leung, D., Law, R., & Lee, H. (2015). Social media marketing and tourist behavior: A systematic review. *Journal of Hospitality and Tourism Research, 39*(2), 151–172.
10. Lee, S., & Kim, D. Y. (2015). A meta-analysis of the determinants of consumer behavior in the travel and tourism sector. *Journal of Travel & Tourism Marketing, 32*(1–2), 62–77.
11. Leung, D., Law, R., & Lee, H. A. (2015). Evaluating the effectiveness of social media marketing in the hotel industry: A model for understanding the impact of social media on brand perception and booking intentions. *Journal of Travel & Tourism Marketing, 32*(8), 1011–1027.
12. Jacobsen, J. K. S., & Munar, A. M. (2014). Motivations for sharing tourism experiences through social media. *Tourism Management, 43*, 37–46.
13. Trainor, K. J., Andzulis, J. M., Rapp, A., & Agnihotri, R. (2014). Social media technology usage and customer relationship performance: A capabilities-based examination of social CRM. *Journal of Business Research, 67*(6), 1201–1208.

14. Gretzel, U., & Xiang, Z. (2010). Traveler use of social media in destination selection: search engine visibility of travel-related user-generated content. *Journal of Travel Research, 49*(2), 179–188.
15. Hudson, S., & Thal, K. (2013). The impact of social media on the consumer decision process: Implications for tourism marketing. *Journal of Travel & Tourism Marketing, 30*(1–2), 156–160. https://doi.org/10.1080/10548408.2013.751276
16. Kaplan, A. M., & Haenlein, M. (2010). Users of the world, unite! The challenges and opportunities of Social Media. *Business Horizons, 53*(1), 59–68. https://doi.org/10.1016/j.bushor.2009.09.003
17. Mangold, W. G., & Faulds, D. J. (2009). Social media: The new hybrid element of the promotion mix. *Business Horizons, 52*(4), 357–365. https://doi.org/10.1016/j.bushor.2009.03.002
18. Ajzen, I. (1991). The theory of planned behavior. *Organizational Behavior and Human Decision Processes, 50*(2), 179–211. https://doi.org/10.1016/0749-5978(91)90020-T
19. Katz, E., Blumler, J. G., & Gurevitch, M. (1973). Uses and gratifications research. *Public Opinion Quarterly, 37*(4), 509–523. https://doi.org/10.1086/268109

Stock Market Investments: An Exploration of Investor Perceptions

Kathari Santosh, M. Jahnavi, N. Nagasubba Reddy, B. Sireesha, M. Sandeep Kumar, and G. Ramanjaneyulu

Abstract

"To investigate the association in the investors' perception attributes towards the stock market across various demographic variables" is the stated goal of this research paper. Using a sample of 223 investors' perceptions of the Hyderabad stock market, the paper uses data reduction using the Chi-Square test, CFA, and Reliability test. It condenses a collection of 19 items representing six investor perception qualities. A model of investors' perceptions of the stock market across demographic characteristics is proposed in this study. The study discovered a substantial correlation between the respondents' demographic profile and six investor perception variables. Although the Indian market for stocks is prone to numerous unanticipated swings, it also offers investors who are strategic and forward-thinking significant profits and capital appreciation.

K. Santosh (✉) · M. S. Kumar
Department of MBA, CMR Institute of Technology, Bengaluru, India
e-mail: katarisantoshmba@gmail.com

M. Jahnavi · N. N. Reddy
RV Institute of Management, Bengaluru, India
e-mail: jahnavim.rvim@rvei.edu.in

N. N. Reddy
e-mail: nnsreddy.rvim@rvei.edu.in

B. Sireesha
Cambridge Institute of Technology, Bengaluru, India
e-mail: sireesha.mba@cambridge.edu.in

G. Ramanjaneyulu
Department of MBA, Annamacharya Institute of Technology and Sciences, Kadapa, India

S. Patnaik et al. (eds.), *People, Planet and Profit: Advancing Sustainable Business Practices*, Springer Proceedings in Geography and Sustainability,
https://doi.org/10.1007/978-3-032-11296-5_15

Keywords

Maturity period • Capital increase • ROI • Safety of principle • Tax benefits risk • Investors perception • Stock market

1 Introduction

An important participant in the Indian stock market is the investor. The stock market regulators cannot ignore the actions of individual financial backers as they create a more significant share of initiatives and compensation. The aforementioned research purposes to understand and learn about the elements that influence the behavior and attitude of the retail monetary supporter. The primary focus is on investors' decision-making processes and viewpoints toward investing in the stock market. Increasing compensation and cutting expenses ultimately become the main ideas of venture capitalists. The investors' prudent behavior teaches them to divide their earnings between investment and usage money. The investors' prudent behavior teaches them to divide their earnings between investment and usage money. When the possibility of income and expenses is taken into concern, players in the speculating cycle become more independent. A well-planned and detailed survey is a method used to understand stock market investors' perspectives. A great deal of care is required to focus on the investors' mental capabilities because their stock predilection and character traits are factors that significantly impact the venture decisions. In this volatile market, investors' perceptions and attitudes toward the securities exchange fluctuate periodically. Taking this into consideration, this review was adopted in order to understand investor behaviors and to learn about investors' awareness, preferences, and inclinations regarding various venture roads.

2 Review of Literature

The value disclosure tool is typically feasible for the majority of goods, although it might not be particularly effective for a few. In particular, value changes in goods markets may be hypothesized or hedged against using causality. Although skilled theoretical processes can be calculated whether a change in destinies expenses power adjustments in spot costs, equipped supporting methods can be prepared in the unlikely event that variations in spot costs power variations in fates costs. Additionally, causality may be used to predict item location and fate costs. The majority of Indian investors are unaware of the coordinated item market, and their judgment regarding it ranges from risky to really risky speculation. Many of them harbor misconceptions about the product market. They act absurdly toward the item market as a result. Concerned professionals must take action to improve the method of exchanging products in a way that is simple and painless [1]. The Granger causality, co-joining, and error correction model to research pepper in order to examine the value revelation process. In the pepper options market, there existed a unidirectional causal relationship between futures and spot costs [2]. They saw an effective value disclosure procedure in place. They proposed strengthening the administrative framework of the marketplace. The Forwards Market Commission's (FMC) independence was given significance. Additionally, their focus revealed the need for globally developed warehouses and market connections [3]. The long-run connection among investments and commercial expenses for farming supplies such as maize, chickpeas, sugar, black lentil, pepper, cotton seed, and soybeans. In the futures market and Spot costs, they saw co-joining. They were temporarily associated, and futures exchanges were able to speculate on the spot prices of sugar, castor seeds, soybeans, and chickpeas. In the short term, there existed a reciprocal interaction between peppers, maize, and black lentil [4].

To close the data streams between Spot and Contracts costs in agricultural products used the Granger Causality test. It was shown that futures markets are usually where spot costs may be discovered. They differ in order to combat the uncertainty in business sectors and to provide a basis of sufficient food grain holdings worldwide [5]. Generally speaking, trading horticulture products—particularly food items—have not resulted in value transparency or decreased price volatility. There are no discernible effects on ranchers' ability to get higher prices as a rule across the lookout. Like in other places, future product business sectors in India appear to possess introduced value market dealers to new theoretical approaches. Along with a causal relationship from later to ad pricing for goods for which upcoming valuing numbers was provided, they came across the stark increases in spot costs of major food items [6]. To determine how offering chances for exchange affected the current prices of pulses in India. They discovered that the prices of wheat, urad, and gram were higher during the post-prospects era than they were during the pre-futures period, which is consistent with the destabilizing effect of the fates contract [7]. The financial market for goods in the last few weeks and for goods in the pre-contract expiration weeks was an unadulterated payback of the fortunes market [8]. The level of unpredictability was reduced when the markets of the subordinates were put up as spurious variables. As a result, the exchange of the superiors on file showed a net positive stress. As the smaller business sectors were introduced, the NSE Nifty-50 record's trading volume has also increased. This result is evident even after the capacity exported on the lapsed days was indifferent. Since subsidiaries have a significant impact, their presentation does not pose a difficult for the commercial marketplace. These goals cannot help but run counter to the widely held belief that trading in subsidiary' markets increases volatility and decreases the core market's liquidity. Therefore, he differed with essayists who promote a more comprehensive set of subordinate economic sectors since this set of rules may limit the possible results of ventures [9].

The impact of prospective deals on the volatile nature of India's hidden goods was considered by [10]. He discovered that unexpected growth in future activity, as measured by increases in open revenue and exchanging volumes, has led to an increase in actual money value volatility. He suggested that the influence of prospective exchanging on spot value items was lessened. Examined the relationship between money and future expenses in rural business industries in Brazil, focusing on the outcome of the value revelation tool's exchange activity. Their findings suggested that increased trading movement isn't linked to the presence of a long-standing harmonious association among money and its fates costs, but that short-term or long-term communications are crucial in economic sectors with little trade [11]. Five commodities' price discovery across six Indian commodity marketplaces was studied by [12]. They also empirically tested the futures markets' capacity to effectively assimilate information by calculating the relation of typical errors of present and upcoming taxes. The research has experimentally examined the competence of both commercial and future markets. They looked into the future market's inability to properly integrate information and verified its inefficiency. However, the authors came to the conclusion that the future markets for agricultural goods in India are not yet developed and effective.

The examination of how investor behavior and perception affect volatility in the Indian stock market is the main subject of this research. Three biases serve as a stand-in for assessing investor perception and behavior in this study. These are herding, the disposition effect, and overconfidence. 612 investors who are active participants in the Indian stock market made up the final sample size. The survey was administered online. The findings show that the Indian stock market's volatility is positively impacted by herding and overconfidence bias. Conversely, there is no statistically significant relationship between disposition bias and volatility in the Indian stock market. The study comes to the conclusion that there is a correlation between investor behavior or perception and the volatility of the Indian stock market [13].

The addition of sentiment metrics as a risk factor in asset pricing is investigated in this study. We develop a new sentiment variable and build raw and clean sentiment indices using a principal component analysis of many sentiment proxies recognized in the literature using UK market data from January 1993 to December 2020. In essence, the model estimates fall into two categories: first, the study records how well the conventional pricing models perform on portfolios constructed based on various attributes. Second, by including sentiment variables into the model definition, the study enhances the first category. The results show that when it comes to explaining the excess returns of the portfolios, sentiment-augmented asset pricing models perform better than standard approaches [14].

A variety of psychological elements continually influence stock market movement. Fear of Missing Out, or FOMO, is becoming a significant determinant of investing behavior, especially among individual investors. FOMO, a psychological condition brought on by the fear of missing out on special possibilities, has become a strong argument for investing in particular stock market shares. Investors in the stock market are continuously exposed to messages and ads in the contemporary world that are meant to evoke strong feelings and a desire to make stock market investments. Stock brokerage companies are constantly looking for new and creative methods to attract investors, and FOMO is one such instrument that is becoming more and more well-liked. It is crucial to investigate how stock market participants perceive FOMO as it provides insight into how investor behavior is affected by this anxiety [15].

The study looked at why people want to invest in the stock market and learned how they decide which stocks to buy based on behavioral and economic factors. Investors' self-analysis served as the primary foundation for stock selection. There is also discussion of various stock picking considerations. Informants also made care to diversify their assets and used a variety of techniques to deal with their emotions. They also experienced a variety of behavioral biases. Since individual investors are the most affected and least aware stock market participants, this study offers new information to improve their financial stability. Along with examining several notable behavioral patterns displayed by retail investors, the report compiles practical methods for reducing behavioral biases [16] (Fig. 1).

In order to examine how demographic attributes affect investors' perspectives in the Indian stock market, a conceptual framework is developed drawing insights from the reviewed literature and behavioral finance theories [17–19]. In order to understand their effects on significant investment perception characteristics like ROI, risk, and safety, this framework links traditional demographic profiles with psychological concepts such as herding, overconfidence, and risk aversion. The model also figures out the ways in which these perceptions shape portfolio behavior and investment decisions.

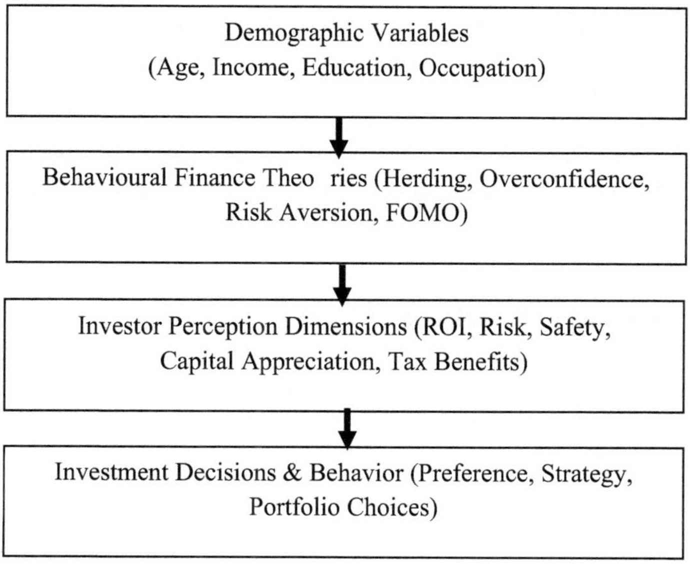

Fig. 1 Conceptual framework illustrating the influence of demographics and behavioral finance on investor perception and investment decisions

3 Research Gap

Although there is a sizable amount of study on investor behavior in the Indian stock market, the majority of that research has concentrated on conventional elements like investing goals, risk tolerance, and demographic variables. Nonetheless, there is still a great deal to learn about how investors' perspectives are altering in light of the quickening pace of financial innovation, digital trading platforms, and shifting market dynamics after COVID-19. Prior research frequently lacked a thorough examination of the behavioral, emotional, and psychological factors that affect investing choices in real time. Furthermore, the ways in which investor opinions change among other investor categories—such as retail traders, millennials, and first-time investors who are increasingly involved in the Indian capital markets have received less attention. By examining the complex attitudes of modern Indian investors and accounting for sentiment-driven decision-making, new behavioral trends, and the influence of technology integration on investing confidence and tactics, this research aims to close this gap. By doing this, it hopes to provide new perspectives and advance a more comprehensive comprehension of investor behavior in the quickly changing Indian financial environment.

4 Research Problem

Investors are now given greater prominence because to technological advancements and the Indian capital market's explosive expansion after the implementation of the economic policy of 1991. Additionally, investor behavior tends to shift from immediate trading of instruments in the capital market to savings to investments. As a result of the capitalist market's liberalized regulations, more brokers joined the market. Depending on the needs of the investors, brokers offer a variety of services under one roof. So, the aim of this research is to determine how investors see these services and how they use them.

5 Research Objectives

1. To determine the elements of investors' perceptions of the stock market.
2. To investigate how investors' perceptions of the stock market relate to various demographic factors.

6 Methodology

6.1 Statistical Tools

This study employed a number of statistical techniques. The reliability of the survey items was tested using a reliability investigation. The validity of the questions was confirmed by confirmatory factor analysis (CFA). To see if there were associations between the various categories, the chi-square test was used.

6.2 Sampling Procedure

Investigating the way demographic characteristics—including age, income, employment, and educational background—influence investors' impressions of the Hyderabad stock market is the goal of the current multistage analytical study. A judgmental (purposive) sample approach was used to select a total of 223 investors, which was deemed suitable because the study required particular, informed respondents who were actively involved in stock market activities. By ensuring that the sample includes individuals with appropriate expertise and familiarity with investing procedures, this non-probability sampling strategy was used to improve the relevance and applicability of the data gathered. Only individuals who met the inclusion requirements—such as having active equity investments and a basic understanding of finance—were retained for the study after the initial respondents were vetted to reduce response error. By gathering information from knowledgeable participants, this strategy offered focused insights into investor attitudes, which aligned with the study's goals. Despite its potential to restrict generalizability, judgmental sampling provides the depth and context necessary to comprehend intricate investor behavior in the selected region.

7 Data Analysis

7.1 Reliability Test

The 19 questionnaire's internal consistency, with a Cronbach's Alpha rating of 0.871, indicates that the data is 87.1% trustworthy (Table 1).

Table 1 Reliability statistics

Cronbach's alpha	Number of items
.871	19

7.2 Analysis of CFA in Investor Perception

As was previously indicated, several studies have already verified the metrics used to evaluate the retention strategies. Therefore, in instruction to determine the correct component structure for stock opinion elements, we have simply carried out a confirmatory factor analysis. The research study was found to meet the suggested six-factor model as shown in Fig. 2). The model has a df = 137 and a chi-square of 472.809 AIC = 837.066, RMSEA = 0.088, TLI = 0.741, and CFI = 0.810. According to Hair et al. (2010), good-fit values for RMSEA must be closer to 0.091 and for CFI and TLI; they should be close to 0.95.

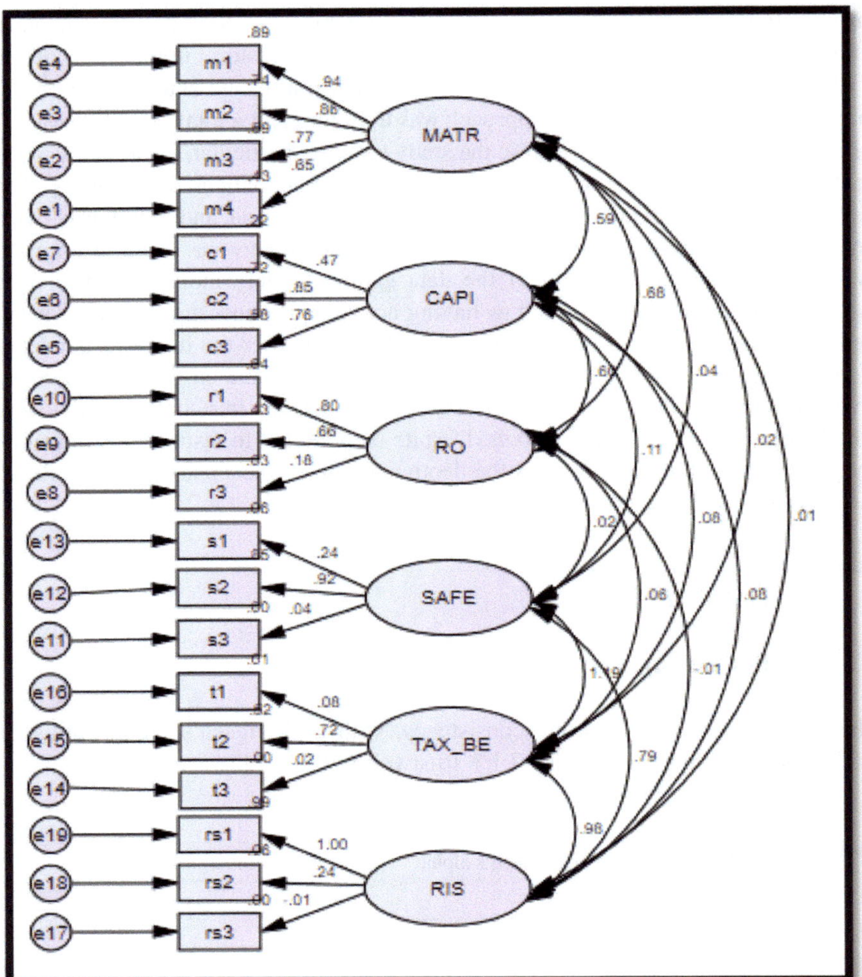

Fig. 2 CFA of investors' perception elements

Table 2 CFA of alternate replicas

Model	χ2	Df	χ2/Df	TLI	CFI	RMSEA
Six factor model	472.809	137	3.451	0.741	0.810	0.091

7.3 Selected Variables Expansion

MATR(Maturity Period), CAPI (Capital Increase), RO (Return on Investment), SAFE (Safety of Principle), TAX_BE (Tax Benefits), and RIS (Risk).

Table 2 displays the findings of the six-factor model's Confirmatory Factor Analysis (CFA), which point to a decent match to the data. A statistically significant chi-square value ($\chi^2 = 472.809$, df $= 137$) usually means that the model is not well fitted. Even with acceptable model fits, chi-square values are frequently significant due to the huge sample size. A moderately satisfactory fit is suggested by the chi-square to degrees of freedom ratio (χ^2/df $= 3.451$), which is marginally over the generally recognized cutoff of 3. Both the Comparative Fit Index (CFI $= 0.810$) and the Tucker-Lewis Index (TLI $= 0.741$) fall below the optimal cut-off of 0.90, showing inadequate incremental fit and raising the possibility that the model does not accurately represent the underlying data structure. Finally, a less-than-ideal model fit is indicated by the Root Mean Square Error of Approximation (RMSEA $= 0.091$), which is greater than the suggested maximum value of 0.08. Together, these fit indices show that although the six-factor model explains part of the data's variance, it does not show a good match. To get a better fit and increase the model's usefulness in practice, extra model modifications can be needed, such as changing factor loadings or correlating error terms.

8 Chi-Square Test

Ho1: There is no significant association in investors perception attributes towards stock market with respect to demographic variables.

With significant variations in preferences for return on investment (ROI), safety of primary, maturity time, and risk, the cross-tabulation results in Table 3 provide light on how investor opinions of the stock market differ among various age groups. The results of the chi-square test, which are shown in Table 4, show that investor opinions of several stock market characteristics and age are statistically significantly correlated ($\chi^2 = 33.610$, df $= 15$, p $= 0.004$). The substantial link is further supported by the likelihood ratio test ($\chi^2 = 34.371$, p $= 0.003$). These results imply that investor attitudes and decision-making criteria are significantly influenced by age. Thus, younger investors (less than 30) seem to be more risk-tolerant and capital growth oriented, whereas investors in the 30- to 40-year-old age range prioritize return on investment and principle safety. Investors over 50 show more balanced choices in every category. The overall chi-square results show significant variation in stock market perceptions across age groups, despite

Table 3 Cross-tabulation of investor perceptions of the stock market by age

		Maturity period	Capital increase	ROI	Safety of principle	Tax benefits	Risk	Total
Age	<30	4	7	6	3	5	8	33
	30–40	7	4	28	22	5	8	74
	41–50	14	14	9	11	12	8	68
	>50	9	6	12	11	5	5	48
Total		34	31	55	47	27	29	223

Table 4 $\chi 2$ test

	Value	Degree of freedom	*p*-value
Pearson $\chi 2$	33.610[a]	15	0.004
Likelihood ratio	34.371	15	0.003
Associate of linear	2.079	1	0.149
Number of valid cases	223		

the non-significant linear-by-linear association statistic ($p = 0.149$). This has significant implications for the development of age-specific investment products and communication strategies.

The distribution of investor perceptions across various educational backgrounds with respect to important investment attributes like risk, return on investment (ROI), safety of principal, tax benefits, maturity period, and capital increase is shown by the cross-tabulation results in Table 5. Postgraduate (PG) investors, in particular, have a greater propensity for return on investment (ROI) and principle safety, indicating a more methodical and knowledgeable approach to investing choices. People with different qualifications, on the other hand, have a more diverse view of various qualities, which could point to a less targeted approach to investing. The findings of the Chi-square test for independence, which looks at the relationship between investor credentials and how they see the stock market, are shown in Table 6. There is a statistically significant correlation between the two variables at the 5% significance level, as indicated by the Pearson Chi-square value of 25.024 with 10 degrees of freedom and a p-value of 0.005. Additionally, the likelihood ratio (26.817, $p = 0.003$) supports this conclusion. But according to the linear-by-linear association statistic (0.069, $p = 0.793$), the connection is not linear, suggesting that judgments vary more complexly between categories rather than only rising or falling with qualification level.

Different occupational groups have distinct priorities, according to the cross-tabulation analysis (Table 7) between investor opinions of the stock market and occupation. For example, return on investment (ROI), capital growth, and principle protection are the top priorities for salaried people, whereas self-employed investors are more inclined toward ROI and capital growth. "Others and students" preferences are more fairly spread, with a relatively smaller focus on tax

Table 5 Cross-tabulation of investor perception of the stock market and qualifications

		Maturity period	Capital increase	ROI	Safety of principle	Tax benefits	Risk	Total
Qualification	UG	8	4	14	3	3	4	36
	PG	9	12	32	28	16	12	109
	Others	17	15	9	16	8	13	78
Total		34	31	55	47	27	29	223

Table 6 $\chi 2$ test

	Value	Degree of freedom	*p*-value
Pearson $\chi 2$	25.024[a]	10	0.005
Ratio of likelihood	26.817	10	0.003
Association of Linear	0.069	1	0.793
Number of valid cases	223		

advantages and return on investment. The existence of a statistically significant correlation between employment and investor perception factors is further supported by the chi-square test findings in Table 8. The likelihood ratio test (23.253, $p = 0.000$) and Pearson chi-square value ($\chi^2 = 23.129$, df $= 15$, $p = 0.001$) both show that it is improbable that the observed distribution of investor impressions across various professional groups happened by accident. The link may not be absolutely linear, though, as the linear-by-linear association ($p = 0.784$) is not statistically significant.

Table 7 Cross-tabulation of investor perceptions of the stock market and occupation

		Maturity period	Capital increase	ROI	Safety of principle	Tax benefits	Risk	Total
Occupation	Student	4	4	2	2	1	2	15
	Self-employed	6	7	24	17	3	10	67
	Salaried	16	16	24	19	20	12	107
	Others	8	4	5	9	3	5	34
Total		34	31	55	47	27	29	223

Table 8 $\chi 2$ test

	Value	Degree of freedom	*p*-value
Pearson $\chi 2$	23.129	15	0.001
Ratio of likelihood	23.253	15	0.000
Association of linear	0.075	1	0.784
Number of valid cases	223		

The cross-tabulation findings in Table 9 examine how investor income levels relate to how they see many aspects of the stock market, including risk, maturity term, capital appreciation, return on investment (ROI), safety of primary, and tax advantages. Across income groups, there is a wide range of responses, with some trends showing up. For example, investors in the ₹25,001–₹50,000 and ₹50,001–₹75,000 income brackets exhibit a greater preference for return on investment and principal safety, indicating a more balanced risk-return outlook in the mid-income segments. The findings of the Chi-square test of independence, which determines whether income level and investor impressions are statistically significantly correlated, are shown in Table 10. At the 0.05 level, there is a substantial correlation between income and investor perception, as indicated by the Pearson Chi-square value of 30.236 with 20 degrees of freedom and a p-value of 0.003. This conclusion is further supported by the Likelihood Ratio value of 32.506 and the p-value of 0.038. However, it appears that the link is not strictly linear based on the linear-by-linear association value (2.467, p = 0.116). These findings suggest that there are substantial differences in how investors see the stock market depending on their income bracket, which emphasizes the need of taking income into account when examining investing patterns and developing focused financial recommendations or legislative measures.

Since the dimensions were drawn from behavioral finance theories and established literature, this research utilized Confirmatory Factor Analysis (CFA) to test the hypothesized six-factor structure of investor perception qualities. The conceptual clarity of these constructs, which were operationalized from validated previous literature, served as the basis for not to conduct exploratory factor analysis (EFA).

Table 9 Cross-tabulation of investor perceptions of the stock market and income

		Maturity period	Capital increase	ROI	Safety of principle	Tax benefits	Risk	Total
Income	<25 K	6	7	10	2	5	2	32
	25 K-50 K	9	5	21	7	6	4	52
	50,001-75 K	5	7	11	13	7	9	52
	75,001–1 Lakh	8	5	11	17	7	9	57
	>1 Lakh	6	7	2	8	2	5	30
Total		34	31	55	47	27	29	223

Table 10 $\chi 2$ test

	Value	Degree of freedom	*p*-value
Pearson $\chi 2$	30.236	20	0.003
Ratio of likelihood	32.506	20	0.038
Association of linear	2.467	1	0.116
Number of valid cases	223		

The marginal fit indices (TLI = 0.741, CFI = 0.810, RMSEA = 0.091) found from the CFA findings, however, proposed that the model could be improved. In view of this, we confirm the value of a two-step EFA–CFA method, especially in studies that are context-specific like this one. We recommend that EFA be used in future studies to explore the latent factor structure, and then CFA be used to confirm its validity in diverse investor population.

9 Suggestions

- In order to make an effective investment decision, investors must talk to share traders and brokers on the state and performance of the market in addition to their friends.
- Local investors will find it easier to make investments in the financial sector if more favorable attitudes and improvement tactics are used.
- The company's window dressing must be examined by the investors.
- The organization's results over the previous several years must be shown to the investors.
- The business must guarantee that investors get dividends on a regular basis.

10 Discussion: Linking Findings to Prior Research

According to the study's findings, a number of factors, including risk tolerance, market expertise, prior investing experiences, and the impact of social and digital media, have a major impact on investors' impressions of the stock market. These observations align with previous studies especially that of [17], who highlighted the psychological foundations of investor behavior and pointed out those biases and heuristics frequently influence market decisions rather than just rational analysis. In a similar vein, [18] emphasized how behavioral finance helps explain why investor choices usually diverge from conventional economic theories. The result is also consistent with more recent research by [19], which indicates that in today's digital world, social sentiment—influenced by peer networks and the media—influences investment decisions more and more. These results support the move in academic literature from simply rational models to those that contain behavioral and psychological components, highlighting the significance of both cognitive and emotional factors in influencing investor perceptions and behavior. Thus, this study adds to the increasing amount of evidence showing that a multidisciplinary approach bridging communication, psychology, and finance is necessary to comprehend investor impressions.

11 Conclusion

Indians have long been recognized for their preference for safe investments and their attitude toward saving. India's savings rates have consistently increased since gaining its independence. In order to appeal to the public, several new investment vehicles have been launched during the past 20 years. Safety and security are the main factors in this investment's selection. Although the Indian stock exchange is prone to numerous unanticipated swings, it also offers investors who are strategic and forward-thinking significant profits and capital appreciation. As the situation in the financial market has changed, investors' attitudes have also changed. Share trading may be a profitable financial avenue for the investor. It is entirely dependent on how investors see stock investments and share trading. Investors might experience a sense of wealth from the stock market after they have access to correct and trustworthy information.

References

1. Defeat, B. S., Das, N. M., & Rao, K. C. (2021). Value disclosure tools and investor perception in Indian commodity markets. *Journal of Commodity Market Analysis, 7*(3), 112–125.
2. Dey, M., & Maitra, D. (2012). Price discovery in Indian pepper market using Granger causality, co-integration, and error correction model. *Agricultural Economics Research Review, 25*(1), 103–110.
3. Sehgal, S., Rajput, N., & Dua, R. (2012). Price discovery in Indian agricultural commodity markets. *International Journal of Economics and Finance, 4*(3), 40–51.
4. Ali, J., & Gupta, K. B. (2011). Examining the long-term relationship between futures and spot prices for agricultural commodities in India. *Journal of Agricultural Economics and Development, 3*(2), 45–56.
5. Baldi, L., Peri, M., & Vandone, D. (2011). Price discovery in agricultural commodities: A Granger causality test between spot and futures markets. *Applied Economics Letters, 18*(2), 107–112.
6. Sen, S., & Paul, M. (2010). Does futures trading destabilize spot prices in agricultural commodity markets? *Evidence from India. Indian Economic Review, 45*(1), 91–110.
7. Nath, G. C., & Lingareddy, T. (2008). Impact of futures trading on spot markets: An empirical analysis of pulses market in India. *Indian Journal of Agricultural Marketing, 22*(3), 23–36.
8. Bose, S. (2007). Commodity futures market in India: A study of price discovery and volatility in Indian commodity markets. Indian Institute of Capital Markets 10th Capital Markets Conference Paper.
9. Mahalwala, R. (2016). Impact of derivatives market on volatility and trading volume in Indian equity markets. *International Journal of Financial Markets and Derivatives, 5*(2), 89–101.
10. Mukherjee, D. (2011). Impact of futures trading on spot market volatility: Evidence from Indian commodity markets. *International Review of Applied Financial Issues and Economics, 3*(2), 327–340.
11. Mattos, F., & Garcia, P. (2004). Price discovery in thinly traded markets: Cash and futures relationships in Brazilian agricultural futures markets. *Journal of Futures Markets, 24*(11), 1021–1041.
12. Kumar, A., & Sunil, R. (2004). Price discovery and market efficiency in Indian commodity futures markets. *Indian Journal of Economics and Business, 3*(1), 145–160.
13. Gupta, P. (2023). How volatility is impacted by investor's perception and behavior in Indian stock market. *Elementary Education Online, 19*(4), 8348–8361.

14. Sakariyahu, R., Paterson, A., Chatzivgeri, E., et al. (2024). Chasing noise in the stock market: An inquiry into the dynamics of investor sentiment and asset pricing. *Rev Quant Finan Acc, 62*, 135–169.
15. Jose, J. (2024). Investor Perception on Stock Market FOMO (Fear Of Missing Out). *International Journal of Research in Commerce and Management Studies., 6*(6), 141–145.
16. Ahuja, S., & Kumar, B. (2025). An elicitation study to understand the equity investment motivation and decisions among Indian millennials. *Qualitative Research in Financial Markets, 17*(1), 66–86.
17. Barberis, N., & Thaler, R. (2003). A survey of behavioral finance. In G. M. Constantinides, M. Harris, & R. M. Stulz (Eds.), *Handbook of the economics of finance* (Vol. 1, pp. 1053–1128). Elsevier. https://doi.org/10.1016/S1574-0102(03)01027-6.
18. Shefrin, H. (2007). Behavioral corporate finance: Decisions that create value. McGraw-Hill/ Irwin.
19. Statman, M. (2019). Behavioral finance: The second generation. *CFA Institute Research Foundation*. https://doi.org/10.2139/ssrn.3240383

Intrinsic Motivation in the Workplace: The Roles of Age, Gender, and Tenure

Swati Mishra, Bamadev Mahapatra, and Vivek Sharma Dhakal

Abstract

Employee motivation is vital to organizational effectiveness and has influenced management practices and industrial organizational psychology. Organizations have escalating challenges stemming from globalization, competitive demands, and the characteristics of a multigenerational workforce. Comprehending the factors that motivate personnel is crucial for augmenting retention and optimizing job effectiveness. Although extrinsic motivators such as salary and incentives have historically been prioritized, recent studies underscore the increasing significance of intrinsic motivation, especially among Generation Y (Gen Y) employees. Intrinsic motivation originates from the inherent gratification of executing a task and encompasses aspects such as fervent involvement, pursuit of novelty, attainment of objectives, and pleasure derived from labor which leads to improved productivity. Gen Y, a substantial and swiftly expanding workforce, prioritizes meaningful employment, work–life equilibrium, and stimulating tasks more than preceding generations. This transition highlights the necessity for firms to prioritize intrinsic motivators to line with Gen Y's tastes and enhance employee engagement. This research examines the influence of age, gender, and organizational tenure on the intrinsic motivation of Gen Y software professionals in India. Results witnessed high inclination of

S. Mishra (✉)
Assistant Professor in HRM, Institute of Management and Information Technology (IMIT), Cuttack, India
e-mail: swatimishra0674@gmil.com

B. Mahapatra
Assistant Professor in Economics, Institute of Management and Information Technology (IMIT), Cuttack, India

V. S. Dhakal
Veltara Consultants Pvt. Ltd., Itahari, Nepal

S. Patnaik et al. (eds.), *People, Planet and Profit: Advancing Sustainable Business Practices*, Springer Proceedings in Geography and Sustainability,
https://doi.org/10.1007/978-3-032-11296-5_16

younger age group toward Intrinsic Motivation in comparison to previous generations. Inclination toward Intrinsic Motivation is equal irrespective of genders and employees with less organizational tenure prefer challenges and sense of choice. This exploration of predictors offers insights on cultivating a more motivated and productive workforce, while also addressing the drawbacks of over dependence on extrinsic motivators in contemporary organizational settings.

Keywords

Industrial organizational psychology • Extrinsic motivation • Intrinsic motivation • Gen Y workforce

1 Introduction

Employee motivation has been one of the most prolific research areas in management theory and practice and is central to work performance [1]. The ability to motivate employees is critical to organizational success and performance [2]. Thus, organizations need to understand different components of employee motivation while considering globalization, economy, and the labor market [3]. Therefore, managers are always on the quest to discover newer and more appropriate motivators for their employees to aid retention and improve work performance. Amidst economic volatility, retaining personnel is extremely challenging for all organizations across the globe. The fillip to research on employee retention was provided through considerable research on employee turnover and identifying factors that force employee turnover and factors that force employees to leave the organization [4, 5]. While research on employee turnover retains importance, there is also a need to recognize the factors that encourage employees to continue in the organization. Retention and employee motivation are directly correlated, as per the work of Gallup, the Work Foundation, and Roffey Park [6]. Though studies on employee motivation are a continually evolving domain, they are broadly classified as intrinsic and extrinsic, as per the different content and process theories of motivation.

Extrinsic motivators, though contingent on the work, may not inherently be a part of it and involve monetary and nonmonetary rewards [7, 8]. In Contrast, intrinsic motivation emerges from the enjoyment of doing the task without expecting any external reward. The predictive capacity of intrinsic motivators on employee productivity has witnessed a paucity of addresses. This paucity is critical, considering intrinsic motivation's association with higher cognitive processes. The emphasis on research of intrinsic motivation and its relationship with retention is deemed as particularly relevant in the Indian context because of a rapidly growing segment of the workforce popularly designated as Generation or Gen Y. While many Western countries grapple with problems associated with an aging population, the average age of Indians will be 29 by 2025. Members of Gen Y were atypical in their pursuit of meaningful work, recognition, work–life balance, sense of achievement, progress, social involvement, and desire to have mental wellbeing at work [9, 10]. Their predilection toward intrinsic motivators, such as challenging tasks and work involving greater responsibility, is higher than in previous generations.

Among the various industrial sectors, the Indian IT industry has experienced phenomenal growth in the past decade and a half. The IT industry has an overwhelmingly large Gen Y populace, with over two million people born after 1980.Intrinsic motivation (IM) can affect employee engagement in several ways. First, IM denotes the internal impetus that compels individuals to engage in activities due to enjoyment, challenge, or personal fulfillment rather than external incentives such as income or promotion. In the Indian IT sector, characterized by problem-solving, innovation, and ongoing learning, intrinsic motivation is a crucial determinant of employee engagement in their positions. Second, employees with IM typically exhibit enhanced ownership of their tasks. In Indian IT companies, where employees frequently engage in intricate coding, software development, or client-specific projects, personal engagement and intrinsic fulfillment are pivotal. When employees perceive their work as meaningful, they are more inclined to exert additional effort, propose ideas, and dedicate themselves to producing high-quality results. Third, the Indian IT sector flourishes via innovation, particularly in response to increasing demands in fields such as artificial intelligence, cybersecurity, and cloud computing. Employees with intrinsic motivation are more inclined to engage in innovative problem-solving and contribute novel insights to their tasks. Their intrinsic motivation to acquire knowledge and develop cultivates a culture of creativity, essential for companies aiming to remain competitive in a rapidly evolving technology landscape. Fourth, extended working hours and deadline constraints are prevalent in the Indian IT sector. People who are intrinsically driven often view problems as opportunities instead of responsibilities. This intrinsic motivation can facilitate a positive perspective, foster perseverance, and enhance resilience in the face of failures, thus augmenting overall engagement and diminishing burnout rates. Finally, IM influences individual performance and enhances positive team dynamics. Engaged employees exhibit enhanced collaboration, supportiveness, and communication. In substantial Indian IT firms where collaboration is crucial, this may result in a more favorable work atmosphere, enhanced morale, and reduced turnover. Therefore, it is appealing to understand the differences between groups of respondents based on age, gender, and organizational tenure, and their views toward IM concerning Indian Software Employees.

Section 2 of the study provides a comprehensive literature review. Section 3 delineates the data series and methods employed in this study. The results and discussions are detailed in Sect. 4. Finally, Sect. 5 provides the concluding remarks of the study.

2 Literature Review

2.1 Overview of Motivation

Employee motivation has dominated management thought theoretically and practically [1, 11], providing substance to industrial and organizational psychology. The interest in employee motivation in management stems from the belief that motivating staff members is essential to the performance and success of an organization [2, 12–14]. The study of motivation is directly associated with other fields of management research, such as employee retention, leadership styles, team and group dynamics, human resource management, and organizational change management. The challenge for any organization is to balance employee productivity and motivation. During the era of the drive/reinforcement theories of motivation and the early content theories, motivation was attributed to extrinsic factors.

2.2 Studies on Extrinsic Motivation

Extrinsic motivation (EM) is the motive that generates individual focus on tasks through outside incentives [15, 16]. Extrinsic motivators include rewards, sanctions, physical work environment, vacation time, health insurance, company policies, quality of supervision, relations with peers' subordinates, and so on [16–18]. A cache of studies contributed to developing varied fixed and differential incentive plan methods, understanding the impact of vacation, fatigue, rest intervals, etc., on performance and motivation to perform. Initial conceptualization of the pay-for-performance plans (one of the most effective extrinsic motivators) was also rooted in the Taylorian management principles. The Society for Human Resource Management, in its annual surveys of 600 employees from 2004–2012, ranked 21 rewards in order of their relative importance and concluded that pay/compensation was ranked first. Incentive effects are directly related to motivation and refer to the level and intensity of individual and aggregate motivation. Though extrinsic motivators are varied, they are usually associated with financial incentives in organizations, hence the presence of a noticeable trend of research on the effect of such rewards and incentives.

2.3 Studies on Intrinsic Motivation

Intrinsic motivation is the motive that builds an individual's focus on a task through internal qualities [16]. Intrinsic motivators include enthusiasm for tasks, openness to novel experiences, a tendency for excellence, goal achievement, and enjoyment in one's work [15, 19]. The earliest known influential contribution to theorization on intrinsic motivation was reflected in White's description of a motive called 'effectance'. He believed in the importance of the effectance motive and regretted its neglect in orthodox drive theories of motivation prevalent in the 1920s–50 s.

A parallel thought on intrinsic motivation concerned with curiosity, uncertainty, incongruity, and discrepancy was found in the theories of Berlyne [17]. A likely answer to these questions could be ascribed to motivational variables specific to a drive termed 'curiosity' appeased by the reception and subsequent rehearsal of knowledge. Finally, theorists such as deCharms, deCharms and Carpenter further augmented the nascent understanding of intrinsic motivation in their studies on a motive called control or self-determination [20–22].

Early thoughts on intrinsic motivation also paved the path for the classic self-determination theory (SDT) of intrinsic motivation by Deci and Ryan [16]. SDT, over the past few decades, has evolved through the growth of four mini theories, each related to specific social phenomena. The first mini theory was cognitive evaluation theory (CET), which described the effects of one's context on intrinsic motivation [23]. Organismic integration theory (OIT) was the second mini theory developed to explain the dynamics of extrinsic motivation and the level of autonomy experienced by individuals while engaging in extrinsically motivated behavior [24]. The third-causality orientations theory (COT) explains individual differences in orientation to the social environment. Finally, basic needs theory (BNT) was developed to describe associations of motivation and goals to individual health and wellbeing [23]. However, SDT of Intrinsic Motivation examines the innate psychological needs and growth tendencies that build employee motivation. IM in activities was enhanced through the possibilities of challenge in the activity itself [21]. Employees derive a sense of IM when they feel that the organization recognizes/appreciates/values their work quality. Employees naturally explore and experiment with different working styles and methods to accomplish their work. When they receive meaningful feedback from essential stakeholders (such as a superior or peer), their level of IM increases, and they feel valued [23]. The concept of choice/autonomy is central to the foundations of SDT. The tendency to derive pleasure or enjoyment in one's own work is core to the ontology of intrinsic motivation [23]. Accomplishment on the job helps buttress feelings of satisfaction, pride, and enjoyment, thus enhancing IM. Encouragement of employee creativity, ideas, and proposals augments the scope of IM. Entrusting responsibilities related to job assignments to employees instigates their level of IM and helps them become managers or team leaders. A sense of progress is the satisfaction that employees are on the correct track and moving in the right direction. Hence, recent trends in generational theory urged further research on discovering intergenerational differences (if any) based on differences in personal demography such as age, gender, race, ethnicity, etc., in the workplace as well as their personal domain [9, 14].

3 Data and Methodology

This research's statement of purpose helped select quantitative research as a suitable and relevant paradigm for this study. Empirical data was collected from five major IT organizations in India. The sample demographic profile is summarized in Table 2. The study adopted a survey as a research methodology, and a sample size of 378 was deemed optimal for the study utilizing scientific calculation procedures [25]. Bearing in mind the requirements of confidentiality by the organizations that provided the researcher with permission to collect data, snowball sampling was considered while identifying respondents and collecting data. Snowball sampling, a nonprobability technique particularly effective for accessing enigmatic or specific populations, was implemented in the study to collect data. Snowball or chain referral sampling, a method frequently employed in social science research, entails the generation of a sample through participant referrals. This method is typically employed among individuals conversing with or sharing the desired characteristics [26]. Participants nominate others after the initial recruitment process, generating a 'snowball effect' [27]. In the current study, this method was implemented. The HR leaders of organizations were sent formal invitation letters and questionnaires, which were subsequently disseminated through a centralized portal accessible to employees. These initial participants, in turn, motivated others in their networks to participate, thereby ensuring a broader audience. Some people do not want to come forward and participate in research studies because they don't want their identity to be exposed. Snowball sampling helps in this situation as they ask for a reference from people known to each other. However, snowball sampling is advantageous in exploratory research, where trust and accessibility are essential, as it allows researchers to gather comprehensive data from individuals who may be challenging to engage in other ways. To ensure that the sample was representative and to address potential limitations, the number of sampling units was intentionally determined in accordance with the research topic. This ensured that essential elements were captured across pertinent demographic characteristics [28].

A close-ended questionnaire with a five-point Likert scale was designed to gather data. The questionnaire design was guided by utilizing a valid extant scale for the variable of the research construct, i.e., intrinsic motivation. This study utilizes the framework of SDT to enhance comprehension of intrinsic motivation (IM) facets. It specifically employs the Intrinsic Motivation Inventory (IMI), which has been extensively utilized in research grounded by SDT. The IMI utilized in this research is modified from the original scale created by McAuley et al. [29]. The present study measured IM through nine dimensions: Challenge, Perceived Choice, Accomplishment, Encouraging Initiatives, Feedback, Recognition, Responsibility, Sense of Progress, and Enjoyment. This study used Cronbach's alpha values to compute scale reliability. IM scale internal reliabilities (Cronbach's coefficient alphas) were compared to the acceptable threshold of 0.70 ($\chi 2 = 6790.586$, df $= 741$, $p < 0.001$). An extensive literature review was conducted to establish the validity, and a well-accepted scale was used in the study. Different professional experts also pre-tested the instruments and selected respondents to assess their adequacy and relevance during the pilot study (Table 1).

Various techniques were used to obtain appropriate and scientific answers to the research objective, such as the requisite descriptive statistics, the t-test, and the analysis of variance (ANOVA). Descriptive statistics furnish a fundamental description of the dataset by elucidating the central tendency, dispersion, and distribution of the examined variables. Statistical measures such as mean, median,

Table 1 Factor analysis scores for IM scale

Factors and items	Factor loadings	Rotated eigenvalues	Percentage of variance	Cronbach's alpha (α)
Factor 1: Challenge I like difficult assignments because I find them more interesting The real reward in working is engaging in stimulating activities I like to go onto a new work that's at a more difficult level	0.570 0.623 0.687	2.644	6.780	0.887
Factor 2: Recognition I get recognition for my work in front of my peers I get credit for what I do I am praised for the work that I do	0.745 0.802 0.764	2.709	6.945	0.852
Factor 3: Feedback The feedback that I receive on how I do my job is highly relevant I have a pretty good idea of how well I am performing my work from my supervisor	0.667 0.693	1.334	3.420	0.801
Factor 4: Perceived Choice I can choose appropriate methods to conduct my work I do my job because I want to I can exercise independent thought and action in my job I felt like I was doing what I wanted to do while I was working on the job	0.577 0.679 0.704 0.563	3.830	9.821	0.743
Factor 5: Enjoyment My job is so interesting that it is a motivation in itself I enjoy doing my job very much	0.590 0.713	1.813	4.648	0.824
Factor 6: Encouraging Initiatives My manager encourages engagement on the job My manager publicly recognizes my initiatives taken to implement my job My manager respects my ability to function independently on the job My manager expects me to undertake initiatives to solve problems on the job	0.629 0.605 0.727 0.654	2.341	6.002	0.706

(continued)

Table 1 (continued)

Factors and items	Factor loadings	Rotated eigenvalues	Percentage of variance	Cronbach's alpha (α)
Factor 7: Accomplishment I experience satisfaction while exceling in my work I feel satisfied when I exceed my own standards of performance I am satisfied when I accomplish difficult tasks in my job	0.614 0.678 0.685	2.463	6.316	0.768
Factor 8: Responsibility I experience a great deal of personal responsibility for the work that I do When I experience a problem in my work, I take decisions to resolve it I feel that whether the job gets done right is clearly my own responsibility	0.617 0.517 0.670	2.126	5.451	0.789
Factor 9: Sense of Progress I prefer difficult jobs so that I can learn a great deal I look for opportunities to develop new skills and knowledge I am satisfied when I meet my set goals on the job I like tasks that force me to think hard I truly enjoy learning for the sake of learning on the job I am willing to undertake difficult jobs so that I can learn a lot while doing them	0.526 0.570 0.746 0.732 0.664 0.571	4.719	12.10	0.707

Source Author's own findings
Notes Extraction Method: Principal Component Analysis; Rotation Method: Varimax, KMO measure of sampling adequacy = 0.894, Bartlett's test of sphericity: ($\chi^2 = 6790.586$, $df = 741$, $p < 0.001$)

mode, standard deviation, and range assist in comprehending the data's overarching patterns and attributes. These statistics are essential for detecting outliers, assessing data normality, and identifying trends, which underpin subsequent inferential research. The t-test is a robust inferential statistical method employed to compare the means of two groups to ascertain if they are statistically distinct. The t-test enables us to derive accurate inferences on population parameters from sample data by assessing the significance of mean differences. It augments the scientific rigor of the study by supplying evidence regarding whether the observed differences result from chance or genuine variance among the groups. Further, the t-test is extended to comparisons of more than two groups using ANOVA. ANOVA is necessary to assess the impact of a categorical independent variable with three or

Table 2 Demographic characteristics of sample

Characteristics	Frequencies	Percentage (%)
Gender		
Male	238	63
Female	140	37
Age		
21–25	159	42.1
26–30	161	42.6
31–35	42	11.1
36–40	16	4.2
Organizational Tenure (years)		
1–4 yrs	193	51.1
5–8 yrs	137	36.2
9–12 yrs	34	9
13–16 yrs	14	3.7
***N* = 378**		

Source Author's findings

more levels on a dependent variable. It examines the null hypothesis that all group means are equal to determine if one group varies significantly. This test controls Type I errors from many pairwise comparisons, improving scientific validity.

4 Results and Discussions

4.1 Gender and Intrinsic Motivation

To ascertain whether male and female respondents exhibited differences in their responses to the chosen dimensions of IM, a *t*-test was conducted at 95% confidence interval level (see Table 2). The resultant findings depict no significant differences in response on the said dimensions of IM among male and female respondents. The mean scores of male respondents were marginally higher than those of their female counterparts in all dimensions of IM except for feedback. It can also be observed from Table 2 that the difference in mean values between male and female respondents on all the Nine IM Dimensions is insignificant (at $p < 0.05$).

Here are several plausible explanations for the potential insignificance of gender variations on intrinsic motivation. First, innate psychological needs are universally familiar. The SDT theory posits that intrinsic motivation emerges from satisfying fundamental psychological needs—autonomy, competence, and relatedness—not contingent upon gender. Consequently, both genders may exhibit equivalent motivation when these requirements are fulfilled, leading to negligible gender disparities. Second, in contemporary culture, men and women are increasingly subjected to analogous social, educational, and professional contexts. Consequently, gender-based disparities in the evolution of motivation are less evident. Increased gender equality in school, employment prospects, and media coverage may foster IM in both men and women, diminishing the probability of substantial disparities. Third, intrinsic motivation is frequently more significantly shaped by individual elements such as personality traits (e.g., curiosity, openness), personal interests, or particular experiences rather than by gender alone. Numerous research indicates that individual variations, such as intrinsic interests, prior experiences, and cognitive styles, may surpass the impact of gender on motivation, potentially elucidating the low or insignificant gender disparities in intrinsic motivation. Finally, the shifting in cultural norms may be another probable reason of such a finding. Cultural assumptions around gender roles have become increasingly variable over time. Traditionally, males and girls were socialized to prioritize distinct pursuits (e.g., STEM disciplines against the arts), but this distinction is progressively diminishing in modern culture. As cultural attitudes progress, the possibilities for both genders to engage in their inherent interests across multiple domains (e.g., academic, professional, or creative pursuits) have become more equitable, resulting in diminished gender disparities in motivation.

4.2 Age and Intrinsic Motivation

Respondents were divided into four categories based on age. Table 3 shows that the maximum respondents were relatively young, with nearly 159 in the age group of 21–25 years and 161 in the age category of 26–30 years. Forty-two (42) respondents belonged to the age group of 31–35 years, and only 16 remained in the age group of 36–40 years. To examine whether respondents across the four age groups differed in their responses to the dimensions of IM, an ANOVA test was carried out at 95% confidence interval level. The results of the ANOVA test are presented in Table 3. Mean scores of respondents in the age group of 31–35 years were the highest on dimensions such as 'Sense of Progress', 'Perceived Choice', 'Accomplishment', 'Encouraging Initiative', 'Responsibility', and 'Enjoyment', while respondents in the age group of 26–30 years recorded the highest mean scores on the remaining dimensions of IM (Table 3). Responses on dimensions such as 'Sense of Progress'($F = 2.196$; $df = 3/374$; $p = 0.088$), 'Perceived Choice'($F = 0.993$; $df = 3/374$; $p = 0.396$), 'Accomplishment'($F = 0.904$; $df = 3/374$; $p = 0.439$), 'Encouraging Initiatives'($F = 0.791$; $df = 3/374$; $p = 0.500$), 'Enjoyment'($F = 2.495$; $df = 3/374$; $p = 0.060$) and 'Feedback'($F = 0.810$; $df = 3/374$; $p = 0.489$) signified insignificant differences in mean scores between respondents belonging to different age groups. But responses on dimensions such as 'Recognition'($F = 2.772$; $df = 3/374$; $p < 0.05$), 'Challenge'($F = 3.974$; $df = 3/374$; $p < 0.01$) and 'Responsibility'($F = 4.461$; $df = 3/374$; $p < 0.01$) indicated differences in the pattern of mean scores across age groups (the significance was declared at $p < 0.05$). Respondents in the age group of 26–30 years exhibited a greater preference for recognition and challenge in their job assignments, while respondents in the age group of 31–35 years recorded the highest mean scores for responsibility in job assignments.

Table 3 Comparative responses of male and female respondents on IM dimensions

IM Dimensions	Gender	*N*	Mean	SD	*t*-Test (p-value)
Sense of Progress	Male Female	238 140	3.6960 3.7178	0.55170 0.55210	−0.371 (0.71)
Perceived choice	Male Female	238 140	3.5588 3.5821	0.59330 0.52410	−0.385 (0.70)
Recognition	Male Female	238 140	3.7226 3.7262	0.61684 0.53556	−0.054 (0.96)
Challenge	Male Female	238 140	3.8570 3.8672	0.53556 0.50230	−0.184 (0.85)
Accomplishment	Male Female	238 140	3.7300 3.7715	0.63133 0.64705	−0.612 (0.54)
Encouraging Initiative	Male Female	238 140	3.6838 3.6964	0.61201 0.59508	−0.195 (0.84)
Responsibility	Male Female	238 140	3.8014 3.8454	0.62573 0.60965	−0.666 (0.51)
Enjoyment	Male Female	238 140	3.7542 3.8071	0.64915 0.51449	−0.824 (0.41)
Feedback	Male Female	238 140	4.0231 3.9643	0.53020 0.67227	0.941 (0.35)

Notes N = 378

The findings suggest that respondents aged 26–30 preferred recognition in their professional positions. Typically, individuals in this age bracket are in their careers' early to mid-phases, and acknowledgment from peers and superiors is a significant source of motivation and reinforcement. This can be attributed to this. Professionals are currently establishing a presence in their respective fields, pursuing validation for their contributions, and enhancing their reputations. Recognition increases confidence and visibility within the organization, which can be essential for career advancement and goal setting during these formative professional years. Similarly, the 26–30 age group's preference for job-related challenges is likely a result of their aspiration to advance and develop in their careers. This period is frequently characterized by high ambition, energy, and enthusiasm, as individuals are eager to challenge their capabilities, take risks, and accept challenging tasks that provide learning opportunities. They can demonstrate their competencies, improve their problem-solving abilities, and acquire experiences crucial for career advancement by participating in challenging work. The desire for meaningful and stimulating work, which professionals frequently pursue during this dynamic period of their careers, may also indicate an appetite for challenges.

However, the mean scores for responsibility were highest among respondents aged 31–35 years, suggesting a greater sense of accountability and ownership in their work. This can be attributed to the fact that individuals in this age group are generally more experienced, are frequently assigned to more senior positions, and are more established in their professional endeavors. They are likely to have acquired a more comprehensive understanding of the broader implications of their actions within an organizational context and an increased sense of confidence and maturity. Moreover, life-stage factors, such as family responsibilities and long-term career planning, may further contribute to a heightened sense of duty and commitment, thereby increasing their likelihood of effectively accepting and managing responsibility.

4.3 Organizational Tenure and Intrinsic Motivation

Respondents were divided into four categories regarding their organizational tenure. Table 4 indicated that the maximum respondents-193 were relatively new in the organization and had been with the organization in the range of 1–4 years. One hundred thirty-seven (137) respondents had been with the organization in the range of 5–8 years, 34 in the range of 9–12 years, and finally, only 14 respondents had continued with their current organization for more than 10 years and were placed in the tenure range of 13–16 years. To examine whether respondents across the four tenure groups differed in their responses to the dimensions of IM, an ANOVA test was carried out at 95% confidence interval level. The results of the ANOVA test are presented in Table 4. Response on dimensions such as 'Sense of Progress'($F = 0.099$; $df = 3/374$; $p = 0.961$), 'Perceived Choice'($F = 1.612$; $df = 3/374$; $p = 0.186$),'Recognition'($F = 0.336$; $df = 3/374$; $p = 0.799$), 'Accomplishment'($F = 0.550$; $df = 3/374$; $p = 0.649$), 'Encouraging Initiative'($F = 0.771$; $df = 3/374$; $p = 0.511$), 'Responsibility'($F = 0.312$; $df = 3/374$; $p = 0.817$), 'Enjoyment'($F = 0.866$; $df = 3/374$; $p = 0.459$) and 'Feedback'($F = 0.179$; $df = 3/374$; $p = 0.911$) signified insignificant differences in mean scores between respondents belonging to different tenure groups (the significance was declared at $p < 0.05$).However, respondents differed in their predilection toward 'Challenge'($F = 3.743$; $df = 3/374$; $p < 0.01$). Respondents who had been with the organization for 1–4 years displayed maximum propensity for challenge, while respondents whose organizational tenure was 9–12 years were least inclined to prefer challenge in their job assignments on a relative continuum (Table 5).

The findings suggest respondents with 1–4 years of tenure demonstrated the most remarkable propensity for challenging job assignments. The probable reasons for such conclusions can be as follows. First, new employees are typically keen to build their reputation and credibility inside the organization. Undertaking demanding tasks is a means to exhibit proficiency, initiative, and potential. Their motivation is fueled by the desire to distinguish themselves, attain recognition, and enhance career opportunities, including promotions or significant positions. Second, New employees generally commence their careers with a robust eagerness to acquire knowledge, develop, and assess their potential. They perceive problems as opportunities for personal and professional growth. Their receptiveness to novel experiences and willingness to investigate the organization's structures and procedures render them more amenable to challenging assignments. Third, Individuals in their professions' early phase frequently exhibit significant ambition and

Table 4 Comparative responses across age groups on IM dimensions

IM dimensions	Age groups	*N*	Mean	SD	*F*
Sense of progress	21–25 years	159	3.6668	0.54825	2.196
	26–30 years	161	3.7502	0.53751	
	31–35 years	42	3.7733	0.44698	
	36–40 years	16	3.4281	0.84291	
Perceived choice	21–25 years	159	3.5566	0.58794	0.993
	26–30 years	161	3.5916	0.53010	
	31–35 years	42	3.6012	0.58711	
	36–40 years	16	3.3438	0.68237	
Recognition	21–25 years	159	3.6542	0.66647	2.772*
	26–30 years	161	3.8095	0.59578	
	31–35 years	42	3.7614	0.51757	
	36–40 years	16	3.4581	0.47061	
Challenge	21–25 years	159	3.8827	0.49616	3.974**
	26–30 years	161	3.8925	0.50501	
	31–35 years	42	3.8174	0.55616	
	36–40 years	16	3.4375	0.70657	
Accomplishment	21–25 years	159	3.7316	0.64005	0.904
	26–30 years	161	3.7622	0.63166	
	31–35 years	42	3.8179	0.60803	
	36–40 years	16	3.5213	0.72967	
Encouraging initiative	21–25 years	159	3.6887	0.59249	0.791
	26–30 years	161	3.6988	0.62819	
	31–35 years	42	3.7321	0.57486	
	36–40 years	16	3.4688	0.57645	
Responsibility	21–25 years	159	3.7633	0.66619	4.461**
	26–30 years	161	3.8678	0.53633	
	31–35 years	42	3.9924	0.54827	
	36–40 years	16	3.3963	0.86260	
Enjoyment	21–25 years	159	3.7673	0.57583	2.495
	26–30 years	161	3.7888	0.62660	
	31–35 years	42	3.8810	0.56103	
	36–40 years	16	3.4063	0.63819	
Feedback	21–25 years	159	3.9591	0.61616	0.810
	26–30 years	161	4.0559	0.55338	
	31–35 years	42	3.9643	0.59871	
	36–40 years	16	3.9688	0.59073	
$N = 378$					

Source Author's findings
Notes F = one-way ANOVA; degrees of freedom (df) = 3/374, *$p < 0.05$ ** $p < 0.01$

a forward-thinking mindset. They contend that embracing demanding assignments might expedite their career advancement and enhance their professional profile. This forward-thinking mentality compels people to do duties beyond their current obligations. Finally, early-tenure employees may perceive that they have less to lose due to the comparatively lower number of responsibilities or expectations

Table 5 Comparative responses across tenure groups on dimensions of IM

Dimensions of IM	Tenure groups	*N*	Mean	SD	*F*
Sense of progress	1–4 years	193	3.7098	0.56379	0.099
	5–8 years	137	3.7101	0.50242	
	9–12 years	34	3.6621	0.68295	
	13–16 years	14	3.6671	0.53406	
Perceived choice	1–4 years	193	3.5777	0.57093	1.612
	5–8 years	137	3.5931	0.54571	
	9–12 years	34	3.5368	0.63405	
	13–16 years	14	3.2500	0.53709	
Recognition	1–4 years	193	3.7168	0.66070	0.336
	5–8 years	137	3.7566	0.57944	
	9–12 years	34	3.6759	0.60658	
	13–16 years	14	3.6193	0.43084	
Challenge	1–4 years	193	3.9276	0.48633	3.743**
	5–8 years	137	3.8396	0.51146	
	9–12 years	34	3.6471	0.65000	
	13–16 years	14	3.6664	0.62771	
Accomplishment	1–4 years	193	3.7756	0.63190	0.550
	5–8 years	137	3.7374	0.64079	
	9–12 years	34	3.6574	0.70356	
	13–16 years	14	3.6193	0.50364	
Encouraging initiative	1–4 years	193	3.7267	0.60065	0.771
	5–8 years	137	3.6387	0.59788	
	9–12 years	34	3.7206	0.70915	
	13–16 years	14	3.5714	0.45392	
Responsibility	1–4 years	193	3.8311	0.62861	0.312
	5–8 years	137	3.8176	0.56395	
	9–12 years	34	3.8047	0.67732	
	13–16 years	14	3.6664	0.87735	
Enjoyment	1–4 years	193	3.8135	0.58757	0.866
	5–8 years	137	3.7226	0.61213	
	9–12 years	34	3.8088	0.61564	
	13–16 years	14	3.6429	0.69139	
Feedback	1–4 years	193	4.0130	0.61330	0.179
	5–8 years	137	3.9891	0.53711	
	9–12 years	34	3.9559	0.64394	
	13–16 years	14	4.0714	0.58366	
N = 378					

Source Author's findings
Notes F = one-way ANOVA; degrees of freedom (df) = 3/374, ** $p < 0.01$

than their more experienced colleagues. This diminished dread of failure enables them to accept challenges and take calculated risks without the psychological burden that frequently accompanies seniority. Additionally, they may regard obstacles as thrilling rather than intimidating.

However, the findings suggest respondents with 9–12 years of tenure were least likely to pursue a challenge. The possible reasons for such findings are as follows. First, Employees who have been employed for 9–12 years may have attained a career plateau, during which their opportunities for rapid advancement decrease and their responsibilities become more routine. This can result in a lack of motivation to pursue new challenges or a perception of professional stagnation. A more settled and complacent approach to work may replace the enthusiasm and motivation that fuel early career years. Second, Employees frequently prioritize stability over novelty during this phase of their careers. They may prioritize predictable work environments and routines that enable them to manage work-life balance effectively, given the potential for increased personal responsibilities, such as family or financial obligations. Challenging assignments may disrupt this equilibrium, necessitating more time and energy. Third, These employees may experience moderate burnout or fatigue due to their near-decade-long tenure in comparable positions or within the same organizational structure. Repetitive tasks, rigid structures, or a lack of recognition can diminish motivation, reducing the inclination to pursue challenging tasks. The initial fervor to investigate and push boundaries may have waned over time. Finally, Mid-tenure employees may experience feelings of demotivation if the organization fails to consistently reward or acknowledge them for undertaking challenging assignments. They may be under the impression that the extra effort necessary to complete complex tasks does not result in tangible benefits, such as promotions, raises, or appreciation, which could result in a decreased inclination to participate in such activities.

5 Conclusion

The research objective sought to examine differences (if any) in the pattern of responses of respondents from varied demographic groups toward dimensions of IM as identified within the scope of the study. While gender did not induce significant differences in responses to varied dimensions of IM, respondents of different age groups revealed significant differences in their responses on three IM dimensions: recognition, challenge, and responsibility. Respondents in age groups 26–30 years revealed an increased preference for recognition and challenge, while respondents in the age of 31–35 years were predilected toward responsibility in their job assignments compared to their counterparts. Similarly, respondents belonging to different tenure groups exhibited similarity in their responses to all dimensions of IM except for one (i.e., challenge) as younger employees who had been with the organization in the tenure range of 1–4 years preferred challenge in their job assignments in comparison to their older counterparts.

The outcome demonstrates that intrinsic job motivation is a significant concern among Generation Y workers in the company. Gen Y population is the highest in number in any organizations today. Therefore, the primary goal of management should be to motivate them and retain talented and skilled workers effectively. Excessive turnover causes issues for clients, employees, and employers alike. A lower turnover rate shows a more positive approach toward clients. In turn, satisfied customers yield higher satisfaction ratings, increasing sales. Now is the moment to consider creatively and concentrate on intrinsic motivational variables, which can also impact employee productivity when employed effectively by management. Generation Y employees will make up most of the workforce in the upcoming years. For intrinsic motivation to become a potent instrument in the hands of the company to keep its talented, innovative, and dynamic Generation Y workers, further research is needed to close the gap and offer fresh perspectives on intrinsic motivation, its forms, applicability, and assessment.

References

1. Zhang, J., Zhang, Y., Song, Y., & Gong, Z. (2016). The different relations of extrinsic, introjected, identified regulation and intrinsic motivation on employees' performance: Empirical studies following self-determination theory. *Manage Decision., 54*(10), 2393–2412.
2. Amabile, T. M. (1993). Motivational synergy: Toward new conceptualizations of intrinsic and extrinsic motivation in the workplace. *Human Resource Management Review, 3*(3), 185–201.
3. Hausknecht, J. P., Rodda, J., & Howard, M. J. (2008). Targeted employee retention: Performance-based and job-related differences in reported reasons for staying. *Human Resource Management, 48*(2), 269–288.
4. Dysvik, A., & Kuvaas, B. (2011). Intrinsic motivation as a moderator on the relationship between perceived job autonomy and work performance. *European Journal of Work and Organizational Psychology, 20*(3), 367–387.
5. Griffeth, R. W., Hom, P. W., & Gaertner, S. (2000). A meta-analysis of antecedents and correlates of employee turnover: Update, moderator tests, and research implications for the next millennium. *Journal of Management, 26*(3), 463–488.
6. Poole, E. (2009). Organizational spirituality–a literature review. *Journal of Business Ethics, 84*(4), 577–588.
7. Manolopoulos, D. (2011). Work motivation in the Hellenic extended public sector: An empirical investigation. *International Journal of Human Resource Management, 19*(9), 1738–1762.
8. Cerasoli, C. P., Nicklin, J. M., & Ford, M. T. (2014). Intrinsic motivation and extrinsic incentives jointly predict performance: A 40-year meta-analysis. *Psychological Bulletin, 140*(4), 980–1008.
9. Catano, V. M., Morrow Hines, H. (2016). The influence of corporate social responsibility, psychologically healthy workplaces, and individual values in attracting millennial job applicants. *Canadian Journal of Behavioural, 48*(2), 142–154.
10. Schwartz, B. (2015). *Why We Work.* Simon & Schuster.
11. Steers, R. M., Mowday, R. T., & Shapiro, D. L. (2004). Introduction to special topic forum: The future of work motivation theory. *Academy of Management Review, 29*(3), 379–387.
12. Imran, H., Arif, I., Cheema, S., & Azeem, M. (2014). Relationship between job satisfaction, job performance, attitude towards work and organizational commitment. *Entre. Innov. Manage. J., 2*(2), 135–144.
13. Le Grand, J. (2006). Equality and choice in public services. *Social Research, 73*(2), 695–710.
14. Manolopoulos, D. (2008). Work motivation in the Hellenic extended public sector: An empirical investigation. *International Journal of Human Resource Management, 19*(9), 1738–1762.
15. Lee, J. Q., McInerney, D. M., Liem, G. A. D., & Ortiga, Y. P. (2010). The relationship between future goals and achievement goal orientations: An intrinsic–extrinsic motivation perspective. *Contemporary Educational Psychology*, *35*(4), 264–279.
16. Ryan, E. L. (2000). Deci, Intrinsic and extrinsic motivations: Classic definitions and new directions. *Contemporary Educational Psychology, 25*(1), 54–67.

17. Berlyne, D. E. (1966). Curiosity and exploration. Science 153(3731), 25–33.
18. Vansteenkiste, M., Lens, W., & Deci, E. L. (2006). Intrinsic versus extrinsic goal contents in self-determination theory: Another look at the quality of academic motivation. *Educational Psychology, 41*(1), 19–31.
19. Reeve, J., Deci, E. L., Ryan, R. M. (2002). Self-determination Theory: A dialectical framework for understanding socio-cultural influences on student motivation. In D. M. McInerney & S. Van Etten (Eds.), *Big Theories Revisited. Greenwich*, CT: Information Age, 31–60
20. Condry, J. (1977). Enemies of exploration: Self-initiated versus other-initiated learning. *Journal of Personality and Social Psychology, 35*(7), 459–477.
21. Deci, E. L. (1975). *Intrinsic motivation.* Plenum.
22. Nuttin, J. R. (1973). Pleasure and reward in human motivation and learning. In: D. E. Berlyne & K. B. Madsen (Eds.), *Pleasure, reward, preference* (pp. 243–274). New York: Academic Press
23. Ryan, R. M., & Deci, E. L. (2000). Self-determination theory and the facilitation of intrinsic motivation, social development, and well-being. *American Psychologist, 55*(1), 68–78.
24. Ryan, R. M., & Connell, J. P. (1989). Perceived locus of causality and internalization: Examining reasons for acting in two domains. *Journal of Personality and Social Psychology, 57*(5), 749–761.
25. Kothari, C. R. (2004). *Research methodology: Methods and techniques* (2nd Edition). New Age International
26. Biernacki, P., & Waldorf, D. (1981). Snowball sampling: Problems and techniques of chain referral sampling. *Sociological Methods and Research, 10*(2), 141–163.
27. Sadler, G. R., Lee, H. C., Lim, R. S. H., & Fullerton, J. (2010). Recruitment of hard-to-reach population subgroups via adaptations of the snowball sampling strategy. *Nursing and Health Sciences Journal, 12*(3), 369–374.
28. Sarangi, P. (2010). *Research methodology* (10th ed.). Taxmann Publications Pvt Ltd.
29. McAuley, E., Duncan, T., & Tammen, V. V. (1989). Psychometric properties of the Intrinsic Motivation Inventory in a competitive sport setting: A confirmatory factor analysis. *Research Quarterly for Exercise and Sport, 60*(1), 48–58.

Impact of Knowledge Management on Employee Engagement in the Information Technology Industry

N. Gangisetty and S. Mohana

Abstract

This study investigates how employee engagement in the information technology (IT) sector is impacted by knowledge management qualities, such as knowledge generation, knowledge capture, and knowledge storage. A structured questionnaire and 317 IT professionals made up the sample size. SPSS was used to analyze the data using statistical procedures including multiple linear regression, Exploratory Factor Analysis (EFA), and reliability testing. The results show a strong and favorable relationship between increased employee engagement and good knowledge management attributes. Employee engagement and involvement are often higher in organizations that support ongoing learning, encourage information exchange through mentoring programs, and have up-to-date knowledge repositories. To maintain a thriving information-driven culture, the report recommends establishing focused efforts like learning workshops, organized coaching, and knowledge audits. It also emphasizes the need for proactive organizational knowledge management to improve employee engagement.

Keywords

IT industry • Knowledge management • Knowledge application • Projects • Management support

N. Gangisetty (✉)
Department of Management Studies, Madanapalle Institute of Technology and Science, Madanapalle, India
e-mail: ngangisetty2006@gmail.com

S. Mohana
Department of Management Studies, Annamacharya Institute of Technology and Sciences, Kadapa, India

S. Patnaik et al. (eds.), *People, Planet and Profit: Advancing Sustainable Business Practices*, Springer Proceedings in Geography and Sustainability,
https://doi.org/10.1007/978-3-032-11296-5_17

1 Introduction

In the Information Technology (IT) industry, effective knowledge management (KM) is pivotal for enhancing employee engagement. Knowledge management is the rational management of a group's information assets to generate value and satisfy strategic needs. By facilitating seamless knowledge sharing and collaboration, KM fosters a philosophy of continuous knowledge and invention, leading to heightened employee engagement.

Research indicates that robust KM practices significantly impact employee engagement. For instance, a study by [1] in the hospitality sector demonstrated that effective KM practices accounted for a disparity in employee engagement of 50.4%, with a unit change in KM leading to a 0.728 increase in engagement levels. Although this study focused on the hospitality industry, its insights apply to the IT sector, where knowledge sharing and continuous learning are equally critical. Moreover, [2] emphasizes that KM enhances employee flexibility and job satisfaction, making them more adaptable to change. In the dynamic IT industry, such adaptability is essential, as it enables employees to keep pace with rapid technological advancements, thereby increasing their engagement and commitment to the organization. In summary, implementing effective KM practices in the IT industry not only streamlines operations but also fosters a more engaged and motivated workforce, driving long-term organizational success.

IT companies greatly depend on the knowledge, innovation, and ongoing education of their employees in the knowledge-driven economy of today. A new strategic instrument for leveraging and disseminating intellectual capital is knowledge management. However, the degree of employee engagement has a significant impact on its performance. Engaged workers are likelier to share ideas, work well with others, and improve performance in knowledge-intensive industries like IT where creativity and adaptability are critical. Thus, maintaining a competitive advantage requires an awareness of the relationship between knowledge management and employee engagement.

2 Review of Literature

KM is crucial in the IT sector, where knowledge is a key asset and innovation is rapid. Studies have shown that effective KM practices can improve IT project success, reduce knowledge loss, and enhance organizational learning [3, 4]. Knowledge management (KM) in the IT sector includes handling explicit knowledge like code and documentation and tacit knowledge like experience and skill [5].

Research has painted the significance of the KM frame, including technology and processes, in supporting KM initiatives. For example, learning by [6] establishes that IT-enabled KM systems can facilitate knowledge sharing and collaboration, improving project outcomes. Similarly, a study by [7] found that KM processes, such as knowledge mapping and documentation, can help to identify and leverage knowledge assets, leading to improved organizational performance.

However, studies have also highlighted the challenges of implementing effective KM practices in the IT industry. For instance, a study by [8] found that IT professionals often resist KM initiatives due to concerns about loss of autonomy and control over their knowledge. [9] Found that KM initiatives can be hindered by cultural and organizational barriers, such as lack of trust and inadequate communication.

To improve an organization's competitive edge, KM entails recognizing and using the group information inside the company. There are several classifications for KMC in the corpus of KM research. KMC is divided into three primary categories [10] knowledge generation, incorporation, and dissemination. Knowledge construction, embodiment, distribution, and utilization are the four KMCs identified by [11]. Sequences such as information generation, storing and recovery, transmission, and claims are examined [12]. Furthermore, KM encompasses the management actions, infrastructure, and skills that support and enhance these knowledge processes [13]. All things considered, the literature generally acknowledges four to six interrelated, cyclical knowledge processes.

Similar to the previously discussed perspectives, this research aims to examine the possible possessions of the KMC frame that was suggested by job satisfaction, organizational commitment, and engagement of employees. The KMC consists of the following five steps: knowledge production, sharing, structuring, using, and auditing. KMC starts with knowledge production, where new information is produced using research, invention, or teamwork. This knowledge is then captured and stored for later use using methods like databases, digital archives, or documentation. The last phase is methodically arranging the acquired knowledge to enhance the distribution and retrieval procedures [14].

After being arranged, information is disseminated to the pertinent parties via the Internet, conferences, and training sessions. Knowledge sharing encourages collaboration and learning, enabling people to gain from a variety of viewpoints and insights. Knowledge is then put to use in real-world situations to resolve problems, make choices, or improve organizational processes. Following application, knowledge is evaluated to determine its efficacy and influence on organizational objectives and results. This evaluation highlights areas in need of development and offers guidance for the next knowledge management initiatives. The cycle ends with knowledge refinement, in which the organization's knowledge management practices are continuously improved by applying the evaluation's findings to improve the information and processes already in place.

Though opinions differ, EE has emerged as a major topic in academic research and human resource (HR) consultancy. It is frequently commended for its capacity to increase profitability by creating favorable financial results through increased sales, productivity, and staff retention. [15] State that when workers exhibit physical, mental, and emotional participation in their work, they are committed to their jobs (Fig. 1).

A spiritual component, characterizing engagement as capturing employees' thoughts, emotions, and inner beings [16], and other researchers define commitment as the psychological thing associated with the individual expenditure

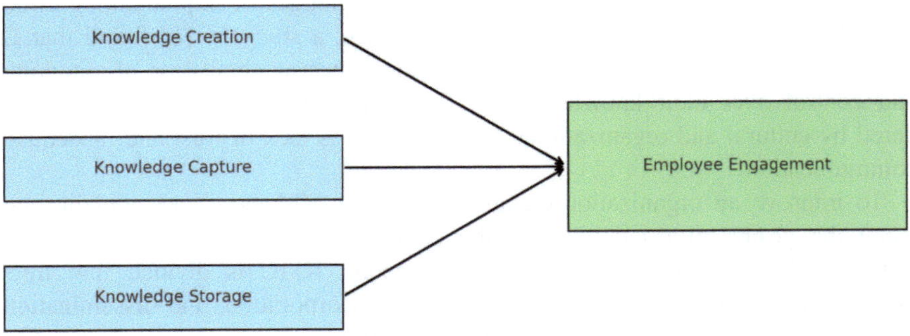

Fig. 1 Conceptual framework—Linking KM dimensions to employee engagement

Table 1 Operational definitions of variables and constructs

Constructs	Definition	References
Knowledge creation	Generating novel ideas or solutions through collaboration, experimentation, and innovation	[5]
Knowledge capture	Systematically documenting and transferring individual expertise and lessons learned	[7]
Knowledge storage	Organizing and maintaining knowledge in centralized repositories for easy access and reuse	[12]
Employee engagement	Emotional and cognitive commitment to work, shown through enthusiasm, involvement, and effort	[18, 19]

of vitality [17]. Three levels of engagement such as engaged, not engaged, and keenly disengaged—are distinguished by [18], and each has a distinct effect on organizational performance. Broadens the definition of engagement to include organizational and work characteristics that are impacted by perceived support, satisfaction, fairness, and job features [19] (Table 1).

Work satisfaction, dedication, intention to leave, and organizational citizenship behavior are examples of engagement outcomes [20]. Increased involvement significantly lowers the inclination to leave, according to studies. Job type, pay, WLB, top-level contacts, job advancement, and lineup subtleties are the six key management practices that [21] identified as having an impact on executives' engagement. In order to create a solid model, this study attempts to thoroughly examine the variables affecting EE.

3 Research Gap

While knowledge management (KM) is becoming more widely acknowledged as a crucial factor in determining organizational performance, less is known about how it directly affects employee engagement, especially in the IT industry. With little focus on how particular KM practices like knowledge sharing, knowledge retention, and organizational learning impact engagement dimensions like commitment, motivation, and job satisfaction, the majority of existing research concentrates on the effects of KM on innovation, decision-making, and productivity. In the context of IT organizations, which function in dynamic, knowledge-intensive contexts where employee engagement is essential for maintaining innovation and agility, this disparity is particularly noteworthy. In an increasingly competitive environment, knowing this link may help IT companies create knowledge management (KM) strategies that not only improve performance but also foster a dedicated and driven team.

4 Research Problem

In the rapidly evolving IT industry, organizations rely on knowledge as a critical asset to drive innovation, enhance productivity, and maintain a competitive edge. Employee engagement is one of the organizational outcomes that are impacted by the difficulty of effectively managing knowledge, which includes its generation, exchange, and utilization. Employee engagement, characterized by enthusiasm, commitment, and discretionary effort, is essential for organizational success, yet many IT companies struggle with disengagement, knowledge silos, and inefficient knowledge-sharing practices. While knowledge management (KM) initiatives such as collaborative platforms, training programs, and mentorship structures are widely implemented, their direct influence on employee engagement is not well understood. It is necessary to investigate if structured knowledge management techniques, organizational knowledge accessibility, and knowledge-sharing cultures lead to increased employee motivation, engagement, and retention. Furthermore, further research is required to fully understand how technology facilitates knowledge transfer and how this affects the engagement of employees in isolated and annoyed task arrangements. To help organizations create information-driven engagement strategies that work, this research will look at how knowledge management affects employee engagement in the IT sector. It will also identify important KM practices that either help or hurt engagement.

5 Objectives of Research

1. To identify the knowledge management attributes in the IT sector.
2. To extend the effect of knowledge management attributes on employee engagement.

6 Statistical Tools

SPSS software was used to examine the gathered data. Cronbach's Alpha was used to assess the scale's reliability, and EFA was used to confirm the build structure. The impact of the knowledge management variables on employee engagement was then investigated using multiple linear regression.

7 Research Methodology

7.1 Research Instrument Design

A structured questionnaire intended to evaluate the connection between IT sector employee engagement and knowledge management techniques served as the main tool for gathering data. To guarantee content relevancy and theoretical compatibility, the questionnaire was created by adapting previously validated measures and doing a thorough assessment of the body of current research. A pre-test was carried out using a pilot sample of thirty IT experts who were not included in the final research sample to guarantee the validity and reliability of the instrument. Uncertain questions were clarified and refined using input from the pilot group. Three academicians and two HR specialists with expertise in knowledge management and employee engagement reviewed the content to ensure its accuracy.

Cronbach's Alpha was used to evaluate internal consistency, and exploratory factor analysis (EFA) was used to show construct validity. All of the constructions showed satisfactory dependability, with Alpha values above the suggested cutoff point of 0.70. By following these procedures, the questionnaire was guaranteed to accurately capture the targeted components and to serve as a firm substance for the primary data collection.

7.2 Sampling Procedure

The influence of knowledge management practices—more especially, knowledge generation, knowledge capture, and knowledge storage—on employee engagement in the information technology (IT) sector was examined in this study using a quantitative research approach. Data from IT professionals at different Indian firms was gathered using a standardized questionnaire. Convenience sampling was the sampling strategy employed in this study, which focused on IT industry experts who were reachable and eager to take part. This group was chosen because of the IT industry's dynamic and knowledge-intensive character, where knowledge management is essential to employee engagement and performance. Based on [22] sample size determination, a minimum of 317 answers was necessary for a population of this scale to achieve statistical significance and representativeness. As a result, more than 450 employees received the poll through email and professional networking sites like LinkedIn. For data analysis, a total of 317 valid responses were collected.

8 Result Analysis

8.1 Reliability and Validity Test

See Tables 2, 3.

The findings of the reliability test show how internally consistent the study's measuring scales were. Knowledge management attributes have a modest but acceptable level of reliability, as shown by the Alpha of 0.782; it's greater than the standard value. This suggests that the knowledge management traits are consistently captured by the items used to measure them. The four items evaluating employee engagement appear to have great internal consistency, as specified by the great dependability of the Alpha for employee engagement, which is 0.830. The scales utilized for both dimensions are statistically sound and appropriate for additional study, according to the reliability scores taken as a whole.

8.1.1 Exploratory Factor Analysis

See Table 4.

A strong level of adequacy for EFA is shown by the KMO Extent of adequacy sample, which stands at 0.841. Given that there is adequate association among the

Table 2 Summary of all variables for case processing

		Frequency	%
Cases	Valid	317	100.0
	Excluded	0	0.0
	Total	317	100.0

Table 3 Dimension-wise reliability test results

Dimension	List of variables	Cronbach's alpha
Knowledge management attributes	13	0.782
Employee engagement	4	0.830

Table 4 Bartlett's test and KMO

Sampling adequacy of KMO		0.841
Bartlett's sphericity test	χ^2 approximately	1625.277
	Degree of freedom	68
	Significance	0.000

variables, a KMO value is more than 0.8 shows that the data is fit for exploratory factor analysis. Furthermore, the Sphericity is significant (χ^2 = 1625.277, df = 68, p = 0.000), indicating that factor analysis is suitable and the matrix of correlation is not an identity matrix. According to the substantial Bartlett's test result, the variables are connected and may be successfully boiled down to underlying causes. Thus, it becomes sense to move further with factor analysis in imperative to investigate the dataset's structure (Table 5).

Rotation of Varimax with Normalization of Kaiser has been utilized to obtain three components. Every element is composed of all the items with factor Eigenvalues > 0.5. Thirteen factors were combined to generate three components. Three factors were found among the 13 variables that were part of the investigation. These three factors accounted for 63.485 percent of the variation in Knowledge Management attributes in the IT industry (Table 6).

8.2 Regressions

Regression analysis was used to determine how knowledge management attributes affected employee engagement, which was the dependent variable.

The model summary shows that among IT professionals, employee engagement (the dependent variable) and knowledge management (the independent variable) have a relatively significant positive association. There is a significant association, as indicated by the R-value of 0.706. A decent model fit is shown by the R Square (0.712), which indicates that knowledge management methods account for around 71.2% of the variation in employee engagement. Even after taking into consideration the number of predictors in the model, the Adjusted R Square (0.684) still shows a high level of explanatory power. The actual and anticipated values appear to differ somewhat, as indicated by the 1.0613 standard deviation from the estimate. The validity of the regression assumptions is supported by the considerable autocorrelation in the residuals, as indicated by the Durbin-Watson statistic of 1.985, which is near 2 (Tables 8, 9).

The regression model's F statistic is statistically at the 0.05 level, indicating that the regression equation fits well (the model is statistically significant), according to ANOVA Table 7.

The findings of the regression coefficient provide light on the connection between IT employees' engagement and knowledge management aspects. The constant (B = 0.712, p = 0.000) indicates the baseline level of employee engagement when all independent variables are zero.

- Among the predictors, knowledge creation (B = 0.306, p = 0.000, β = 0.541) has the highest standardized beta coefficient, suggesting that it has the strongest positive impact on employee engagement. This implies that fostering knowledge creation significantly enhances engagement levels.

Table 5 Explained total variance

Component	Eigenvalues at initial state			Sums of extracted squared loadings			Rotation totals of squared weights		
	Total	percentage of variance	Cumulative percentage	Total	Percentage of variance	Cumulative percentage	Total	Percentage of variance	Cumulative percentage
1	5.162	39.705	39.705	5.162	39.705	39.705	3.359	25.838	25.838
2	1.734	13.335	53.041	1.734	13.335	53.041	2.750	21.152	46.990
3	1.358	10.444	63.485	1.358	10.444	63.485	2.144	16.495	63.485
4	.894	6.879	70.363						
5	.762	5.861	76.224						
6	.537	4.129	80.353						
7	.495	3.806	84.159						
8	.447	3.437	87.596						
9	.433	3.333	90.929						
10	.350	2.692	93.621						
11	.333	2.561	96.182						
12	.290	2.234	98.416						
13	.206	1.584	100.000						

Table 6 Component matrix rotation

Component matrix			
Factor name	Factor loading		
	1	2	3
F1: Knowledge creation			
My organization encourages employees to generate new ideas and solutions	0.714		
Employees have opportunities to collaborate and brainstorm innovative solutions	0.722		
My organization provides resources and tools to facilitate knowledge creation	0.782		
Knowledge-sharing sessions and discussions are regularly conducted	0.642		
Employees are encouraged to experiment and take calculated risks to generate new knowledge	0.610		
F2: Knowledge capture			
My organization has systems in place to document and capture employee knowledge		0.692	
Employees are encouraged to document and share their experiences and expertise		0.541	
Best practices and lessons learned are systematically recorded and made accessible		0.587	
My organization conducts regular debriefings to capture knowledge from completed projects		.531	
Knowledge transfer programs, such as mentorship or training, are actively promoted		0.522	
F3: Knowledge storage			
My organization has a centralized knowledge repository for storing important information			0.711
Employees can easily access stored knowledge when needed			0.543
The organization ensures that stored knowledge is regularly updated and maintained			0.621

Table 7 Model summary

Model	R	R square	Adjusted R square	SD from the estimate	Durbin-Watson
1	0.706	0.712	0.684	1.0613	1.719

Table 8 ANOVA[a]

Model	Sum of squares	DF	Mean square	F	P-value
1	Regression	36.054	1	35.062	0.000[b]
	Residual	312.462	326	1.114	
	Total	382.741	327		

Table 9 Regression-Coefficient

KM attributes		Un-std. coefficients		Std. coefficients	T	P-value
		B	Std. error	Beta		
1	(Constant)	0.712	0.149		7.251	0.000
	Knowledge creation	0.306	0.079	0.541	6.472	0.000
	Knowledge capture	0.241	0.026	0.365	4.258	0.000
	Knowledge storage	0.192	0.015	0.269	3.259	0.001

a. Dependent variable: employee engagement

- Knowledge capture ($B = 0.241$, $p = 0.000$, $\beta = 0.365$) also has a progressive and statistically significant impact, indicating that effective documentation and transfer of knowledge contribute meaningfully to employee engagement.
- Knowledge storage ($B = 0.192$, $p = 0.001$, $\beta = 0.269$) is also a significant predictor, though its impact is comparatively lower. This suggests that while efficient storage and retrieval of knowledge are important, their influence on engagement is relatively weaker than knowledge creation and capture.

Overall, the model confirms that all three dimensions of knowledge management significantly influence employee engagement, with knowledge creation playing the most dominant role. These findings emphasize the need for IT organizations to focus on creating a knowledge-sharing culture to enhance employee engagement.

9 Discussion: Linking Findings to Prior Research

Effective knowledge management (KM) practices and employee engagement in the information technology (IT) sector are strongly positively correlated, according to the study's findings. This is in line with earlier studies that emphasize how KM can improve employee happiness and organizational effectiveness. For example, earlier research by [23, 24] highlighted how KM systems support a culture of information sharing, which gives workers a feeling of purpose and belonging. Knowledge management (KM) initiatives like collaborative platforms, knowledge repositories, and mentoring programs not only promote effective information flow but also empower staff members by allowing them to feel appreciated and included in decision-making processes. This is especially important in the IT industry, where rapid innovation and knowledge turnover are common. These results corroborate those of more recent studies, including the one by [25], which discovered that companies with strong knowledge management (KM) strategies have greater employee engagement and motivation levels. Therefore, this study supports the idea that incorporating knowledge management (KM) into corporate culture is a strategic strategy to developing a more engaged and dedicated workforce in the fast-paced IT sector, in addition to being a technology investment.

10 Suggestions

- Conduct continuous learning programs, workshops, and mentorship initiatives to upskill employees and encourage knowledge generation.
- Encourage senior employees to share expertise through structured mentorship and coaching programs.
- Keep stored knowledge up to date by assigning dedicated knowledge managers to review and refresh content periodically.

11 Conclusion

The research findings highlight the significant effect of knowledge management dimensions such as the creation, capture, and storage of knowledge on employee engagement in the IT industry. Among these dimensions, knowledge creation emerged as the most influential factor, indicating that fostering an environment that encourages idea generation, innovation, and collaborative problem-solving leads to higher employee engagement. Knowledge capture, which involves systematically documenting and transferring expertise, also plays a crucial role in certifying that professionals feel valued and connected to organizational knowledge. Additionally, knowledge storage, while slightly less impactful than the other two dimensions, remains an essential factor in ensuring the accessibility and usability of organizational knowledge, contributing to employee confidence and efficiency. Overall, the results suggest that effective knowledge management strategies enhance employee engagement by promoting a culture of learning, collaboration, and accessibility to information. IT organizations should focus on strengthening these knowledge management practices to drive employee motivation, commitment, and productivity. By investing in structured knowledge-sharing platforms, mentorship programs, and digital knowledge repositories, IT firms can create an engaging work environment that fosters innovation and long-term organizational achievement.

References

1. Onyango, R. O., Egessa, R., & Ojera, P. (2022). Knowledge management and employee engagement in the hospitality industry. *International Journal of Research in Business and Social Science, 11*(6), 209–217.
2. Becerra-Fernandez, I., & Sabherwal, R. (2010). Knowledge Management: Systems and Processes. M.E. Sharpe.
3. Davenport, T. H., & Prusak, L. (1998). Working knowledge: How organizations manage what they know. *Harvard Business School Press.*

4. Lee, J., & Kim, Y. (2001). A stage model of organizational knowledge management: A latent content analysis. *Journal of Knowledge Management, 5*(4), 401–415.
5. Nonaka, I., & Takeuchi, H. (1995). The knowledge-creating company: How Japanese companies create the dynamics of innovation. *Oxford University Press.*
6. Kulkarni, U. R., Ravindran, S., & Freeze, R. D. (2006). A knowledge management success model: Theoretical development and empirical validation. *Journal of Management Information Systems, 23*(3), 309–347.
7. Chaudhry, A. S. (2005). Knowledge sharing and organizational learning: A case study of a software development company. *Journal of Knowledge Management, 9*(3), 114–127.
8. Desouza, K. C. (2003). Barriers to effective use of knowledge management systems in software engineering. *Software Quality Journal, 11*(2), 143–155.
9. Panahi, S., Watson, J., & Partridge, H. (2012). Towards a model for evaluating the effectiveness of knowledge management systems. *Journal of Knowledge Management, 16*(4), 521–535.
10. Nonaka, L., Takeuchi, H., & Umemoto, K. (1996). A theory of organizational knowledge creation. *International Journal of Technology Management, 11*(7–8), 833–845.
11. Demarest, M. (1997). Understanding knowledge management. *Long Range Planning, 30*(3), 374–384.
12. Alavi, M., & Leidner, D. E. (2001). Knowledge management and knowledge management systems: Conceptual foundations and research issues. MIS Quarterly, 107–136.
13. Othman, A. A. E., & ElKady, M. M. (2023). A knowledge management-based framework for enhancing the learning culture in architectural design firms in developing countries. *Journal of Engineering, Design and Technology, 21*(1), 23–57.
14. Sağsan, M. (2006). A new life cycle model for processing of knowledge management. *In 2nd International Congress of Business, Management and Economics* (pp. 15–18).
15. Boccoli, G., Gastaldi, L., & Corso, M. (2023). The evolution of employee engagement: Towards a social and contextual construct for balancing individual performance and wellbeing dynamically. *International Journal of Management Reviews, 25*(1), 75–98.
16. Fleming, J. H., & Asplund, J. (2007). Human sigma: managing the employee-customer encounter. Simon and Schuster.
17. Schaufeli, W. (2021). Engaging leadership: How to promote work engagement. *Frontiers in Psychology, 12*, Article 754556.
18. Gallup. (2024, April 3). The high cost of disengaged employees. *Gallup Business Journal.* https://www.gallup.com/workplace/393497/world-trillion-workplace-problem.aspx.
19. Saks, A. M. (2006). Antecedents and consequences of employee engagement. *Journal of Managerial Psychology, 21*(6), 600–619.
20. Alnehabi, M., & Al-Mekhlafi, A. B. A. (2023). The association between corporate social responsibility, employee performance, and turnover intention is moderated by organizational identification and commitment. *Sustainability, 15*(14202). https://doi.org/10.3390/su15014202.
21. Joshi, R. J., & Sodhi, J. S. (2011). Drivers of employee engagement in Indian organizations. *Indian Journal of Industrial Relations, 47*(2), 162–182.
22. Krejcie, R. V., & Morgan, D. W. (1970). Determining sample size for research activities. *Educational and Psychological Measurement, 30*(3), 607–610.
23. Alavi, M., & Leidner, D. E. (2001). Review: Knowledge management and knowledge management systems: Conceptual foundations and research issues. *MIS Quarterly, 25*(1), 107–136. https://doi.org/10.2307/3250961
24. Davenport, T. H., & Prusak, L. (1998). Working knowledge: How organizations manage what they know. *Harvard Business Press.*
25. Hislop, D., Bosua, R., & Helms, R. (2018). Knowledge management in organizations: A critical introduction (4th ed.). Oxford University Press.
26. Let me know if you need these in a different referencing style or if you'd like to add more sources.

Impact of Hybrid Work Model on Job Satisfaction of Tech Professionals in Information Technology Industry

Sabitha Sunkesula and S. Venkata Subba Reddy

Abstract

The adoption of HWM in the IT trade has generated considerable interest and discussion regarding its impact on job satisfaction among tech professionals. This research aims to measure the implications of hybrid work arrangements on the satisfaction of job of techies in the IT industry. By synthesizing existing literature and conducting empirical research, this research explores various attributes effects satisfaction of job, such as flexibility, work–life balance, communication effectiveness, collaboration, and productivity, within the context of hybrid work environments. Through surveys and interviews with tech professionals, this research seeks to find the perceived advantages and experiments related with hybrid work modes and their influence on overall job satisfaction. Understanding theses dynamics is crucial for organizations and policymakers in optimizing hybrid work preparations to improve job satisfaction, retention, and managerial enactment in the rapidly evolving landscape of the IT industry.

Keywords

Hybrid-Work-Model • Satisfaction of job • Flexibility • WLB • Communication effectiveness • Collaboration and productivity

S. Sunkesula (✉) · S. V. Subba Reddy
Department of Business Administration, Yogi Vemana University, Kadapa, India
e-mail: sabitha.sunkesula@gmail.com

S. Patnaik et al. (eds.), *People, Planet and Profit: Advancing Sustainable Business Practices*, Springer Proceedings in Geography and Sustainability,
https://doi.org/10.1007/978-3-032-11296-5_18

1 Introduction

The emergence of the hybrid work paradigm, which blends in-office and remote work, has drastically changed the nature of the contemporary workplace, especially in the information technology (IT) sector. During and after the COVID-19 pandemic, this adaptable strategy became widely used as businesses looked to preserve output while protecting the health and safety of their workforce. One of the first industries to use this paradigm was the IT sector, which is renowned for its agility and dependence on digital infrastructure. By providing workers with more autonomy, a better work–life balance, and less stress from commuting, hybrid work arrangements have been demonstrated to improve job satisfaction [1]. But there are drawbacks to this strategy as well, such preserving team unity, bridging communication barriers, and guaranteeing equitable chances for professional growth. Designing successful workplace strategies that encourage engagement, performance, and retention in the IT industry requires an awareness of how hybrid work affects employee happiness as firms negotiate this shift [2].

2 Literature Review

The effects of remote and hybrid work arrangements on IT workers' happiness and productivity. The study discovered that by providing flexibility and autonomy, particularly in creative and knowledge-intensive activities, hybrid work arrangements greatly increase job satisfaction. Because of shorter commutes and better focus at home, IT professionals reported greater engagement and a better work–life balance [1]. A comprehensive study conducted by a number of multinational IT companies to examine how employee satisfaction is affected by hybrid employment. According to their research, hybrid work improves workplace satisfaction and lowers attrition, especially for high-achieving IT workers. It was believed that flexibility was a major factor in higher retention and morale [3]. The following the epidemic, 73% of workers in the IT sector choose hybrid work arrangements. According to the survey, IT workers place a high importance on autonomy, flexibility, and mental health, all of which are closely linked to increased organizational commitment and job satisfaction [4]. the advantages and difficulties of switching to hybrid work models in IT companies during the COVID-19 epidemic. Higher employee happiness, more inclusive cooperation, and a deeper feeling of trust between employers and tech experts were among the long-term benefits, according to the research, even if the first transfers presented coordination issues [5].

Working remotely can either peacefully replace or enhance outdated work–life balance. Even so, there has been a preposterous tendency toward "work-from-home" nation, and the graph indicated that remote or distant working was quite inconsistent before the Novel Corona Virus (SARS CoV2) pandemic. If remote or distant work is done after conventional business hours, it might worsen a poor work–life balance. The author's research indicates that a brief rise in job satisfaction might be attributed to a variety of variables. Because of the obvious disadvantages of less in-person encounters and more interactions through telecommuting, which has weakened the bonds between coworkers, it has resulted in lower labor costs and decreased productivity [6].

The largest risk of "work-from-home" is a reduction in efficiency when employees are unable to engage with one another face-to-face, according to research [7]. Depending on the particulars of the profession and the home office that supports it, this could result in a number of different results. Revealing negative chronic work results in high demands and expectations with little control, which raises turnover, presenters, and job satisfaction [8]. The disadvantages of remote work are the main topic of [9], but he also points out that it directly reduces productivity and makes employee communication more difficult at the expense of speed and quality. On the other hand, significant results from [10] show that 76% of workers agree with this attitude and voluntarily choose to work from home on average at least twice a week. However, because of the disruption of team networking and low job motivation, the remaining 24% think this is totally ineffective and that they should immediately return to the workforce. Employees are reportedly under emotional pressure at work and expected to perform more than they usually do. Because their mental health is under stress and they feel compelled to contribute against their will out of fear of losing their jobs, 51% of workers report feeling stressed out. In return, the employees expect to receive a variety of other amenities, including advanced dining services, childcare, healthcare, and well-being services.

Work-from-home (WFH) workers are at a moderate to high risk of experiencing stress, tension, sadness, and physical and mental tiredness, according to Jodi Oakman (2020). Additionally, it significantly affects employees' job happiness and motivation. According to the primary study by [11], the WFH culture's greatest benefit is time and money savings, while its biggest drawback is the absence of face-to-face relationships. In addition to internet speed, the survey found that children's noise in the house is a main source of interruption for salaried women. Marital women and men were clearly more interested in hybrid work than the single population, according to the results.

For several years, remote work has fully increased rapidly owed to the rapid improvement of tools. According to Microsoft's 2021 prediction, hybrid work—which blends the finest aspects of remote and on-premises work—will be the next big thing in the workplace. Businesses must prioritize the needs of their workers in light of the COVID the growing popularity of work hybrid arrangements. WFH has advantages and disadvantages. One benefit is that task plans may be changed with greater flexibility. During the working, employees can attend to personal and family affairs.

Freeing up time that would have been spent traveling to participate in leisure activities or volunteer work also results in a more enjoyable lifestyle. Mungkasa (2020), these include the difficulty of creating restrictions among professional and personal responsibilities, the ability to manage both, a lack of information with superiors that hinders career advancement, and a feeling of estrangement from the workplace social network [13]. 2,877 knowledge workers in the US took part in a study on work-from-home options (Slack, 2020) [14]. Just 10% of research participants said that working from home made them unhappy with their jobs, and only 13% said that WFH decreased productivity. Due to the fact that several staffs were WFH for the paramount time without any previous planning, 2020 and 2021 were difficult years for both the company and the person. Additionally, Microsoft's 2021 survey, which assessed over 30,000 employees across 31 nations, discovered that working of hybrid is quickly increasing popularity.

Looked at inspiring and prime factors that affect work–family fulfillment and efficiency when WFH. While the ideal flow experience during WFH and the reconciliation of work and home life are the driving forces, ICT anxiety and disruptions are the intervening factors. The results and the study indicate that both traits and WFH production and pleasure are significantly correlated. However, the one factor that strongly correlates with employees' intention to continue working in mixed employment is contentment [15]. Workplace adaptability, family interactions, home office setting, and organizational support. According to the study, employees' job happiness and supposed efficiency when WFH are correlated with organizational support and employment flexibility [16].

Organizations can gain from working in hybrid mode in a variety of ways, such as guaranteeing that a steadiness of business strategy is in habitation and will function well in emergency situations, speeding up the digitalization and adoption of new technologies, introducing creative ways to promote cost effectiveness, and creating opportunities for global cooperation [17]. To attract outstanding talent, companies have spent a great deal of time, money, and energy into making the finest cultures. But, WFH guidelines have been around for a while in a variety of sectors, they now demand a standard shift and the expansion of strong and supportable schemes if businesses are to thrive for a much-extended period.

This study blends motivational elements like flow experience and reconciliation with elements that concentrate connected the interaction among workers and organizations, like job flexibility and managerial backing [17]. Finding the elements that affect WFH enjoyment and productivity is the goal. Businesses can realize the assistances of working from home while also overcoming the fear of productivity by putting into practice elements that might boost worker happiness and output. The end of 2019 pandemic has drastically altered many facets of life throughout the world, and WFH with flexible times and internet access has emerged as a viable option to slow the COVID-19 virus's spread. Following the end of the pandemic and the onset of the new normal, many businesses have continued to use remote work or flexible working spaces (FWS). Since employees no longer have to drive to the office, these techniques are now employed to enhance time or cost efficiency rather than to avoid COVID-19 viral infection [18]. Work-from-home or work-for-any place has developed very popular work styles in the modern workplace, even if the epidemic has ended and job duties may once again be completed in the office.

According to the findings of a poll performed by Inside Public Accounting in 2021 among 540 US professionals, 48% of respondents preferred a totally remote work paradigm, while 44% preferred mixed employment. However, just 5% of businesses provide a totally remote work model, whereas 51% support mixed work models. It is undeniable, nevertheless, that working remotely enables one to spend more time with their loved ones. Hybrid work incorporates both regular and remote working methods by setting up work hours. Everybody responds to FWS in a different way, with varying experiences and perspectives. When it comes to the impact of work environment on the FWS scenario, many researches have produced varying conclusions. Despite being in place since the twenty-first century, FWS has only now come to be acknowledged by a number of sources as beneficial to both companies and employees. It has been demonstrated to improve commitment, satisfaction of job, and WLB, as well as attendance, morale, and quality of life. However, studies as well show that millennial workers' intentions to stay do not considerably rise when they work remotely.

The potential for hybrid work—the option to work remotely or in an office—to boost job satisfaction is among the most compelling arguments for additional research into this gap (WFA). Traffic jams and air pollution are caused by Jakarta's high population. This is how pollution and congestion affect resource waste and pollution in the environment. If it is feasible to reduce traffic by reducing the number of employees who have to commute, the option of hybrid employment becomes crucial. The fact that works processes may now be completed online at certain times and from any location is also connected to the dynamic developments in the corporate environment. Gender equity, health and welfare, working conditions, and performance are all improved by hybrid work, making it a strong choice for constructing sustainability.

The Job Demands-Resources (JD-R) Model provides a robust theoretical foundation for examining how workplace variables impact employee outcomes like job satisfaction. The JD-R framework (Bakker & Demerouti, 2007) describes that job resources such as flexibility, effective communication, and work–life balance enhance job satisfaction and boost employee motivation and lower burnout. These are mostly relevant to the IT industry since it frequently practices hybrid work arrangements.

There are growing employee demands for hybrid workspace and resources are reallocated to cater to the demand with increased autonomy, digital connection, and schedule flexibility (more job resources) and physical commuting and strict scheduling reducing (lower job demands). Five significant dimensions are operationalized in this study as job resources that possibly impact IT workers' job satisfaction: flexibility, communication effectiveness, productivity, work–life balance, and collaboration (Fig. 1).

The study assumes that these dimensions directly and positively impact employee job satisfaction in hybrid mode by adopting the JD-R framework.

3 Research Gap

There are still several important research gaps about how hybrid work arrangements affect technological experts' job satisfaction in the IT sector. Hybrid work's long-term impacts are still little understood, despite its short-term effects being

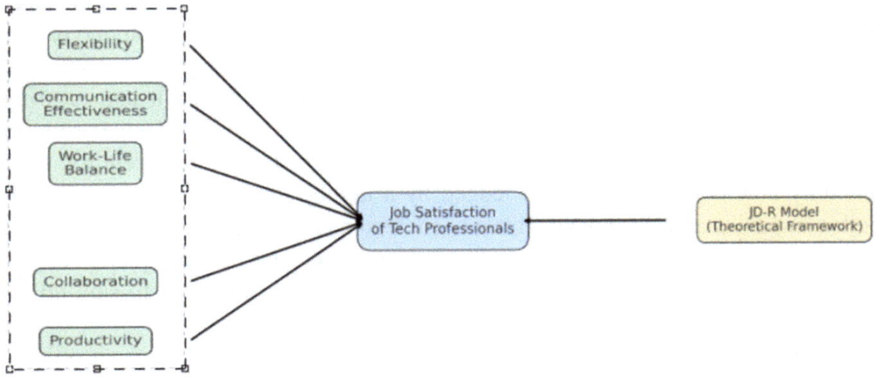

Fig. 1 Conceptual Framework of the Impact of Hybrid Work Model Attributes on Job Satisfaction among Tech Professionals in the IT Industry, Anchored in the Job Demands-Resources (JD-R) Model

extensively recorded. Furthermore, the research currently in publication offers scant understanding of the ways in which hybrid work arrangements affect various demographic groups, including those based on age, gender, and ethnicity. The generalizability of the results is further limited by the conspicuous absence of comparison research across nations or regions. Furthermore, little study has been done on the implications of hybrid work on other job activities within IT, such as project management, software development, and IT support. Despite the fact that work–life balance is often mentioned, more focused research is required to determine how hybrid work models directly influence this balance and, consequently, impact job satisfaction. Similarly, it is yet unknown how perceived productivity, real performance, and work satisfaction relate to one another in hybrid environments. The relationship between company culture and hybrid work arrangements and how it affects employee happiness is another neglected topic.

4 Research Problem

The IT business has changed as a result of the move to hybrid work arrangements, in which workers split their time among employed in the workplace and remote working. However, there is no information on how it impacts the job satisfaction of IT professionals. Flexibility and convenience are two benefits of hybrid employment, but there are drawbacks as well, such loneliness, communication difficulties, and a blurring of work–life boundaries. Depending on how these models are used, tech workers—who are essential to innovation and upkeep of the digital infrastructure may have differing levels of job satisfaction. In-depth research on how different hybrid work arrangements affect job satisfaction across a range of jobs, demographic groups, and organizational cultures in the IT sector is sparse, despite the fact that hybrid work is widely used. Companies must comprehend these factors in order to create hybrid work practices that improve worker happiness, output, and general well-being. By examining the complex consequences of hybrid work on tech professionals' job satisfaction, this study seeks to close this knowledge gap and offer insights that can inform organizational strategies in the rapidly changing workplace.

5 Research Objectives

1. To identify the Hybrid Work Model attributes in IT industry.
2. To measure the impact of Hybrid Work Model attributes on Job Satisfaction of Tech professionals.

6 Research Hypothesis

H01: There is no significant relationship between Hybrid Work Model attributes and Job Satisfaction.

7 Statistical Tools

- Cronbach's Alpha test
- CFA (Confirmatory Factor Analysis)
- Structural Model fit.

8 Sampling Procedure

The study uses data collected from 228 respondents working from three of Bengaluru's top IT companies—Infosys, Cognizant, and IBM using purposing sampling to ensure proper representation from organizations well- known for their hybrid work model. There respondents were accessible and had the experience of working in hybrid work patterns. 228 of the roughly 400 respondents contacted who provided useful responses, resulting a total of 57% response rate. The multi-stage sampling technique was especially useful to maintain practical practicality while ensuring industry relevance. Convenience sampling techniques are likely to limit generalizability, the inclusion of diverse job roles (developers, support, analysts, project managers) and demographic variables (age, gender, education) provides adequate heterogeneity for exploratory analysis. Future studies can be made on these insights by researching in additional areas or stratifying by organization size or specialization.

9 Data Synthesis

9.1 Demographic Breakdown of Sample (228 Respondents)

32% of the 228 IT employees in the sample are female, 3% identify as other or prefer not to reveal their gender, and the bulk of respondents (65%) are male. Only 5% of respondents are 56 years of age or older, with the majority of respondents (40%) being between the ages of 26 and 35, followed by those between the ages of 18 and 25 (25%). Out of all job functions, software developers make up the largest group (35%), followed by IT support and helpdesk positions (25%), project managers (15%), systems analysts (10%), and other IT positions such as quality assurance and network engineers (15%). According to the respondents' experience level, mid-career professionals make up the majority; 40% have 4–7 years of experience, and 25% have 8–12 years. Additionally, 20% of the sample has 0–3 years of experience, while 15% has 13 + years. The vast majority of responders (80%) live in cities, while 20% are concentrated in rural regions. Regarding education, 10% of the respondents have a Ph.D. or equivalent credentials, 40% have a master's degree, and 50% have a bachelor's degree (Table 1).

10 Cronbach's Alpha Test

Table 2 shows that the coefficient of Cronbach's alpha for an 18 scale items is 0.899, suggesting a relatively high level of core reliability. This indicates that the items consistently evaluate a common underlying idea or feature. A number of 0.899 means that your scale has sufficient core reliability, since values of Cronbach's alpha values more than 0.70 are often considered satisfactory for research purposes (Table 3).

11 Hybrid Work Model—Confirmatory Factor Analysis

11.1 Factor Loadings

Factor loadings, which vary from -1 to 1, show how closely an underlying variable and an observable variable are related. A loading of 1 implies a strong advanced connotation, whereas a loading of -1 shows a perfect pessimistic connection. There is no association between the variables when the loading is 0. Hair et al. (1998) state that loadings above 0.5 often recommend a modest to tough association, whereas loadings <0.5 could suggest a less connection, perhaps pointing to problems such as well-defined component or an observable variable that does not adequately reflect the underlying notion. This criterion serves as a standard by which to measure the degree of correlation between experiential and latent variables. A route model is used to assess the factor loadings of variables in relation to production problems and company performance, as illustrated in Fig. 2

Table 1 Demographic Characteristics of the IT employees

Demographic category	Category	Percentage (%)	Number of respondents
Gender	Male	65	148
	Female	32	73
	Other/Prefer not to disclose	3	7
Age	18–25 years	25	57
	26–35 years	40	91
	36–45 years	20	46
	46–55 years	10	23
	56 + years	5	11
Job function	Software Developers	35	79
	IT Support/Helpdesk	25	57
	Project Managers	15	34
	Systems Analysts	10	23
	Other IT roles (Network Engineers, QA)	15	35
Experience level	0–3 years	20	46
	4–7 years	40	91
	8–12 years	25	57
	13 + years	15	34
Location	Urban	80	182
	Rural	20	46
Educational background	Bachelor's Degree	50	114
	Master's Degree	40	91
	PhD/Other	10	23

Table 2 Summary of case processing

		N	%
Cases	Valid	228	100.0
	Excluded[a]	0	0.0
	Total	228	100.0

Table 3 Reliability statistics

Alpha	Number of items
.899	18

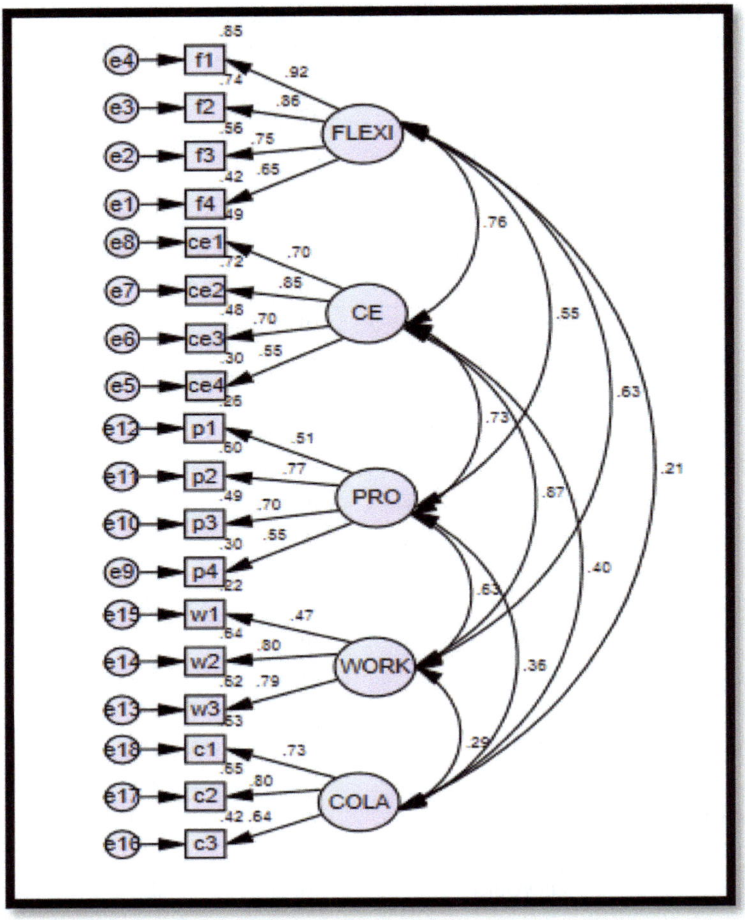

Fig. 2 Path Model- HWM coefficient

12 Expansion of Selected Variables

FLEXI (Flexibility), CE (Communication Effectiveness), PRO (Productivity), WORK (Work–Life Balance) and COLA (Collaboration) (Table 4).

Degrees of Freedom and Chi-Square: An important value of $\chi2$ ($p < 0.001$) advises that the model may not be a perfect match to the data. But, $\chi2$ is profound to sample size, and when the sample size is large, even little variations can significantly affect the findings. The degrees of freedom (125), however, imply a somewhat complex model. The $\chi2$/Df: Given that values around 2 or less are frequently suggestive of a solid match, a score of 4.044 suggests that the model has significant space for improvement. This is to be expected, though, given the complexity of the models and the magnitude of the samples. A model's fit to a

Table 4 Standardized weights of regression

Code	Statements	Estimate	Removed statement	Factor name
f4	Hybrid work model allows me to manage my work schedule more effectively	0.648		Flexibility
f3	The hybrid work model gives me more control over my workplace, in my opinion	0.749		
f2	The flexibility provided by the hybrid work model improves my overall job satisfaction	0.862		
f1	The hybrid work model accommodates my personal needs better than a fully in-office model	0.924		
ce4	The hybrid work model helps me maintain a healthy work–life balance	0.548		Communication Effectiveness
ce3	With the hybrid work paradigm, I can more easily manage my personal and professional obligations	0.695		
ce2	Working in a hybrid model reduces my stress levels related to commuting	.849		
ce1	Because of the hybrid work arrangement, I am able to spend more time with my loved ones	0.700		
p4	My productivity has increased since adopting the hybrid work model	0.549		Productivity
p3	The hybrid work model allows me to focus better on my tasks	0.703		
p2	I am able to meet my deadlines more efficiently with the hybrid work model	0.773		
p1	I feel more motivated to perform my tasks in hybrid work environment	0.512		

(continued)

Table 4 (continued)

Code	Statements	Estimate	Removed statement	Factor name
w3	I am able to keep a good work–life balance because to the hybrid work paradigm	.786		Work–Life Balance
w2	In my experience, the hybrid work paradigm makes it simpler to manage my personal and professional obligations	0.798		
w1	I'm able to spend more time with my loved ones thanks to the hybrid job approach	0.467	Yes	
c3	I find it easy to communicate with my colleagues while working in a hybrid model	0.645		Collaboration
c2	The hybrid work model does not negatively impact my ability to stay informed about work-related matters	0.804		
c1	The hybrid work model enhances my ability to participate in team meetings and discussions	0.725		

baseline model is evaluated using the TLI and the CFI; values closer to 1 indicate a better match. The TLI of 0.773 and CFI of 0.815 show a decent match, but not extraordinary. RMSEA score of 0.116 specifies an excellent match, whereas numbers higher than 0.10 often imply that the model needs to be improved. Overall, there is potential for development even though the Five Factors Model fits the data somewhat. The model is statistically significant, according to the chi-square test, suggesting that adjustments or modifications might enhance its fit. Depending on the situation and particular objectives of the study, the TLI, CFI, and RMSEA indicate that even though the fit is not perfect, it could still be deemed acceptable. It might be helpful to investigate different models or develop the current model more.

13 Structured Model Assessment

Deciphering the complex interactions between hidden qualities within a theoretical framework is the aim of structural equation modeling (SEM), and evaluating the structural model is a crucial stage in this process. The exploration of pathways or coefficients—channels that convey the causal linkages between different latent variables—is the main emphasis of this research. These coefficients not only indicate the direction of a latent variable's impact on another, but also its degree of influence. Researchers may better grasp the difficult web of straight and subsidiary things between hidden variables by using this in-depth study, which aids in their comprehension of the dynamics at work (Fig. 3; Table 6).

Based on the SEM fit indices displayed in Table 5, there are contradictory results about the model's overall fit. The CMIN/df value of 3.745, which falls within the recommended range of 3–5, indicates an acceptable match for this criterion. The p-value of 0.000, however, is below the recommended cutoff of 0.05, indicating that the model lacks statistical significance. The GFI and CFI both meet or surpass the suggested value of 0.90 with perfect scores of 1.000, implying a very strong match. Furthermore, the positive fit indices are supported by the NFI, which likewise scores 1.000, above the recommended limit. In contrast, the Adjusted Goodness of Match Index (AGFI) shows a good match in this region with a score of 0.228, significantly below the suggested criterion of 0.80. The RMSEA of 0.040, which is significantly less than the 0.08 maximum suggested values, suggests that this criterion is well met. Overall, the weak AGFI and statistically insignificant p-value point to possible areas of concern, indicating that more model refinement may be required, even while several indices (such as the GFI, CFI, NFI, and RMSEA) show a satisfactorily model fit (Table 7).

- There is a 0.032 chance of obtaining a critical ratio as high as 1.868 in absolute value. In other words, the regression weight for flexibility in predicting job satisfaction deviates significantly from zero at the two-tailed 0.05 level.

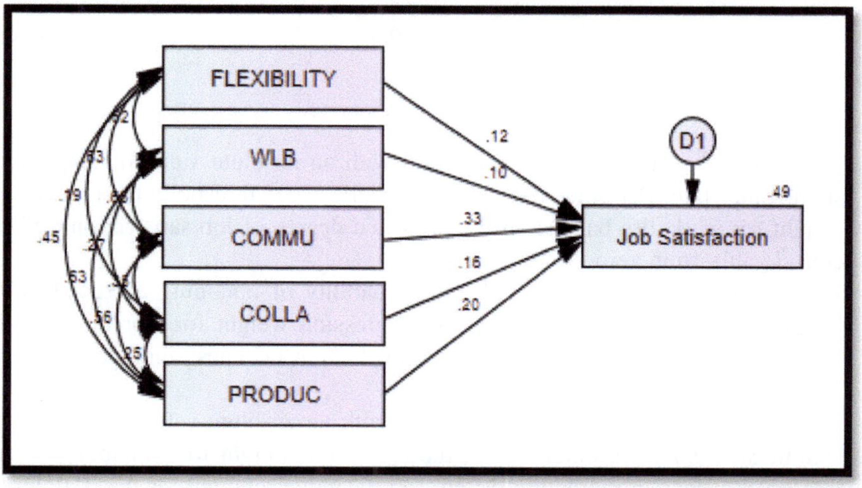

Fig. 3 A SEM that illustrates how the characteristics of the HWM impact IT employees' job happiness

Table 5 CFA of alternate model

Model	Chi-Square	DF	Chi-Square /DF	TLI	CFI	RMSEA
Five Factors	5.589	125	4.044	0.773	0.815	0.116

Source Primary data through questionnaire

Table 6 Estimating the fit of a SEM

Indices	Suggested value	Model fit indices
CMIN/Df	3–5	3.745
P-value	>0.05	0.000
GFI	>0.90	1.000
AGFI	>0.80	0.228
NFI	>0.90	1.000
CFI	>0.90	1.000
RMSEA	>0.08	0.040

Source Primary data through questionnaire

Table 7 Weights of regression

Dependent Variable		Independent Variables	Estimate	Std. Error	Critical Ratio	Sig	Remark
Job Satisfaction	<---	FLEXIBILITY	0.093	0.050	1.868	0.032	Rejected
	<---	WLB	0.081	0.055	1.454	0.046	Rejected
	<---	COMMU	0.267	0.059	4.558	0.000	Rejected
	<---	COLLA	0.131	0.042	3.102	0.002	Rejected
	<---	PRODUC	0.170	0.051	3.307	0.000	Rejected

Source Primary data through questionnaire

- The potential of reaching a crucial ratio with an absolute value of 1.454 and significance 046. Put otherwise, at the two-tailed 0.05 threshold, the regression weight for work–life balance in the projected degree of job satisfaction differs considerably from zero.
- With an absolute value of 4.558, the probability of obtaining a crucial ratio is less than 0.001. In other words, the regression weight for Communication Effectiveness in the prediction of Job Satisfaction deviates significantly from zero at the 0.001 level (two-tailed).
- The probability of getting a critical ratio with an absolute value of 3.102 and significance 0.002. In other words, the regression weight for Collaboration in the Job Satisfaction prediction differs significantly from zero at the 0.01 level (two-tailed).
- The probability of reaching a critical ratio with an absolute value as high as 3.303 is less than 0.001. In other words, the regression weight for productivity

in the prediction of work satisfaction deviates significantly from zero at the 0.001 level (two-tailed).

While some SEM fit indices (e.g., GFI, CFI, NFI = 1.000; RMSEA = 0.040) show a **strong or excellent model fit**, other indices such as AGFI (0.228) and a significant chi-square ($p < 0.001$) advocate that the model may require further refinement. It is significant to note that **chi-square is highly sensitive to sample size**, and minor discrepancies can unnecessarily result in statistical significance in large datasets [24].

Moreover, the low AGFI score indicates that model parsimony may be an issue. Therefore, we interpret the overall model fit as **marginally acceptable**, with strong localized relationships (as shown by significant regression weights), but scope for improving structural consistency. These limitations are acknowledged, and future iterations may incorporate latent interaction terms or refine item loadings for improved model precision.

14 Practical Implications

- Reduces commuting time and costs, providing the option to work from preferred environments.
- Balancing professional and personal responsibilities becomes easier with flexible work arrangements.
- More intentional planning of meetings can lead to more focused and productive discussions.
- Combination of in-person and virtual meetings can ensure inclusive participation from all team members.
- Strong time management and self-discipline are essential for remote work in order to prevent procrastination and sustain productivity.

15 Discussion: Linking Findings to Prior Research

In line with earlier research by [19], which emphasized increased employee satisfaction and productivity with flexible work arrangements, the study's findings indicate that the hybrid work model has a significantly positive impact on the job satisfaction of tech professionals in the IT industry. Respondents cited improved autonomy, work–life balance, and less stress from commuting as major sources of pleasure; these elements are also highlighted in research by [20, 21], which discovered that aspects of remote work have a favorable impact on employee attitudes and performance. Additionally, our results are consistent with recent post-pandemic research [22], which suggests that by encouraging involvement and lowering burnout, the hybrid model overcomes the drawbacks of both totally remote and in-office arrangements. It's interesting to note that this study deviates from previous worries [23] that working remotely could impede teamwork and career progression, indicating that modern digital collaboration tools and agile management techniques have reduced these dangers in the tech industry. Overall, the findings support the growing belief that, in knowledge-intensive businesses, hybrid work arrangements can be deliberately used to increase employee satisfaction.

16 Conclusion

The IT industry's IT professionals' job happiness has been greatly influenced by the hybrid work paradigm. One important advantage is flexibility, which enables workers to modify their schedule and work environment to better suit their individual requirements and raises job satisfaction levels overall. Because professionals are better able to steadiness their job and individual obligations, this flexibility helps to promote work–life balance. However, communication effectiveness can be challenged in a hybrid setting, requiring robust tools and strategies to maintain clear and consistent interaction among team members. Despite these challenges, collaboration can be enhanced through the combination of in-person and remote work, leveraging the strengths of both environments. Productivity has shown a positive trend, as employees can choose their most productive work settings. Overall, the hybrid work model supports higher job satisfaction by offering greater flexibility, improving work–life balance, and maintaining effective collaboration and productivity, albeit with a need for enhanced communication strategies.

References

1. Choudhury, P., Foroughi, C., & Larson, B. Z. (2021). Work-from-anywhere: The productivity effects of geographic flexibility. *Strategic Management Journal, 42*(4), 655–683.
2. Gallup. (2022). State of the global workplace: 2022 Report. Gallup, Inc.
3. Bloom, N., Han, R., & Liang, J. (2022). How hybrid working from home works out. *National Bureau of Economic Research*, Working Paper No. 30292.
4. Microsoft. (2022). The work trend index: annual report. Retrieved from https://www.microsoft.com/en-us/worklab/work-trend-index.
5. Waizenegger, L., McKenna, B., Cai, W., & Bendz, T. (2020). An affordance perspective of team collaboration and enforced working from home during COVID-19. *European Journal of Information Systems, 29*(4), 429–442.
6. Bellmann, L., & Hübler, O. (2020). *Job satisfaction and work-life balance: Differences between homework and work at the workplace of the company* (IZA Discussion Paper No. 13504). Institute of Labor Economics (IZA).
7. Haridas, P., Rahul, P. R., & Subha, K. (2021). Impact of work from home model on the productivity of employees in the IT industry. *International Journal of Innovative Research in Technology, 8*(2), 662–666.
8. Ravalier, J. M., & Boichat, C. (2018). UK social workers: Working conditions and wellbeing. British Association of Social Workers.
9. Singh, R. (2019). The impact of telecommuting on employees' productivity and professionalism. *IOSR Journal of Business and Management, 19*(5), 17–19.

10. Lister, K., & Kamouri, A. (2020). Global work-from-home experience survey. Global Workplace Analytics.
11. Bellmann, L., & Hübler, O. (2021). Working from home, job satisfaction and work–life balance—robust or heterogeneous links? *International Journal of Manpower, 42*(3), 424–441.
12. Lister, K., & Kamouri, A. (2020). Global work-from-home experience survey report. Global Workplace Analytics.
13. Mungkasa, O. (2020). Bekerja dari Rumah (Working From Home/WFH): Menuju Tatanan Baru Era Pandemi COVID-19. *Jurnal Perencanaan Pembangunan: The Indonesian Journal of Development Planning, 4*(2), 126–150.
14. Prodanova, J., & Kocarev, L. (2022). The role of hybrid-working in improving employees' satisfaction, perceived productivity, and organizations' capabilities. *Journal of Theoretical and Applied Management, 208*, 1–15.
15. Yang, E., Kim, Y., & Hong, S. (2023). Does working from home work? Experience of working from home and the value of hybrid workplace post-COVID-19. *Journal of Corporate Real Estate, 25*(1), 50–76.
16. Jaiswal, A., & Arun, C. J. (2024). Working from home during COVID-19 and its impact on Indian employees' stress and creativity. *Asian Business & Management, 23*(2), 312–336.
17. Yang, J., Kim, H., & Hong, S. (2023). Workplace adaptability and organizational support in remote work settings. *International Journal of Human Resource Studies, 13*(1), 101–118.
18. Anderson, A. J., Kaplan, S. A., & Vega, R. P. (2015). The impact of telework on emotional experience: When, and for whom, does telework improve daily affective well-being? *European Journal of Work and Organizational Psychology, 24*(6), 882–897.
19. Allen, T. D., Golden, T. D., & Shockley, K. M. (2013). How effective is telecommuting? Assessing the status of our scientific findings. *Psychological Science in the Public Interest, 16*(2), 40–68. https://doi.org/10.1177/1529100615593273
20. Bailey, D. E., & Kurland, N. B. (2002). A review of telework research: Findings, new directions, and lessons for the study of modern work. *Journal of Organizational Behavior, 23*(4), 383–400. https://doi.org/10.1002/job.144
21. Bloom, N., Liang, J., Roberts, J., & Ying, Z. J. (2015). Does working from home work? Evidence from a Chinese experiment. *The Quarterly Journal of Economics, 130*(1), 165–218. https://doi.org/10.1093/qje/qju032
22. Gajendran, R. S., & Harrison, D. A. (2007). The good, the bad, and the unknown about telecommuting: Meta-analysis of psychological mediators and individual consequences. *Journal of Applied Psychology, 92*(6), 1524–1541. https://doi.org/10.1037/0021-9010.92.6.1524
23. Wang, B., Liu, Y., Qian, J., & Parker, S. K. (2021). Achieving effective remote working during the COVID-19 pandemic: A work design perspective. *Applied Psychology, 70*(1), 16–59. https://doi.org/10.1111/apps.12290
24. Hair, J. F., Black, W. C., Babin, B. J., & Anderson, R. E. (2010). Multivariate data analysis (7th ed.). Pearson Education.

HR Analytics in Predicting Employee Turnover in IT Industry: A Structural Equation Modeling Approach

G. Ramanjaneyulu, Somasekhar Donthu, B. Venkata Lakshmi, V. Mouneswari, Y. Mallikarjuna Achari, and S. Md. Ershad

Abstract

This study examines at how HR analytics may be used to forecast employee turnover in the IT sector. It focuses on five important factors: leadership support, HRM competency, data availability, technological infrastructure, and change management. Data was gathered from 255 HR professionals and IT company workers using a standardized questionnaire using a 5-point Likert scale format. AMOS was used to analyze the data in order to evaluate the predicted correlations among the determinants and employee turnover as well as test the model fit. According to the results, all five criteria have a considerable impact on turnover, but the effects of Leadership Support and Data Availability are more noticeable. With indices like CMIN/df, GFI, AGFI, CFI, and RMSEA falling within suggested bounds, the model showed a satisfactory match. The findings highlight how crucial it is to include data-driven tactics into HR procedures in order to accurately predict and lower employee turnover. For IT

G. Ramanjaneyulu (✉)
Department of MBA, Annamacharya Institute of Technology and Sciences, Kadapa, India
e-mail: g.ramanjaneyulu.mba@gmail.com

S. Donthu
School of Business, GITAM University Bangalore, Bangalore, India

B. Venkata Lakshmi
Department of Humanities and Sciences, KSRM College of Engineering (A), Kadapa, India

V. Mouneswari
School of Management, Annamacharya University, Rajampeta, India

Y. Mallikarjuna Achari
Department of MBA, RGM College of Engineering and Technology, Nandyal, India

S. Md. Ershad
Department of MBA, Santhiram Engineering College, Nandyal, India

S. Patnaik et al. (eds.), *People, Planet and Profit: Advancing Sustainable Business Practices*, Springer Proceedings in Geography and Sustainability,
https://doi.org/10.1007/978-3-032-11296-5_19

companies looking to improve employee retention through the strategic use of HR analytics, the report provides useful takeaways.

Keywords

Data availability • Technological infrastructure • Leadership support • HRM competency • Change management

1 Introduction

Human resource analytics, or HR analytics, has become a crucial strategic tool to handle important personnel concerns in today's dynamic economic environment. Employee turnover is one of the biggest issues facing businesses since it can result in expensive hiring and training expenses, the loss of organizational knowledge, and a decline in morale. Conventional attrition management techniques are frequently impulsive and lack the ability to predict needed to implement changes on time. HR analytics, sometimes referred to as workforce analytics or people analytics, employs data-driven methods to guide HR choices. Organizations can identify risk indicators, anticipate staff attrition, and put proactive retention plans into place by using statistical models and historical employee data. Patterns in tenure, performance, pay, satisfaction with work, and other important factors that affect turnover can be found using predictive analytics. Organizations that successfully apply HR analytics have a major advantage in keeping valuable people, cutting expenses, and enhancing organizational effectiveness as the race for talent heats up. This study investigates the use of HR analytics methods and technologies to improve HR decision-making and forecast employee turnover.

2 Literature Review

HR analytics has advanced from just monitoring HR indicators to producing actionable insights via predictive modeling. Their study highlighted the tactical benefit of employing predictive analytics as opposed to more conventional descriptive techniques. They contend that in order to identify which employees are most likely to quit and why, HR departments need to use forecasting tools and algorithms rather than just historical trend research. In order to enable prompt and focused responses, their methodology urges HR practitioners to concentrate on predictive criteria including manager evaluations, engagement scores, and global job market trends [1].

The elements that contribute to turnover into two categories: organizational and individual. At the individual level, factors including perceived fairness, career advancement, organizational commitment, and work happiness are important. Workplace culture, HR procedures, and leadership caliber are examples of organizational variables. According to the authors, turnover is frequently the result of several minor dissatisfactions compounding over time, therefore prediction models must take a multifactorial approach. Several HR analytics approaches that currently use organizational data and employee input to forecast attrition are based on their work [2].

The idea of work embeddedness as a predictor of employee turnover was first presented by [3]. Job embeddedness, in contrast to more conventional indicators like commitment and satisfaction, measures how much workers experience a sense of belonging to their coworkers, their job duties, and the larger community. In order to create prediction models that might accurately identify "at-risk" personnel and combined demographic information, performance at work, and organizational tenure. According to their research, employees who scored poorly on embeddedness were far more likely to quit, hence it's imperative that HR analytics frameworks include these characteristics.

In order to forecast employee turnover in Korean companies, [4] conducted an empirical research using logistical regression and data mining methods (such as support vector machine learning and decision trees). When compared to conventional HR procedures, the scientists discovered that data-driven techniques greatly improved prediction accuracy. The salary, duration of assistance, and promotion history, and absence rates were important indicators. Crucially, they emphasized that models that were personalized to the sector and organizational context had the best predictive accuracy, highlighting the necessity of bespoke analytics solutions above generic models.

Artificial intelligence in HR analytics assessed the efficacy of many algorithms, including decision trees, random forests, and naïve Bayes, in predicting turnover. Their results demonstrated that these models were capable of predicting retention with reliability rates of more than 85% when trained on complete, high-quality personnel datasets. For real-time monitoring, the research suggested incorporating such algorithms into information systems for human resources. Additionally, they pointed out that elements like management evaluations, peer reviews, and staff satisfaction surveys were crucial in improving model performance [5].

Trends in the expanding significance of employee analytics in contemporary enterprises were highlighted in the Deloitte study. According to the study, 71% of businesses ranked people analytics as their top HR importance, with turnover prediction as their main area of concentration. According to the survey, conventional HR roles are giving way to more strategic, data-centric roles where HR specialists are required to analyze complicated data and counsel executives on personnel risks. The research also covered the incorporation of powered by AI technologies into HR systems, which allow for automatic and ongoing attrition risk monitoring [6].

The use of HR analytics in the Indian IT sector, particularly has traditionally seen significant attrition rates, was examined in a recent paper by [7]. By examining factors like tardiness, the degree of internal grievances, and performance declines, Jain and Singh discovered that machine learning models assisted in the early identification of employees who were disengaged. Their results also shown that prediction results were enhanced by integrating external sources (such as Glass Door reviews and LinkedIn activity) with internal data (such as HRIS, PMS, and LMS). Interestingly, they discovered that companies who implemented analytics-based retention tactics had a 15–20% decrease in yearly turnover rates.

The growing quantities of workforce information and improvements in analytical tools have made predictive HR analytics a potent organizational capacity. Their research demonstrates how predictive models use historical trends in employee behavior, including tenure, performance reviews, and promotions, to find patterns linked to voluntary departures. By doing this, HR managers may identify workers who are more likely to leave and put focused retention plans in place [8].

The significance of including important HR variables into predictive models, including employee engagement ratings, absence rates, performance reviews, and training attendance, is emphasized by [9]. Their study emphasizes how important these indicators are to comprehending the root reasons of turnover. Poor levels of engagement or frequent absences, for example, may serve as early warning indicators that, when examined over time, might provide useful information to stop the loss of talent.

In order to improve the prediction power of turnover models, [10] investigates the combination of algorithms for machine learning with HR data. Their results show that by capturing intricate, non-linear interactions across HR variables, methods like random forests, decision trees, and neural networks may much outperform conventional statistical methods. This leads to more precise forecasts and enables HR managers to react with tailored interventions—from workload modifications to individualized career development plans—before attrition happens.

Implementing data-driven HR initiatives is crucial for improving workforce stability, according to [11]. According to their research, voluntary leave rates are lower in companies that use predictive analytics as a component of their talent management strategy. In order to enhance evidence-based decision-making, these businesses frequently utilize real-time dashboards and ongoing monitoring of predictive indicators. In addition to reducing turnover expenses, this proactive strategy boosts organizational commitment and employee morale.

3 Research Gap

Even if HR analytics is becoming more and more popular in businesses, little is known about how to use it to forecast employee churn, especially in the fast-paced world of IT firms. Without providing insights particular to a given business, the majority of current research concentrates on generic workforce analytics or staff retention tactics. Additionally, although a number of models and methods have been put out to anticipate turnover, little is known about their efficacy and practical use in the fiercely competitive and fast-paced IT industry. Empirical studies that look at how HR departments might employ sophisticated analytics, such as machine learning, modeling for prediction, and big data, to anticipate turnover trends and create proactive retention strategies are scarce. In the environment of IT organizations, where high staff turnover rates continue to pose a barrier to organizational development and stability, this research gap highlights the need for targeted studies that evaluate the predictive capacity of HR analytics.

4 Research Problem

For IT firms, employee turnover has become a major problem that frequently leads to large losses in organizational expertise, productivity, and hiring expenses. Conventional approaches to turnover management have mostly been reactive, neither anticipating future attrition nor adequately addressing the root reasons. There is an increasing chance to switch from proactive to predictive techniques with the introduction of sophisticated HR analytics. Nevertheless, a lot of IT companies find it difficult to use data-driven insights to promptly identify high-risk individuals and predict attrition patterns. By looking at important workforce data, finding trends and predictors, and evaluating the precision and usefulness of prediction models, this study aims to investigate how HR analytics may be used to accurately forecast employee turnover in IT companies. The goal of the project is to close the knowledge gap on how strategic HR decision-making may be supported by predictive analytics technologies to improve employee retention and lower turnover costs.

5 Research Objectives

1. To identify the HR analytics determinants in IT industry.
2. To measure the effect of HR analytics determinants on employee turnover in IT industry.

6 Statistical Tools

The research will use a variety of statistical approaches to guarantee the quality and robustness of the data gathered. The internal validity of the survey responses will first be evaluated using a reliability test, especially Cronbach's Alpha, to make sure the scale is trustworthy for gauging important concepts linked to employee turnover and HR analytics. The measurement model will then be validated using Confirmatory Factor Analysis (CFA), which verifies the factor structure and assesses how well the observed variables reflect the underlying latent constructs, including turnover intention, the use of predictive analytics, and the alignment of HR strategy. Finally, the proposed model of how HR analytics affects forecasting and supervision of employee turnover in IT organizations will be tested, and the structural links between the components will be examined using Structural Equation Modeling (SEM). Together, these resources strengthen the validity of the study's conclusions and offer a thorough framework for analyzing data.

7 Sampling Procedure

The focus of this study is IT companies in Bengaluru City, one of India's most well-known and internationally renowned information technology hubs. The purpose of the study is to investigate how these firms use Human Resource (HR) analytics to forecast and control employee turnover. A stratified random sample technique will be used to guarantee the representativeness and dependability of the results because IT organizations vary widely in terms of size, resources, and organizational procedures.

In order to account for differences in organizational dynamics and HR strategies, all IT firms will be divided into three groups: small, medium, and large businesses. The inherent variations in scale, structure, and technology use across businesses of different sizes are recognized by this classification. Specific respondent groups—workforce analytics specialists, HR department personnel, and employee engagement specialists—will be identified from each stratum and chosen at random to take part in the survey. These groups were selected because they have firsthand experience developing and implementing HR strategies, are knowledgeable about employee turnover patterns, and are proficient with predictive HR technologies.

The goal sample size will be 256 responders in total. About 85–86 people would be selected from each firm category to provide equal representation, reducing sample bias and improving the results' applicability to Bengaluru's larger IT industry. The study can successfully compare the procedures and results of small, medium, and large enterprises because of this even distribution across strata.

8 Data Synthesis

8.1 Reliability Test

The study's measuring scale has a high degree of internal consistency and reliability, as evidenced by its Cronbach's Alpha rating of 0.825 for the 15 items. While a number above 0.80 indicates high dependability, a value above 0.70 is often regarded as satisfactory. This shows that the questionnaire's components assess the underlying construct—in this example, the factors that influence HR analytics' ability to forecast employee turnover—consistently and with good correlation. As a result, the scale is considered trustworthy for additional statistical interpretation and analysis (Table 1).

Table 1 Reliability statistics

Cronbach's alpha	Number of items
0.825	15

9 CFA of HR Analytics Determinants

See Fig. 1.

10 Variables Expansion

Data Availability (DAT), Technological Infrastructures (TEC), Leadership Support (LEAD), HRM Competency (HR), and Change Management (CHAN) (Table 2).

The five-factor model shows a satisfactory fit to the data according to the model fit indices that are presented. With 80 degrees of freedom, the χ^2 value is 257.884,

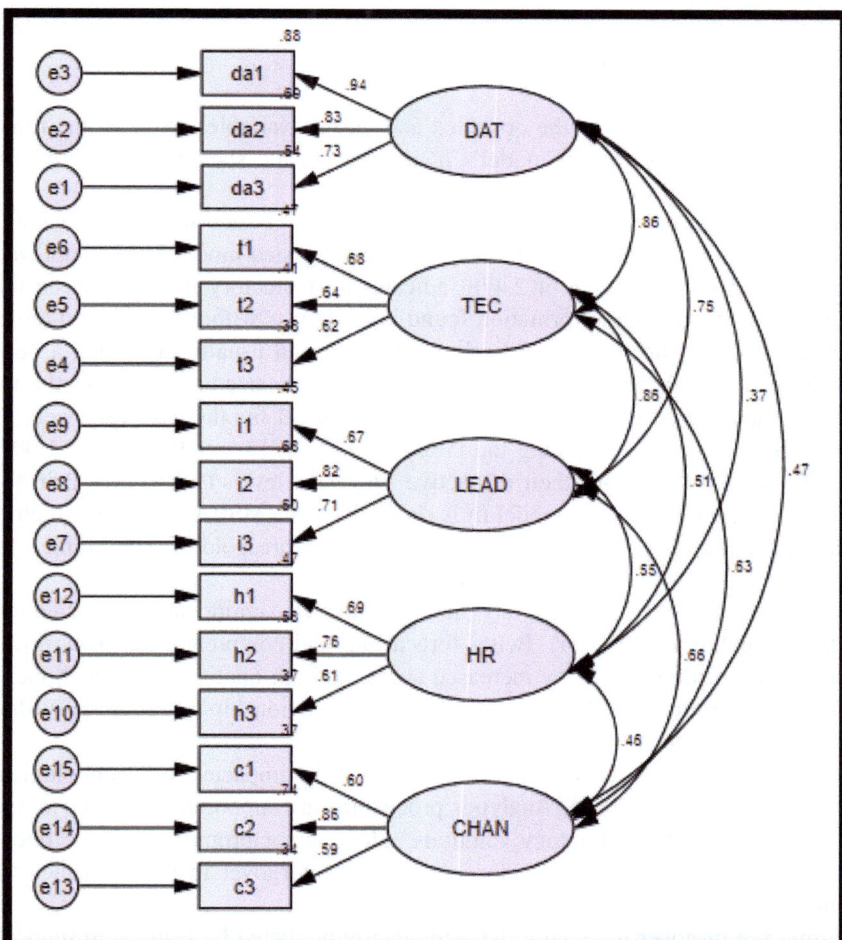

Fig. 1 Measurement model of HR analytics determinants

Table 2 CFA of alternative models

Model	χ2	Df	χ2/Df	TLI	CFI	RMSEA
Five-factor	257.884	80	3.223	0.912	0.895	0.082

yielding a χ^2/df of 3.223 that is within the permissible range (<5), suggesting a reasonably excellent fit. The model's fit may be enhanced, as shown by the TLI of 0.912 and the CFI of 0.895, which are both above the generally recognized cutoff of 0.90. Furthermore, a moderate fit is indicated by the RMSEA of 0.082, which is just higher than the suggested maximum of 0.08. Overall, although though the five-factor model approximates the data quite well, it can still be improved and future model re-specification or refinement might increase the model's fit (Table 3).

11 Structural Model Fit

The graphic illustrates how the observed and unseen variables relate to the route diagram.2. The measurement model's fit indices are the standardized regression coefficients (Fig. 2; Table 4).

The majorities of the indicators reach or surpass suggested thresholds, according to the model fit indices, which show that the suggested model fits the data well overall. The CMIN/df value of 2.478 indicates a satisfactory match between the suggested model and the information found because it is within the allowed range (less than 3). Despite being statistically significant and usually indicating a poor fit, the p-value of 0.000 is extremely variable to sample size and is frequently not the only metric used to evaluate the model. The model fits the data very well, as seen by key fit indices including the GFI = 1.000, AGFI = 0.93, NFI = 1.000, and CFI = 1.000 all above their respective suggested levels (>0.90 or > 0.80 for AGFI). A decent estimate of model fit is shown by the RMSEA of 0.063, which is within a suitable range but more beyond the optimal threshold of 0.08 (Table 5).

- Employee turnover is positively and statistically significantly impacted by data availability ($p < 0.05$). Better forecasting and comprehension of turnover trends are made possible by increased access to high-quality data, which facilitates more efficient HR decision-making. The relationship's strength is further supported by the strong CR value.
- Employee turnover is strongly and significantly impacted by technological infrastructure ($p < 0.001$). Analytics procedures are supported by effective and contemporary HR technology solutions, allowing for prompt employee retention actions. It is a crucial factor in controlling turnover in IT companies, as indicated by the high estimate and CR value.
- Employee turnover is found to be significantly predicted by leadership support. With the greatest estimate of any variable, it demonstrates how strongly staff retention tactics are impacted by leadership's active participation and support of

Table 3 Standardized regression weights

Code	Item	Estimate	Factor name
da3	The data available for HR analytics is accurate and reliable	0.734	Data Availability
da2	Historical HR data is easily accessible when needed	0.831	
da1	There is a clear process for collecting and storing employee performance data	0.941	
t3	Our organization has advanced HR information systems to support analytics	0.619	Technological infrastructures
t2	The tools used for HR analytics are user-friendly and efficient	0.644	
t1	Real-time data visualization tools (e.g., dashboards) are available to HR staff	0.682	
i3	Top management supports the use of HR analytics in decision-making	0.707	Leadership support
i2	Leadership encourages a data-driven culture within the HR function	0.823	
i1	HR analytics insights are considered in strategic business decisions	0.674	
h3	Our HR team has adequate skills to interpret and use analytical data	0.610	HRM competency
h2	HR professionals are regularly trained in analytics tools and techniques	0.759	
h1	HR staffs are confident in applying analytics for workforce planning	0.689	
c3	Employees are receptive to adopting data-driven HR practices	0.586	Change management
c2	There is clear communication during changes related to HR systems or processes	0.862	
c1	Resistance to using HR analytics tools is minimal among HR professionals	0.605	

the use of analytics. This demonstrates how senior management plays a strategic role in advancing analytics-based HR procedures.

- Employee turnover is highly impacted by HRM competency ($p < 0.01$). Competent HR specialists who can successfully use and understand analytics are better equipped to recognize turnover issues and implement remedial measures. Therefore, one of the main factors influencing the use of analytics to lower attrition is the proficiency of HR professionals.
- Additionally, there is a substantial and statistically significant correlation between change management and employee turnover. Talent retention is more successful for companies that facilitate transitions and help staff members adjust

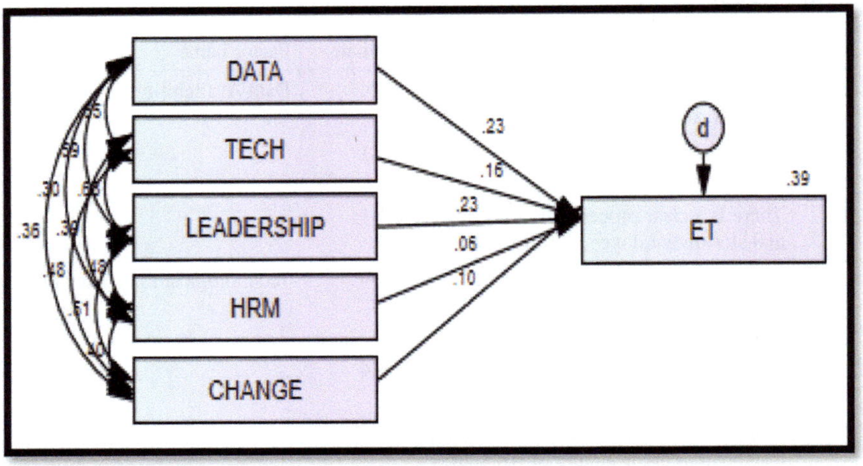

Fig. 2 SEM representing HR analytics determinants on employee turnover

Table 4 Structural model fit estimation

Indices	Recommended value	Model fit indices
CMIN/Df	<3	2.478
P-value	>0.05	0.000
GFI	>0.90	1.000
AGFI	>0.80	0.93
NFI	>0.90	1.000
CFI	>0.90	1.000
RMSEA	>0.08	0.063

Table 5 Variances'

Dependent	HR analytics	Estimate	S.E	CR	P	Label
Employee turnover	Data availability	1.317	0.117	11.269	0.021	Rejected
	Technological infrastructures	1.262	0.112	8.627	0.000	Rejected
	Leadership support	1.325	0.118	9.235	0.000	Rejected
	HRM competency	1.244	0.110	7.246	0.001	Rejected
	Change management	1.299	0.115	7.541	0.000	Rejected

to change. HR analytics may be used successfully and without opposition if change management is done well.

11.1 Discussion: Linking Findings to Prior Research

The current study's findings highlight how important HR data are for forecasting employee churn in the IT sector, especially when examined using a Structural Equation Modeling (SEM) approach. Our results confirm that data-driven HR strategies help to better predict turnover trends, which is in line with other studies [12, 13]. This study backs up the claim made by [14] that companies may proactively handle turnover risks rather than reactively control outcomes by incorporating advanced analytics into HR decision-making. Notably, the characteristics emphasized in the models put forth by [15] and expanded upon in subsequent work by [16] are echoed by the found strong connections between variables including job satisfaction, organizational commitment, and turnover intent. Furthermore, in line with findings from [17], who highlighted that business size frequently corresponds with analytics maturity, our stratified research of small, medium, and large IT firms in Bengaluru shows that larger firms typically have more complex HR analytics systems. Intriguingly, the study also shows that smaller businesses are becoming more conscious of the strategic advantages of predictive analytics, indicating a move away from traditional HR reporting and toward more proactive worker management, in keeping with [18]. These findings add to the expanding corpus of research supporting the strategic use of HR analytics into talent retention and workforce planning plans at all levels of IT organizations.

12 Suggestions

- Make certain that thorough and organized employee data is gathered via departure, engagement, performance, and attendance interviews.
- Use machine learning and artificial intelligence to estimate attrition risk by analyzing past employee data.
- Promote senior management participation in HR analytics projects by tying them to corporate objectives (e.g., lowering expensive turnover).
- Promote cross-functional cooperation between the data science, IT, and HR departments.
- Create a change management plan to progressively deploy HR analytics tools.

13 Conclusion

The study comes to the conclusion that anticipating and controlling employee turnover in the IT sector is greatly aided by HR analytics, which includes elements like data availability, technological infrastructure, leadership support, HRM competency, and change management. The results demonstrate how firms may see early warning indicators of attrition when they successfully gather, integrate, and use employee data, backed by strong IT systems. The strategic application of analytics in workforce scheduling is further strengthened by leadership's dedication to data-driven HR practices. Furthermore, the successful use of analytics-based approaches to retention is guaranteed by the proficiency of HR experts in data interpretation and transition management. Together, these factors enable IT firms to take a proactive stance in reducing voluntary departures, enhancing worker engagement, and promoting organizational stability by making well-informed, fact-based decisions.

References

1. Fitz-enz, J., & Mattox, J. R. (2014). *Predictive analytics for human resources*. Wiley.
2. Hausknecht, J. P., & Trevor, C. O. (2011). Collective turnover at the group, unit, and organizational levels: Evidence, issues, and implications. *Journal of Management, 37*(1), 352–388.
3. Ramesh, A., & Gelfand, M. J. (2010). Will they stay or will they go? The role of job embeddedness in predicting turnover in individualistic and collectivistic cultures. *Journal of Applied Psychology, 95*(5), 807–823.
4. Min, J., Kim, M. J., & Lee, H. (2015). Employee turnover prediction using data mining techniques: A case study of a Korean company. *Journal of Intelligent Information Systems, 44*(3), 619–634.
5. Kaur, P., & Mehta, G. (2018). Machine learning algorithms for predicting employee turnover: A case study. *International Journal of Advanced Research in Computer Science, 9*(2), 101–105.
6. Deloitte. (2020). Global Human Capital Trends: The social enterprise at work—Paradox as a path forward. Deloitte Insights.
7. Jain, R., & Singh, R. (2021). Adoption of HR analytics in Indian IT sector: A study on turnover prediction. *International Journal of Management Studies, 8*(1), 45–55.
8. Banerjee, R., & Kulkarni, S. (2022). Data-driven HR strategies for reducing voluntary employee turnover in the IT sector. *Journal of Human Resource Management, 10*(2), 55–63.
9. Jain, A., & Singh, P. (2023). Predictive HR analytics: Forecasting employee attrition using historical data trends. *International Journal of Human Capital and Information Technology Professionals, 14*(1), 12–25.
10. Sharma, M., & Rao, K. (2022). Leveraging HR metrics to build predictive turnover models: A strategic approach. *Journal of Organizational Analytics, 9*(3), 77–88.
11. Thomas, L., D'Souza, R., & Menon, A. (2023). Machine learning integration in HR analytics for employee turnover prediction. *International Journal of Data Science and HRM, 5*(4), 101–115.
12. Bassi, L. (2011). Raging debates in HR analytics. *People and Strategy, 34*(2), 14–18.
13. Davenport, T. H., Harris, J. G., & Shapiro, J. (2010). Competing on talent analytics. *Harvard Business Review, 88*(10), 52–58.
14. Hausknecht, J. P., & Trevor, C. O. (2011). Collective turnover at the group, unit, and organizational levels: Evidence, issues, and implications. *Journal of Management, 37*(1), 352–388. https://doi.org/10.1177/0149206310383910
15. Levenson, A. (2011). Using targeted analytics to improve talent decisions. *People and Strategy, 34*(2), 34–43.
16. Marler, J. H., & Boudreau, J. W. (2017). An evidence-based review of HR Analytics. *The International Journal of Human Resource Management, 28*(1), 3–26. https://doi.org/10.1080/09585192.2016.1244699

17. Mobley, W. H. (1977). Intermediate linkages in the relationship between job satisfaction and employee turnover. *Journal of Applied Psychology, 62*(2), 237–240. https://doi.org/10.1037/0021-9010.62.2.237
18. Tett, R. P., & Meyer, J. P. (1993). Job satisfaction, organizational commitment, turnover intention, and turnover: Path analyses based on meta-analytic findings. *Personnel Psychology, 46*(2), 259–293. https://doi.org/10.1111/j.1744-6570.1993.tb00874.x

Toward Sustainable Work-Life Balance in the Indian IT Industry: A Predictive Analysis of Challenges, Organizational Initiatives, and Technology-Enabled Interventions

E. Kusuma and A. Amruth Prasad Reddy

Abstract

The dynamics and difficulties of work-life balance (WLB) among workers in the Indian IT industry, which is renowned for its high-performance requirements, international operations, and fast-paced atmosphere, are examined in this study. The study's objectives are to determine the main variables that affect WLB, analyze and contrast the tactics used by top IT companies like Infosys, TCS, and Wipro, and assess how WLB affects employee well-being, organizational commitment, and work satisfaction. The study uses secondary research, drawing information from academic literature, industry reports, NASSCOM and ASSOCHAM polls, and documented HR procedures. The results show that although businesses have implemented remote work policies, flexible scheduling, and wellness programs, their efficacy varies. For instance, Wipro makes investments in digital wellness and management assistance, Infosys concentrates on employee well-being initiatives, while TCS prioritizes structured remote work frameworks. Notwithstanding these initiatives, enduring problems including long workdays, digital exhaustion, and the blurring of work-life balance especially in hybrid environments—remain common. There are also gender differences, with women experiencing more stress from playing two roles. Although this research relies on secondary data, yet critically synthesizes existing literature and provides strategic insights for enhancing work-life balance in

E. Kusuma (✉) · A. A. P. Reddy
Department of Business Administration, Yogi Vemana University, Kadapa, India
e-mail: Kusuma.mba85@gmail.com

A. A. P. Reddy
e-mail: apsreddy14@yvu.edu.in

S. Patnaik et al. (eds.), *People, Planet and Profit: Advancing Sustainable Business Practices*, Springer Proceedings in Geography and Sustainability,
https://doi.org/10.1007/978-3-032-11296-5_20

the Indian IT industry. Further studies may be conducted to enrich our understanding and validate proposed interventions by obtaining primary empirical data via surveys or interviews.

Keywords

Work-life-balance • IT employees • Business process management • Infosys • TCS • Wipro • IBM

1 Introduction

A major contributor to the nation's economic expansion and job creation, the Indian information technology (IT) sector has become a global powerhouse. Over 5.4 million professionals are employed in the IT sector, which also accounts for almost 8% of India's GDP. Although this expansion has created many possibilities, it has also made it more difficult for workers to maintain a good work-life balance (WLB) [1]. The harmony between one's personal and professional obligations that enables one to function well in both areas is known as work-life balance. Due to international customer expectations, workers in the fast-paced and cutthroat IT sector frequently put in long hours, meet deadlines, and provide round-the-clock support. The lines between work and family life have become hazier due to the "always-on" attitude, particularly in companies that serve clients from across the world [2]. With remote and hybrid work becoming commonplace, the COVID-19 epidemic has further changed the nature of the workplace. Although working remotely allows for greater freedom, it has also led to increased digital weariness, longer workdays, and difficulties shutting off from work [3]. It has been more challenging for many IT workers, particularly women and mid-level managers, to strike a balance between work obligations and their personal and family lives.

According to studies, a lack of work-life balance increases stress, burnout, job discontent, and attrition. On the other hand, companies with a balanced work culture report increased productivity, retention, and employee engagement [4]. Given these circumstances, it is both pertinent and crucial to look at the work-life balance policies in the Indian IT industry. In addition to evaluating organizational tactics for fostering work-life balance (WLB) and its effects on employee performance and well-being, this study investigates the factors that influence WLB among Indian IT workers.

2 Background

Over the past few decades, the Indian information technology (IT) sector has expanded rapidly, making a substantial contribution to the GDP, exports, and employment of the country. According to the National Association of Software and

Service Companies [1], the IT and Business Process Management (BPM) industry employed more than 5.4 million people and brought in more than USD 245 billion in sales as of 2023. Notwithstanding this impressive expansion, the sector has also had issues with worker welfare, particularly with regard to work-life balance (WLB). Tight project deadlines, long workdays, coordinating with clients worldwide (across time zones), and the need to continuously provide high-quality services are all common aspects of the IT industry's work environment. Due to the blurring of the lines between personal and professional life, especially after COVID-19, as a result of hybrid and remote work models, WLB has become a significant problem [3]. Employees in many Indian IT organizations are required to work weekends, attend late-night calls, and be available after usual business hours, particularly when assisting clients who are located abroad. The issue has been made worse by the pandemic. According to a 2021 Times of India study, 70% of Indian IT workers said that their work-life balance had gotten worse while they were working remotely since there was no physical boundary between work and home [5]. Workplace stress has also been exacerbated by elements including high attrition rates, fast technological development, industry competition, and job instability. IT workers reported greater levels of stress and burnout than those in other industries, according research published in the Indian Journal of Occupational and Environmental Medicine [6].

The employees' demographic face influences how WLB is seen as well. While mid-career workers and those with families find it more difficult to find a balance, younger professionals frequently place a higher priority on job advancement than on their personal lives. Gender inequalities are also apparent; women in IT are under more pressure to balance their professional and home commitments, which increases female employee churn [7]. Many Indian IT companies, including Infosys, TCS, Wipro, and IBM India, have implemented wellness and flexibility programs, such as paid time off, telecommuting, flexible work schedules, and mental health counselling, in recognition of these difficulties. However, corporate culture, workload allocation, and managerial attitudes frequently affect how effective such measures are.

3 Literature Review

India has gained international recognition thanks in part to the Indian IT sector. Female employment has increased in tandem with the sector's growth. All genders are impacted by the issue of work-life balance, although women are more affected than males. Due to cultural factors and the fact that professional women in India still perform the majority of household chores, if not all of them, work-life balance is essential for women working in the Indian IT sector. In order to determine the work-life balance issues that top female IT workers have, particularly while working from home, the present study examines their mental health. While most respondents acknowledged that women's career perspectives are influenced by parental recognition and domestic economic conditions, they disputed that women who prioritize job advancement are less likely to be competent mothers, wives, or daughters-in-law or to take on additional family responsibilities. The respondents also concurred that ambition and self-inclination are linked to work-life balance; some of them thought that a woman who desires it would look for answers. For Indian women, juggling work and personal obligations is a challenging undertaking [8].

This research examines the issues and level of work-life balance faced by IT professionals employed by Bengaluru-based IT organizations. By employing random and snowballing procedures to examine the population, the study has combined quantitative and qualitative analysis. The data is gathered via an online survey form with a structured question. The connection PLS method and bootstrapping for quantitative analysis have been tested using IBM SPSS and Smart PLS. We employed the one-on-one interview approach for qualitative analysis. We discovered that women professionals experienced greater work-life balance and less pressure than men, and that the WLB model could account for 63% of responses on work-life balance. The results of the study show that female employees outperformed male employees. This research has significance for both new developments in the field of organizational development and growth in general and for academics and practitioners in the field of human resource management in particular [9].

Work-life balance and job happiness have emerged as essential variables in maintaining worker loyalty, making employee retention a major problem for firms in the digital era. This study examines the connection between job happiness, work-life balance, and employee retention in the IT industry, particularly in light of technology developments that make flexible working models like remote work and hybrid structures possible. Retention rates have been directly impacted by the shift to digital work environments, which has brought with it both possibilities and problems for managing employee happiness and maintaining an effective work-life balance. Employee retention and work satisfaction were shown to be strongly positively correlated. The IT industry must concentrate on establishing a rewarding work environment and offering growth chances if they want to retain talent, according to regression research, which revealed a high positive correlation between job satisfaction and employee retention. It is anticipated that the research's conclusions would help firms create more successful retention plans suited to the digital age, emphasizing employee happiness and well-being as key factors in sustained business performance [10].

Empirical research on the difficulties workers in the Indian IT industry have in maintaining a work-life balance was carried out by Kaur and Kaur [10]. The study underlined how work-life happiness is greatly impacted by lengthy workdays, stress at work, and a lack of organizational support. Nonetheless, it was discovered that the implementation of remote working choices and flexible work schedules improved employees' capacity to strike a good balance between their personal and professional lives [11].

The consequences of remote and hybrid work models implemented during the COVID-19 pandemic were investigated. Their research showed that although working remotely improved family time and decreased commuter stress, it also led to a blurring of work-life boundaries, which had a negative effect on mental health [12].

Investigated how work-life balance is experienced differently by gender in the Indian IT sector. According to the survey, the combined weight of work and home obligations caused more stress for female employees. The problem was made worse by a lack of supportive employment practices including flexible leave, daycare facilities, and maternity benefits [13].

The relationship between work-life balance and company culture among IT workers. Their research found that companies that promoted a friendly, open, and inclusive work atmosphere were better at keeping staff members and guaranteeing their general contentment, particularly when it came to handling work-life demands [14].

Technology developments promoted employee burnout even if they were helpful in allowing for flexible work schedules. The research noted the emergence of the "always-on" culture, in which workers felt pressured to be available at all times, which had a detrimental effect on their personal time [15].

Work-life balance and well-being in important Indian IT centers including Hyderabad, Pune, and Bengaluru. They discovered that work-life balance and productivity were significantly improved by programs including mindfulness training, wellness seminars, and counselling services [16].

The connection between job satisfaction and work-life conflict. According to their findings, there was a significant inverse relationship between job satisfaction and work-life conflict. To lessen this tension, the research suggested HR practices including flexible leave, less overtime, and employee health initiatives [17].

4 Literature Gap

Notwithstanding the expanding body of research on WLB in the Indian IT industry, substantial limitations persist. Existing research mostly focuses on traditional HR practices, neglecting integration of innovative technology solutions like AI-based task management and digital well-being instruments. Comparative assessments across leading IT firms are infrequent, since the majority of research works focus on individual examples rather than providing cross-organizational perspectives. Moreover, few research systematically utilizes integrated theoretical frameworks such as the JD-R Model and Work-Family Border Theory in conjunction. Although gender-specific WLB problems are recognized, effective organizational solutions are still little investigated. Furthermore, an important portion of the present material predates the COVID-19 pandemic, missing contemporary insights on hybrid work, digital weariness, and post-pandemic firm tactics. This research purpose is to bridge these gaps by rigorously synthesizing secondary data, comparing corporate practices like technology-driven interventions, and offering practical recommendations for augmenting WLB in the dynamic Indian IT sector.

5 Objective of the Study

1. to identify key factors influencing work-life balance among Indian IT employees,
2. to critically assess the initiative undertaken by leading IT companies,
3. to evaluate the impacts of work-life balance practices on employee well-being, satisfaction, and retention,
4. To recommend appropriate interventions for improving the quality of work life of employees in the IT industry.

6 Methods

This study uses a qualitative, exploratory, and secondary research approach to critically investigate work-life balance (WLB) difficulties, organizational initiatives and developing technology solutions within the Indian IT industry. Secondary data were collected from peer-reviewed academic publications, industry reports (NASSCOM, ASSOCHAM), company sustainability reports (Infosys, TCS, Wipro), and business media outlets. Critical content analysis ad comparative assessment approaches were utilized to identify pertinent elements, assess organizational tactics, and analyze variations in effectiveness. The investigation was informed by established theoretical frameworks, namely the Job Demands-Resources (JD-R) Model and the Work-Family Border Theory. Technological interventions, like AI-driven task management and digital well-being platforms, were incorporated to improve traditional work-life balance models. A conceptual framework connecting corporate efforts, moderators, and employee outcomes was established based on the findings. The study acknowledges that the results are drawn from secondary sources like literature reviews, reports, and industry case analyses, and may not represent real-time employee sentiments and future study can explore these obtaining primary data.

7 Findings

a. **High Workload and Long Working Hours Remain a Major Challenge**
   - A sizable portion of IT workers put in more than the recommended 8 h a day, sometimes up to 10 or 12 h, especially in positions involving direct customer interaction.
   - The peculiar working hours required by offshore and global delivery models interfere with daily routines and family time.
   - An ASSOCHAM [1] research found that 78% of Indian IT workers felt overworked and found it difficult to manage their personal and professional life.

b. **Flexible Work Policies Exist but Are Inconsistently Implemented**
   - Flexible work arrangements, such as remote work, flexi-time, and hybrid work, have been adopted by major IT firms like Infosys, TCS, and Wipro; however, adoption varies by department and management discretion.
   - Why a lot of workers believe that demands for continuous availability still exist and that flexibility is only "on paper."

c. **Increased Digital Fatigue and Burnout Post-COVID**
   - The rise in screen time brought about by the move to remote and hybrid work led to mental burnout, eye strain, and digital fatigue.
   - The lines between work and home have become hazier due to constant interaction via virtual platforms like Teams, Zoom, and Slack.
   - According to a 2022 NASSCOM survey, 62% of workers suffered from moderate to severe digital weariness.

d. **Gender Differences in Work-Life Balance**
   - Due to their twin duties (caregiving and career), female employees report more difficulties juggling work and life.
   - Stress is exacerbated by a lack of strong support networks, including as flexible maternity return plans, gender-sensitive policies, and child care.
   - Research also shown that in competitive settings, males feel pressured to put in more hours or shoulder more work.

e. **Organizational Support Influences WLB Satisfaction**
   - Employee satisfaction ratings were higher for businesses that provided counselling services, regular breaks, wellness initiatives, and employee assistance programs (EAP).
   - For instance, Infosys' iCare mental health support program and Wipro's "Wellbeing@Wipro" campaign were cited as excellent practices.

f. **Work-Life Balance Is a Key Driver of Retention**
   - Employees with low WLB expressed a greater desire to quit during the upcoming year.
   - Teams with organized WLB assistance saw 23% reduced turnover than other teams, according to TCS's internal assessment from 2023.

g. **Organizational Culture and Leadership Play a Crucial Role**
   - The way policies are implemented is significantly impacted by the leadership's attitude toward WLB.
   - Regardless of workload, supportive managers and compassionate leadership enhance work-life balance.

h. **Pandemic Accelerated WLB Awareness but Also Added New Challenges**
   - Businesses reviewed their human resources policies as a result of the COVID-19 pandemic's emphasis on mental health and wellbeing.
   - On the other hand, remote work raised demands for continuous responsiveness, which in some companies resulted in an "always-on" mentality.

Table 1 Impact of poor WLB

Impact	Description
Burnout	Chronic physical and emotional exhaustion caused by prolonged work stress, leading to fatigue, detachment, and reduced job efficacy. Often worsened by lack of rest and continuous digital connectivity
Decreased productivity	Employees facing poor WLB often exhibit reduced concentration, increased errors, lower creativity, and frequent delays. Lack of rest and mental clarity undermines performance and innovation
Attrition	Poor WLB is a strong predictor of employee turnover, especially among experienced mid-career professionals who seek better balance elsewhere. Companies with weak WLB policies see higher exit rates
Health issues	Stress-related illnesses like hypertension, migraines, anxiety, depression, cardiovascular diseases, and sleep disorders are commonly reported among overworked employees
Family conflict	Employees may face tension and conflicts at home due to long hours, emotional unavailability, or missing important family events, leading to strained spousal and parent–child relationships
Reduced employee engagement	Workers with poor WLB tend to feel disconnected, unmotivated, and less committed to organizational goals, affecting team morale and workplace culture
Loss of organizational reputation	Companies with a reputation for overworking employees or lacking WLB support may struggle to attract top talent and may face public criticism or poor employer branding
Legal and compliance risks	In some cases, consistent overwork may violate labor laws regarding work hours, leading to compliance issues, employee grievances, or legal challenges

8 Impact of Poor WLB

See Table 1.

9 Work-Life Balance Issues in IT Industry

The Job Demands-Resources (JD-R) Model and the Work-Family Border Theory are useful frameworks for comprehending employee experiences while analyzing work-life balance (WLB) concerns in the Indian IT industry. "The Job Demands-Resources (JD-R) Model explains how employee well-being and performance are shaped by the balance between job demands that cause stress and job resources that support motivation and resilience" [5]. In the IT sector, where long hours, remote work, and international scheduling frequently cause these boundaries to become blurred, the Work-Family Border Theory provides insight into how people negotiate and maintain the borders between the work and family spheres. "The Work-Family Border Theory explains how individuals manage and negotiate the boundaries between work and family roles to maintain balance and reduce conflict between the two domains" [4]. In contrast, the JD-R Model places a strong emphasis on striking a balance between the resources that are available (such as flexible policies and managerial support) and the demands of the job (such as workload and time pressure). By using these frameworks, it becomes clearer how high demands without sufficient support can result in burnout, while well-placed resources like counseling, wellness programs, and flexible work schedules can improve employee happiness, engagement, and general well-being. These theories offer a strong framework for examining the difficulties and successful strategies associated with WLB in the IT sector.

10 Best Practices in IT Companies

See Tables 2 and 3 and Fig. 1.

11 Recommendations

"The study's findings are consistent with the JD-R Model, where excessive job demands such as long working hours, digital fatigue, and customer-driven schedules increase employee stress, while organizational resources like flexible work policies, AI-driven workload management, and wellness programs help buffer these demands. Companies like TCS and Wipro that effectively strengthened these resources reported better employee well-being and lower attrition, validating the JD-R Model's central proposition.'

Integrating the review of literature and the findings, the following solutions are suggested to address ongoing WLB problems in the Indian IT sector:

a. **Implement and Uphold Flexible Work Arrangements Based on Research and Literature**

Although remote and flexible work norms are in place, their implementation is uneven [11, 12]. The corporate houses must institutionalize flexible work arrangements across the necessary divisions and guarantee management responsibility for their uniform implementation.

Implementable Measure: The companies should integrate flexible work arrangements into HR policy and incorporate compliance metrices into management key performance indicators.

Table 2 Best practices in IT companies

Company	WLB initiative	Description
TCS	AI-based workload analysis to avoid burnout	Uses machine learning to monitor employee workloads and flag overburdened teams for intervention; promotes balanced task allocation
Wipro	"Wellbeing@Wipro"—holistic health programs	Includes emotional counseling, fitness sessions, stress management workshops, and wellness apps
Infosys	Remote work, counseling support, work-hour tracking	Offers flexible schedules and remote working; iCare mental health platform provides 24/7 employee counseling and tracks weekly work hours to prevent overworking
IBM India	"Work from anywhere" and mindfulness programs	Enables location flexibility and incorporates daily virtual mindfulness sessions, yoga, and digital detox reminders
HCL technologies	Flexi-Time and Career Sabbatical Options	Offers adjustable shifts and extended unpaid leaves (up to 12 months) for personal commitments, study, or care giving
Accenture India	Mental Wellness Support and "Truly Human" approach	Provides resilience training, wellness coaches, and emphasizes emotional, mental, physical, and spiritual well-being
Tech Mahindra	Happiness@Work and Gamified Wellness Platforms	Runs employee engagement programs, mobile-based health apps, and mindfulness gamification to build consistent well-being habits
Capgemini India	"Reboot@Capgemini"—structured time-off initiatives	Promotes planned time-off for rejuvenation, discourages after-hours emails, and provides support for work-life reset
Cognizant	Return-to-Work and Parental Support Programs	Provides extended maternity/paternity leave, daycare support, and phased re-entry plans for returning parents
Mindtree	"Work-Life Harmony" initiative	Encourages team-level planning for flexible hours, shared goals, and social bonding activities to strengthen personal-professional boundaries

b. **Incorporate AI-Enhanced Workload Management and Digital Well-being Instruments Based on Research and Literature**

Digital weariness and perpetual availability are very pertinent problems [14].

Table 3 Comparison of work-life balance initiatives among leading IT companies

Company	Key WLB initiatives	Unique features
TCS	Structured remote work, AI workload monitoring	AI-based burnout detection
Wipro	Wellbeing programs, digital wellness apps	Gamified mental health engagement
Infosys	Flexible scheduling, mental health counselling	iCare 24/7 mental health platform
IBM India	Work-from-anywhere model, mindfulness programs	Virtual detox programs, no-meeting zones

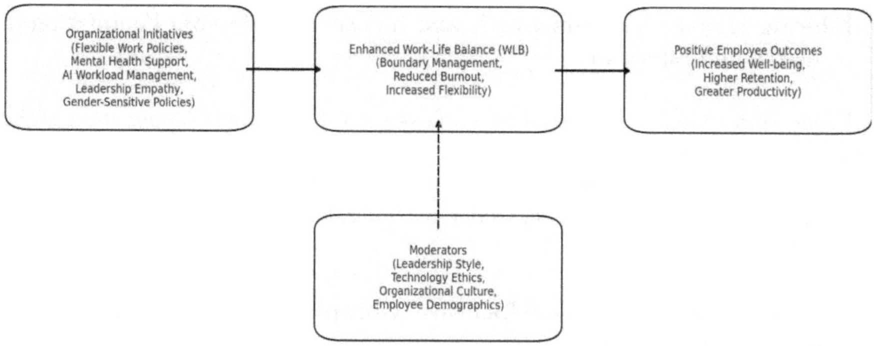

Fig. 1 Conceptual framework linking organizational initiatives, moderators, and WLB outcomes

Practical Measure: The companies should implement AI-driven tools to oversee task allocation (e.g., TCS's model) and add mindfulness and digital detox applications (e.g., IBM's Mindfulness Program) with consistent employee feedback mechanisms.

c. **Advocate for Mental Health and Well-Being Initiatives Foundation in Research and Literature**

Mental health assistance markedly improves retention and engagement [14]; Wipro's Wellbeing@Wipro program).

Practical Measure: The companies should establish counseling services, stress management sessions, and routine mental health evaluations as standard offerings rather than optional benefits.

d. **Reassess Performance Metrics to Emphasize Outcomes Rather than Hours**

Increased working hours are inversely related to productivity and work-life balance [5, 11].

Table 4 Work-life balance issues vs. solutions

WLB challenge	Recommended intervention
Long working hours	Redefine performance metrics, flexible scheduling
Digital fatigue	Digital well-being apps, scheduled screen breaks
Gender disparities	Childcare support, gender-sensitive leave policies
Managerial unsupportiveness	Empathetic leadership training

Practical Measure: The companies should redirect the assessment emphasis from "hours worked" to "outcomes achieved" to mitigate presenteeism and incentivize efficiency.

e. **Educate Managers in Empathetic and Inclusive Leadership Foundation in Research and Literature**

Leadership style most importantly impacts work-life balance (application of the JD-R Model; [14]).

Implementable Measure: The companies should mandate leadership training focused on empathic management, active listening, and the facilitation of work-life balance techniques.

f. **Cultivate a Supportive and Inclusive Workplace Culture Foundation in Research and Literature**

The business should inculcate supportive business cultures markedly to improve work-life balance and employee retention [14]. Implementable Measure: The business should encourage transparent communication, standardize the allocation of personal time, and publicly acknowledge managers who advocate for employee welfare (Table 4).

g. **Implement Gender-Sensitive Work-Life Policies Foundation in Research and Literature**

Gender differences in work-life balance are well reported in literature [13].

Implementable Measure: The business should increase childcare assistance, implement flexible maternity and paternity return activities, and set up gradual job reintegration interventions for caregivers.

h. **Establish Work-Life Balance as a Standard Practice Metrics and Periodic Surveys** Foundation in Research and Literature

Data-driven methodologies provide continuous improvement [12, 17].

Implementable Measure: The business should conduct quarterly WLB satisfaction surveys, monitor stress and engagement metrics, and modify HR interventions according to data analysis.

i. **Increase Collaboration Between Industry and Government Foundation in Research and Literature**: Comprehensive sectoral transformation necessitates extensive alignment [18].

Practical Measure:
The business can engage with NASSCOM, ASSOCHAM, and labor authorities to develop standards and certification programs for best practices in WLB.

Accomplishing sustainable WLB in the Indian IT sector necessitates a deliberate transition from ad hoc wellness initiatives to embedded business practices, powered by empathetic leadership, responsible technology use, and inclusive policies. By operationalizing these solutions, the corporate houses can foster a future-ready workforce that is resilient, engaged, and aligned with the requirements of the evolving digital economy."

12 Conclusion

The Indian IT industry is one of the crucial drivers of economic development in India. The global competitiveness poses tenacious challenges to sustainable work-life balance (WLB) for its workforce. This research through the critical synthesis of secondary data reveals that despite the implementation of flexible work models, wellness initiatives, and digital technologies, systemic problems such as excessive workloads, uneven policy enforcement, digital weariness, and gender inequities still continues.

The combination of literary sources and business case studies reveals that WLB programs are effective when paired with empathic leadership, a conducive company culture, and judicious integration of technology. The corporate houses like Infosys, TCS, and Wipro exhibit commendable practices; yet, comprehensive sectoral transformations are required to elevate work-life balance to the level of strategic necessity rather than an optional advantage. Implementable strategies like AI-enhanced workload management, established wellness initiatives, redefined performance metrics, and gender-sensitive policies are indispensable for converting WLB from a mere policy declaration to tangible business practice. Furthermore, ongoing data-driven oversight and leadership responsibility must encourage all WLB tactics to assure lasting impact. Ultimately, sustaining WLB in the advancing digital workplace is not only the employee's obligation; it is a vital organizational responsibility that directly impacts productivity, retention, mental well-being, and company's reputation. A steadfast commitment to human-centered work design, ethical use of technology, and inclusive practices will shape the future resilience and success of the Indian IT sector. Since authors have relied this study on secondary data, it is recommended recommend that future research may be conducted using primary data using empirical techniques (e.g., surveys, interviews) to validate and improve generalizability across diverse IT contexts.

References

1. ASSOCHAM. (2021). Survey on work-life balance in Indian IT sector. Retrieved from https://assocham.org
2. Bhumika, J. (2021). Challenges to work-life balance in the Indian IT sector. *Journal of Human Resource and Sustainability Studies, 9*(1), 34–45.
3. ASSOCHAM (2021). Work-life balance in the Indian IT industry: An analytical report. Retrieved from https://www.assocham.org
4. Clark, S. C. (2000). Work/family border theory: A new theory of work/family balance. *Human Relations, 53*(6), 747–770. https://doi.org/10.1177/0018726700536001
5. Demerouti, E., Bakker, A. B., Nachreiner, F., & Schaufeli, W. B. (2001). The job demands-resources model of burnout. *Journal of Applied Psychology, 86*(3), 499–512. https://doi.org/10.1037/0021-9010.86.3.499
6. Economic Times. (2023). Burnout in Indian IT: A rising concern. Retrieved from https://economictimes.indiatimes.com
7. Hungund, S. S., Gujanal, P. U., & Suresh, S. D. (2024). A study of the work-life balance of IT professionals in Bengaluru: An empirical investigation. *SDMIMD Journal of Management, 15*, 1–14.
8. Infosys Ltd. (2023). Annual report 2022–23. Retrieved from https://www.infosys.com
9. Joshi, V., & Singh, A. (2021). Role of organizational culture in work-life balance of IT professionals. *South Asian Journal of HRM, 14*(1), 77–89.
10. Kaur, A., & Kaur, S. (2023). Work-life balance challenges in the Indian IT sector: An empirical study. *Journal of Human Resource Development, 12*(1), 45–57.
11. Kumar, A., & Raj, S. (2019). Work-life conflict and job satisfaction among IT employees in India. *International Journal of Human Capital, 11*(3), 33–45.
12. Mehta, P., & Thomas, J. (2020). Impact of technological advancement on work-life balance in IT sector. *Indian Journal of Management Studies, 18*(2), 120–132.
13. Mishra, P., & Jha, R. (2021). The impact of work-life balance on employee engagement in the Indian IT sector. *Journal of Human Resource and Sustainability Studies, 9*(2), 112–123.
14. Narayanan, V., & Krishnan, R. (2020). Employee well-being and work-life balance in Indian IT hubs. *Journal of Contemporary HR Practices, 7*(4), 89–101.
15. NASSCOM (2022). IT industry workforce insights report. Retrieved from https://www.nasscom.in
16. NASSCOM (2023). Strategic review—Technology sector in India. Retrieved from https://nasscom.in
17. Rajadhyaksha, U., & Smita, R. (2020). Occupational stress among Indian IT professionals: A study on work-life imbalance. *Indian Journal of Occupational and Environmental Medicine, 24*(2), 80–85.
18. Singh, R., & Khanna, M. (2020). Work-life balance practices in Indian IT companies: A case of increasing burnout. *Indian Journal of Industrial Relations, 56*(1), 85–95.
19. Rao, M., & Iyer, K. (2021). Gender differences in work-life balance in the Indian IT industry. *Journal of Organizational Behavior and Gender Studies, 9*(2), 56–68.
20. Saxena, R., & Dey, S. (2022). Gender differences in work-life balance among Indian IT employees. *Journal of Human Resource Management, 10*(3), 47–56.
21. Sharma, R., & Singh, A. (2020). Gender differences in work-life balance among IT professionals in India. *Indian Journal of Industrial Relations, 56*(2), 210–223.

22. Sharma, R., & Verma, P. (2022). Remote work and work-life integration in Indian IT firms post-COVID-19. *International Journal of Business and Management Research, 20*(3), 101–110.
23. Sheshadri, T. (2024). Employee retention in the digital age: The role of work-life balance and job satisfaction with reference to IT sector. *Journal of Informatics Education and Research, 4*(3), 1340–1350.
24. TCS (2023). Integrated report 2022–23. Retrieved from https://www.tcs.com
25. Times of India (2021). Remote working blurs boundaries, IT employees struggle with work-life balance. Retrieved from https://timesofindia.indiatimes.com
26. Times of India (2023). How hybrid work is affecting employees' work-life balance in Indian IT. Retrieved from https://timesofindia.indiatimes.com
27. Vijayalakshmi, S., Nirmala, T., & Subasree, R. (2023). An analysis of work-life balance and mental health of women in India during work from home with special reference to IT sector. *Asian Review of Social Sciences, 12*(1), 29–33.
28. Wipro Ltd. (2023). *Sustainability and well-being report*. Retrieved from https://www.wipro.com

Impact of Compensation Management Practices on Job Satisfaction in Indian Information Technology Industry

M. Jahnavi, N. Nagasubba Reddy, N. Chandrika, B. Sireesha, M. Sandeep Kumar, and V. Lakshmi Prasanna Kumari

Abstract

This research papers examines the impacts compensation management practices (CMP) on employee job satisfaction in the Indian information technology (IT) sector. A structured questionnaire was used to obtain data from 278 IT professional from six leading corporate houses in Bengaluru, and the findings were evaluated using exploratory factor analysis (EFA) and Multiple Regression Analysis. EFA identified four main elements: compensation, incentives, leave rules, and fringe benefits. These account for 62.18% of the overall variation. Regression analysis showed that compensation practices substantially impact job satisfaction ($R^2 = 0.703$, Adjusted $R^2 = 0.592$), with fringe benefits ($\beta = 0.337$, $p < 0.001$) and incentives ($\beta = 0.602$, $p = 0.003$) having the most pronounced effects, followed by leave policies ($\beta = 0.131$, $p < 0.001$) and wage and

M. Jahnavi (✉) · N. N. Reddy
RV Institute of Management, Bengaluru, India
e-mail: jahnavim.rvim@rvei.edu.in

N. N. Reddy
e-mail: nnsreddy.rvim@rvei.edu.in

N. Chandrika · M. S. Kumar
Department of MBA, CMR Institute of Technology, Bengaluru, India
e-mail: chandrikaraju02@gmail.com

M. S. Kumar
e-mail: sandeepmachavolu@gmail.com

B. Sireesha
Cambridge Institute of Technology, Bengaluru, India
e-mail: sireesha.mba@cambridge.edu.in

V. L. P. Kumari
Department of MBA, CMR Institute of Technology, Hyderabad, India
e-mail: vlprasanna527@gmail.com

S. Patnaik et al. (eds.), *People, Planet and Profit: Advancing Sustainable Business Practices*, Springer Proceedings in Geography and Sustainability,
https://doi.org/10.1007/978-3-032-11296-5_21

salary ($\beta = 0.037$, $p < 0.001$). The survey reveals that IT businesses should prioritize on clear incentive programs and extensive fringe benefit offers to increase employee retention and satisfaction. Moreover, formulating employee-centric leave policies can improve staff loyalty. The results offer practical insights on IT companies seeking to establish sustainable HR policies that align with the changing demands of India's knowledge-based economy.

Keywords

Compensation • Job satisfaction • Wages and salary • Leave policy • Incentives and fringe benefits

1 Introduction

Businesses provide their employees' pay for the work they do. Despite the widespread misconception that pay and compensation are interchangeable; compensation encompasses far more than just the monetary perks that a business provides. Compensation is distinct as "all forms of financial returns and tangible services and benefits employees receive as part of an employment relationship." "Financial rewards" encompass short- and long-term benefits in addition to an individual's base pay. The following are an example of "tangible services and benefits": pension plans, insurance, paid time off for illness and vacation, and employee discounts. In his book Human Resource Management, Gary Dessler states that "employee compensation refers to all forms of pay going to employees and arising from their employment. The phrase "all forms of pay" in the description refers to all monetary payments, which includes direct and indirect, and excludes other forms of compensation [1].

Simply put, employee happiness, also known as satisfaction of job, is the grade to which employees are happy or content with their occupations. An employee satisfaction survey is usually used to gauge employee satisfaction. These surveys cover a wide range of subjects, containing salary, assignment, executive opinions, rigidity, cooperation, and possessions. All of these factors are critical for businesses observing to maintain employee satisfaction and lower attrition, but employee satisfaction is only one component of the total picture. In reality, there are some companies that would be better off without contented employees. On the surface, employee engagement and employee happiness are related ideas, and many people use them interchangeably. The fundamental requirements and concerns of employees are covered by employee satisfaction. Although it is a nice place to start, it typically misses the mark. Over the past 20 years, the idea of employee satisfaction. It is regarded as a crucial component of organizational performance. As much as customer satisfaction, a number of academics and organization "gurus" emphasized the significance of worker fulfillment and how it affects company performance. [1]

2 Literature Review

Job satisfaction is a pleasant or upbeat emotional state brought on by an evaluation of one's job and their activities at work. According to this, employee happiness is determined by the "perception of connection among what one desires from a position and the benefits that one feels it as offering."[2]. On the other hand, turnover, absences, and perceived stress are negatively correlated with employee satisfaction, whereas motivation, job participation, behavior regarding commitment to the organization, personal fulfilment, and job performance are positively correlated with it. Employee satisfaction is the extent to which an individual is happy with their job [3]. Employee happiness is influenced by a number of different factors. Therefore, overall employee satisfaction refers to an individual's whole emotional reaction to the array of work and work-related features, whereas the elements of job satisfaction are concerned with workers' attitudes toward specific parts of their occupations and work settings [4].

The current study, pay management has an impact on employees' organizational commitment and job satisfaction [5]. Motivations are costs made to workforces based on their contributions to the organization. These payments might be monetary, non-monetary, or psychological [6]. Another aspect of the working relationship is reward, wherein employees receive all material benefits and provisions [7]. The company's value for each position is known as the work rate (a); compensation (b); personal or specialized allowances (c); and fringe benefits (d), which include paid time off, pensions, and other benefits. Employees are encouraged by compensation, which is based on their performance [8].

Money remains the most powerful motivating tool and scientific management associate stated that the main motivator for industrial employees to boost output was money [9]. A lot of managers use cash incentives or sanctions to deal with employees. This is accomplished by rewarding employees for increased output by creating a fear of losing their jobs or facing other related consequences (such as not receiving a raise or promotion after a year of reduced enactment). Employees may also be inspired by the prospect of promotion and increased pay. An additional line of inquiry into compensation emphasizes the internal direction toward an individual's response to remuneration [10]. Industrial organizational psychologists are leading this study direction, which has greatly influenced the field of human resource management. The key driving force behind this research has been the theory that compensation has an influence on employees' primary work habits or motivation as well as their overall job satisfaction [11].

Herzberg's Two-Factor Theory (1959) states that pay elements such as salary, benefits, and leave rules most importantly play the roles as hygiene factors, justifying employee unhappiness. However, a well-structured performance-based incentives and fringe benefits program can function as motivators to enhance overall work satisfaction and employee engagement. Equity Theory [1] states that workers have a tendency to evaluate the equity of their remuneration by combining their input–output ratio with that of comparable individuals. Perceived disparities

Table 1 Theories linked to the present study

Theories	Factors	Descriptions
Herzberg's two-factor theory	Job satisfaction influenced by hygiene and motivator factors	Compensation and incentives act as both hygiene factors and motivators
Adams' equity theory	Perceived fairness of compensation affects satisfaction	Employees compare pay/ benefits; perceived inequity reduces satisfaction

in payroll, perks, or prospects for career progression may decrease work satisfaction and increase turnover, especially in highly competitive and dynamic industries like IT (Table 1).

Less payment can vex and discourage people, but this does not imply that having extra payment will make people happier or more motivated. It makes sense to suppose that if someone's take-home salary was halved, their morale would deteriorate to the point where performance would be compromised. However, that doesn't mean that better work would follow if that person's income were doubled [12]. When looked at the influence of pay on performance of employee, they found a strong correlation between performance and the results that valence pay and expectations provided. Lifetime earnings and earnings related to employee salary [13]. Organizational wage structures and pay scales affect financial performance, patient care results, and resource quality. They also examined the connection between job performance and organizational remuneration decisions [14].

Compensation management is recognized as an essential aspect of human resource strategy which includes all forms of financial returns and tangible benefits that workers receive through their employment relationship [15]. Compensation does not refer to just direct wages and salary that an employee receives but also includes indirect benefits such as bonuses, healthcare, pensions, paid leave, and employee recognition initiatives. In firms, efficient compensation schemes are vital for recruiting, motivating, and retaining talent, particularly in knowledge-driven IT sector.

Employee work satisfaction, marked by the degree of good or bad feelings, people have towards their employment [16], is intricately integrated with payroll practice. Researchers like [17] emphasized the importance of extrinsic elements, such as compensation and perks, as hygiene factors that lessen employee unhappiness but do not inherently improve employee motivation. Conversely, [5] states that people are motivated when they perceive a distinct connection between effort, performance, and rewards. These theoretical perspectives highlight the complex impact of compensation on employee attitudes.

Empirical research studies have repeatedly established a favorable correlation between pay management strategies and job satisfaction. Proved that equitable pay structure, performance-driven incentives, and compensation benefit programs significantly improve employee morale and retention [18]. Likewise, research findings of [19, 20] demonstrated both monetary and non-monetary rewards impact job satisfaction outcomes. However, some scholars also warn that lack of adequate compensation can dissuade employees, although just increasing financial incentives does not inherently result in improved job satisfaction or performance [21].

Despite having a good body of relevant research available in the area of compensation management, the most of the studies have made a very generic approach across sectors, failing to adequately consider sector-specific dynamics. The Indian IT sector, characterized by fast changing technological shifts, intense global competition, and increased employee attrition experienced distinct compensation challenges that markedly differ significantly from traditional industries like manufacturing or banking [22, 23]. IT professionals often offer flexible benefits, stock options, remote work option, and possibilities for professional development prospects above fixed salary components increases.

Besides, fresh changes in work patterns after COVID-19 like hybrid models, mental health programs, and digital skill premium shave meaningfully changed employee expectations for compensation [24]. Nevertheless, very little research has empirically investigated the total effect of these emerging remuneration variables on work satisfaction in the Indian IT sector.

Moreover, prior studies often focus on single compensation component (e.g., pay or bonuses) instead of embracing a comprehensive perspective to include other variables like pay equity, incentives, leave policies, and fringe benefits. Methodologically, several studies used basic correlational designs, failing to include multivariate approaches like exploratory factor analysis (EFA) and regression modeling to explain the relative importance of several other compensation components (Fig. 1).

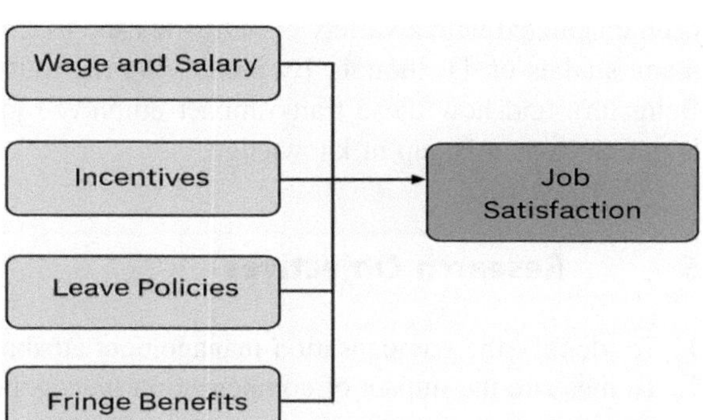

Fig. 1 Conceptual framework

This research empirically evaluates the impact of four essential compensation management practices such as wages and salaries, incentives, leave policies, and fringe benefits on job satisfaction among IT professionals in Bengaluru. The research uses a multivariate analytical technique to offer detailed insights into the collective and individual impacts of pay variables, offer specific suggestions for HR professionals in the IT industry.

3 Research Gap

The relationship between compensation management practices and employee satisfaction in a variety of industries is well covered in the literature currently in publication, but little is known about how these practices specifically affect job satisfaction in the Indian information technology (IT) sector. Although studies have looked at aspects of compensation like pay, perks, and incentives, little attention has been paid to the particular circumstances of the Indian IT industry, where employment expectations and compensation structures vary from those of other sectors. Moreover, the dynamic and quickly changing character of the IT sector, which poses unique issues in compensation management, is not taken into consideration in the majority of research. By investigating the effects of different pay management strategies on job satisfaction among Indian IT workers, this study seeks to close this knowledge gap and provide a targeted understanding of the unique requirements and practices of this industry.

4 Research Problem

As part of their pay for their services to the company, the employees receive welfare benefits in addition to direct and indirect incentives. The advantages of having a successful career include health insurance, disability financial security, retirement benefits, childcare, reimbursement for fees, paid and unpaid vacations, community help, convenient and alternative work schedules, and more. Most of the research mentioned above came to the conclusion that, in any firm, pay management procedures have the biggest influence on employees' job satisfaction. Many studies have been conducted with a variety of industries and in a variety of sectors. There aren't many studies on IT, though. By identifying the traits of the best IT companies in Bengaluru and how these traits impact employee job satisfaction, the researcher hopes to close this gap in knowledge.

5 Research Objectives

1. To identify the compensation management attributes in IT industry.
2. To measure the impact of compensation management attributes on job satisfaction.

3. To recommend actionable compensation strategies for IT organizations.

6 Research Methodology

This research employs a quantitative, cross-sectional, and descriptive research method to examine the impact of compensation management practices on job satisfaction among IT professionals in Bengaluru, India. Primary data were obtained through a structured questionnaire, comprising two parts: The first part obtained demographic data like gender, age, education, work experience, and job level, while the second part collected the respondents' perceptions of compensation management attributes and job satisfaction using a 5-point Likert scale. A convenience sampling technique was used to select respondents, enabling convenient access to IT employees; however, a convenience sampling technique adopted in this study may limit the generalizability of findings beyond the Bengaluru IT sector. Seven hundred surveys were given away through physical and digital channels, yielding 278 completed filled-in responses for analysis. Pilot research was also made with a small sample prior to the main survey to check the clarity and relevance of the questionnaire items. The obtained responses were coded and used input into Microsoft Excel 2010, followed by analysis using SPSS Version 20.0. The data analysis included reliability testing to check internal consistency of the collected responses, followed by EFA to explore the underlying insights of perceived compensation and satisfaction. Multiple linear regression analysis was used to explore the correlation between the identified payroll factors and employee job satisfaction. Ethical considerations were rigorously upheld throughout the research, with informed consent collected from all the respondents, ensuring confidentiality and voluntary participation throughout the research process.

7 Data Analysis

The following section presents the analysis of data, obtained through primary source.

7.1 Demographic Profile

According to the demographic profile of the 278 IT employees, women made up 41.7% of the sample, while men made up the majority of respondents (58.3%). The highest percentage (45.7%) was made up of people aged 31 to 40, followed by those aged 20 to 30 (33.8%). 35.3% of respondents had a bachelor's degree, while the majority (56.5%) had a master's degree. Thirty-six percent had between six and ten years of job experience, while the largest group (34.9%) had three to five years. In terms of job level, more over half of the respondents (51.1%) held mid-level positions, while equal numbers (24.5%) held entry-level and senior/

Table 2 Demographic profile of IT employees

Variable	Category	Frequency (n)	Percentage (%)
Gender	Male	162	58.3
	Female	116	41.7
Age (in years)	20–30	94	33.8
	31–40	127	45.7
	41–50	44	15.8
	Above 50	13	4.7
Educational level	Bachelor's degree	98	35.3
	Master's degree	157	56.5
	Other/Professional	23	8.2
Work experience	6 months–2 years	61	21.9
	3–5 years	97	34.9
	6–10 years	85	30.6
	Above 10 years	35	12.6
Job level	Entry level	68	24.5
	Mid-Level	142	51.1
	Senior Level/Managerial	68	24.5

Table 3 Summary of case processing

		N	%
Cases	Valid	278	100.0
	Excluded[a]	0	0
	Total	278	100.0

managerial positions. In the IT industry, this varied profile guarantees equitable representation across age, experience, and job hierarchy (Tables 2 and 3).

7.2 Reliability Test

The reliability data for the measurement scale used in the investigation are shown in Table 4. The 16 items in the questionnaire have a high degree of internal consistency, as indicated by the Cronbach's Alpha rating of 0.836. This implies that the data gathered by the instrument is trustworthy for additional analysis and that the items accurately assess the underlying constructs. The acquired number validates the reliability of the scale used in this study, and a Cronbach's Alpha value above 0.7 is generally regarded as satisfactory.

Table 4 Reliability statistics

Cronbach's alpha	Number of items
.836	16

Table 5 KMO and Bartlett's test

KMO measure of sampling adequacy		0.871
Bartlett's Test of Sphericity	Approx. Chi-Square	2177.194
	Df	136
	Sig	0.000

7.3 Exploratory Factor Analysis

The outcomes of the Bartlett's Test of Sphericity and the Kaiser-Meyer-Olkin (KMO) Measure of Sampling Adequacy are displayed in Table 5. Given those values more than 0.8 are regarded as meritorious, the sample's KMO score of 0.871 suggests that it is ideal for factor analysis. Furthermore, the results of Bartlett's Test of Sphericity are significant (Chi-square = 2177.194, df = 136, $p < 0.001$), indicating that the variables have sufficient correlations to move forward with factor analysis. All of these findings support the notion that exploratory factor analysis is a suitable method for the dataset.

Four components have been retrieved using Kaiser Normalization and Varimax Rotation. Sixteen variables were combined into 4 components, each of which is composed of all the variables with factor loadings more than 0.5. Sixty-two percent of the variability was explained by these four recovered values (Table 7).

7.4 Regression

A multiple linear regression approach was used to explore the effect of compensation management attributes on job satisfaction.

The regression analysis's model summary is shown in Table 8. There is a high positive correlation between the independent and dependent variables, as indicated by the R-value of 0.750. The R2 value of 0.703 indicates that the model can account for roughly 70.3% of the variance in the dependent variable, which is regarded as significant. Even after taking into consideration the number of predictors in the model, the adjusted R2 value of 0.592 still shows a good match. The average separation between the observed values and the regression line is indicated by the standard error of 1.266. Furthermore, the regression model's assumption of independence is validated because the Durbin-Watson statistic of 1.793 is near to 2, suggesting that there is no significant autocorrelation problem in the residuals.

The findings of the regression model's Analysis of Variance (ANOVA) are shown in Table 9. The regression model is statistically significant, as indicated by the F-value of 18.968 and the significance level (p-value) of 0.000. This indicates

Table 6 Total variance explained

Component	Initial Eigenvalues			Extraction sums of squared Loadings			Rotation sums of squared loadings		
	Total	% of variance	Cumulative %	Total	% of variance	Cumulative %	Total	% of variance	Cumulative %
1	6.041	35.536	35.536	6.041	35.536	35.536	3.452	20.307	20.307
2	2.017	11.863	47.399	2.017	11.863	47.399	2.678	15.750	36.058
3	1.314	7.727	55.126	1.314	7.727	55.126	2.365	13.910	49.968
4	1.200	7.058	62.184	1.200	7.058	62.184	2.077	12.216	62.184
5	1.083	6.369	68.553						
6	.821	4.828	73.381						
7	.585	3.438	76.820						
8	.550	3.233	80.052						
9	.500	2.941	82.993						
10	.486	2.858	85.851						
11	.455	2.677	88.528						
12	.411	2.415	90.943						
13	.382	2.248	93.191						
14	.343	2.020	95.211						
15	.310	1.824	97.036						
16	.286	1.682	98.718						

Table 7 Rotated component matrix

Items	Component			
	1	2	3	4
The wage and salary practices of the company are internally equitable	.792			
Fair amount of bonus is provided in the company	.781			
Leave provision of the company support the psychological wellbeing of the employees	.690			
The company provides medical and healthcare facilities or reimburses the expenses for it to its employees	.680			
Leave provision of the company support the sociological wellbeing of the employees	.603			
Incentives are fairly given to the employees in the organization		.823		
Retirement benefits (pensions, provident fund and gratuity) are fairly appreciated in the company		.770		
The employee promotion in the company is based on fairness and justice		.764		
The company extends adequate transport facilities to the employees		.513		
My company provides gifts on special events or occasions			.831	
My supervisor appreciates and praises me for my work achievements			.760	
Adequate welfare measures are provided to the employees in the company			.712	
Salary practices I received as per industry norms in the company				.768
Adequate fringe benefits are provided to the employees in the company				.732
My company provides opportunities for career development and promotion				.561

Table 8 Model summary

Model	R	R^2	Adjusted R^2	Std. Error	Durbin-Watson
1	0.750[a]	0.703	0.592	1.266	1.793

that the dependent variable is significantly impacted by the independent variables taken together. The validity of the model employed in this study is supported by the high F-ratio and low p-value, which verify that the variation described by the regression model is substantially larger than the unexplained variance (residual).

Table 10, stands for standardized regression coefficients, which show the influence's extent and direction. In order to support the formulation of the measured hypothesis, it additionally includes t and significant values. Considerable effects on IT employees' job satisfaction from aspects of CMP attributes. The equation of

Table 9 Analysis of variance

Model		Sum of squares	Df	Mean square	F	Sig
1	Regression	121.590	4	30.398	18.968	.000[b]
	Residual	477.565	298	1.603		
	Total	599.155	302			

Table 10 Coefficients

Model		Unstandardized coefficients		Std. coefficients	T	Sig
		B	Std. Error	Beta		
1	(Constant)	0.352	0.311		3.132	0.258
	Wage and salary	0.009	0.086	0.037	2.099	0.000
	Leave policy	0.165	0.093	0.131	1.771	0.000
	Incentives	0.075	0.083	0.602	3.901	0.003
	Fringe benefits	0.525	0.095	0.337	5.516	0.000

regression of this model is: $Y = MX + C$

$$\begin{aligned} \text{Y (Job Satisfaction)} = {} & 0.037(\text{Wage and Salary}) \\ & +0.131\ (\text{Leave Policy}) + 0.602\ (\text{Incentives}) \\ & +0.062\ (\text{Fringe Benefits}) + 0.352\ (\text{Constant}) \end{aligned}$$

7.4.1 Wage and Salary

Table 9's beta value of 0.037 shows that wage and salary and job satisfaction are positively correlated. Given that the t. value of 0.000, or less than 0.05, the value and significance of 2.099 indicate that wage and salaries have a considerable impact on the job satisfaction of IT employees.

7.4.2 Leave Policy

The leave policy positively affects job satisfaction, as shown by Table 9's beta value of 0.131. IT employees' job satisfaction is significantly impacted by leave policies, as seen by the value and sig of 1.771 and the t. value of 0.000, which is less than 0.05.

7.4.3 Adaptability to Incentives

Table 9's beta value of 0.602 suggests that job satisfaction is positively impacted by the ability to adapt to incentives. Given that the t. value is 0.003, or less than 0.05, the value and sig have been 3.901, suggesting that incentives significantly affect how content IT employees are with their positions.

7.4.4 Fringe Benefits

With a beta value of 0.182, Table 9 demonstrates that fringe benefits positively affect job satisfaction. Given that the t. value is 0.000, which is less than 0.05, the value and sig have been 5.516, indicating that fringe benefits significantly affect IT employees' job satisfaction.

8 Linking Literature with Findings

The correlation between pay management practices and job satisfaction has been extensively examined in previous studies, particularly in the lens [19, 20]. Herzberg described wage and benefits as hygiene variables that prevent unhappiness but do not inherently improve motivation, while Vroom emphasized that workers' perceived link with effort, performance, and rewards propels motivation and contentment. The findings of present research validate these theoretical perspectives, affirming that pay components like wages, salaries, incentives, leave rules, and fringe benefits, substantially impact job satisfaction among Indian IT employees.

Prior empirical studies like [21, 25], recommended that fair compensation and equitable bonus structure are significant drivers of boosting employee satisfaction and lessening employee turnover. In accordance with these findings, the present research noted that pay and salary practices, nevertheless their moderate impact ($\beta = 0.037$), considerably boost job satisfaction among IT personnel. This shows that, despite increasing importance on intrinsic motivations in the dynamic IT industry, financial equity continues to be an essential driver of pleasure.

Highlighted that employee incentive are important motivators in performance-driven business firms. The research found that incentives had a significant positive impact ($\beta = 0.602$) on employee satisfaction, and found that performance-driven recognition practices remain highly valued among IT workers. This corroborates Vroom's assertion that perceived incentives significantly impact individual engagement and organizational commitment [22].

Leave policies, although is underexamined in earlier studies, have emerged as relevant in present literature, particularly in post-pandemic work landscape where work-life balance has received high attention [26]. The study's findings strengthen this emerging perspective, revealing that leave policies significantly improve job satisfaction ($\beta = 0.131$), implying that flexible and employee-focused leave provisions are important for sustaining employee morale and retention in the IT industry.

Recognized fringe benefits like medical insurance, wellness programs, retirement plans, and other non-monetary incentives, as important for long-term employee commitment. The present study found fringe benefits as a significant predictor of job satisfaction ($\beta = 0.337$), ranking just behind incentives. This shows a change in employee expectations within the Indian IT sector, where non-monetary rewards are increasingly valued as equally important, if not more than direct salary increments [23].

The researches confirm previous theoretical and empirical results and add to the body of literature by illustrating the collective and comparative impacts of several other compensation variables on job satisfaction within the Indian IT sector. The findings highlight that there is a need for IT companies to develop comprehensive, equitable, and industry-specific pay policy that cater to both monetary and non-monetary employee need to improve overall satisfaction and organizational efficacy.

9 Policy Implications

This study's findings have significant policy implications for HRM strategies in the Indian IT sector. The firms should focus on the establishment of comprehensive and transparent payroll management system that can cater to both monetary and non-monetary employee need. While competitive compensation systems are significant for improving fundamental employee happiness, further focus should be directed into establishing performance-based incentive program to recognize individual and team accomplishments. The firms need to reevaluate and improve their leave policies to foster work-life balance, recognizing the fact that the adaptable and employee-centric leave frameworks significantly improve employee morale and retention. Moreover, comprehensive fringe benefits such as health insurance, retirement plans, fitness programs, and childcare assistance, may significantly improve employee loyalty and engagement. In an industry characterized by increased employee turnover and tough competition for retaining competent personnel, a comprehensive compensation strategy should incorporate equitable pay, meaningful incentives, supportive leave policy framework, and attractive fringe benefits is indispensable. HR Head should adopt a strategic, employee-centric approach to payroll management, positioning it not mere a cost factor, but as a crucial lever for sustaining organizational performance, talent retention, and long-term competitiveness in the ever- evolving IT industry (Fig. 2).

Fig. 2 Policy implications for improving employee retention and job satisfaction

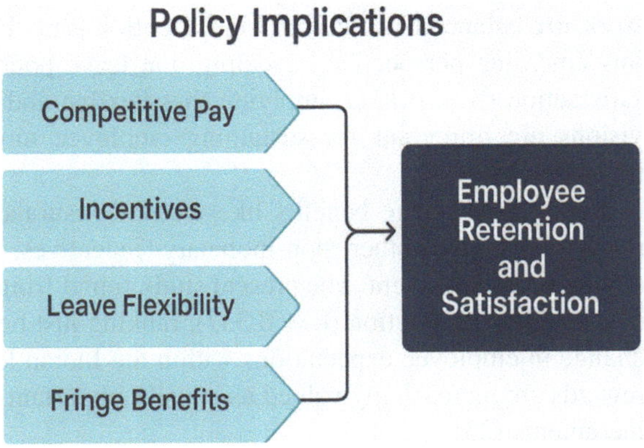

10 Direction for Future Research

This study provides important insights further research possibilities into how compensation practices impact job satisfaction in the Indian IT industry. The present study used a cross-sectional strategy and convenience sampling technique in Bangaluru city only, therefore, future research many be conducted across several Indian cities to investigate temporal differences in compensation-satisfaction dynamics. Studies on sector-specific differences in employee satisfaction trends and salary expectations may be conducted to compare between the IT and non-IT sectors.

Besides, including mediating or moderating factors such as leadership style, corporate culture, or flexibility in remote work set up may improve understanding of the mechanisms through which employee remuneration impacts happiness. Further research might also look at the increasing significance of non-traditional perks, especially in the post-pandemic context like flexible work schedules, sponsorship for professional development and mental health support programs. Understanding employee perceptions and motivations would be improved by utilizing mixed-methods approaches, which combine quantitative surveys with qualitative interviews. In the ever-changing Indian employment scenario, such research would enable a broader generalization of results and offer thorough insights for bettering employee engagement and retention.

Furthermore, given how quickly workplace processes are digitalized, further studies should look at how AI-driven technologies impact employee wellbeing and stress management. In order to balance work-life commitment, the IT industry has been using tools like AI-based job allocation, predictive workload balancing, smart scheduling software, and employee sentiment analysis platforms. Researching the uptake and effectiveness of these technologies may explore new factors impacting job satisfaction and can open the way for HR policy frameworks, enabled by technology.

Future research should also investigate the aspects of pay and work-life balance that are unique to both male and female employees. This study's gender representation (41.7% female, 58.3% male) highlights the need of understanding how flexible benefits, leave and pay policies impact men and women in IT companies. Integrating gender as a moderating variable, along with required qualitative interviews, could offer deeper insights to address equity gaps and formulating inclusive HR policies, sensitive to diverse employee experiences.

11 Conclusion

This research examined the impact of compensation management practices on job satisfaction among IT professional in Bengaluru. This study fills a significant gap in the existing body of literature that often neglects sector-specific employee compensation dynamics. The study utilizes exploratory factor analysis and regression modeling to identify four primary pay elements: wage and salary, incentives, leave regulations, and fringe benefits as major determinants of job satisfaction. The findings corroborate established theories, including Herzberg's Two-Factor Theory and Vroom's Expectancy Theory, while significantly focusing on the increasing significance of non-monetary benefits and work life balance initiatives in influencing employees' perceptions. Among the identified attributes, incentives and fringe benefits emerged as the most significant drivers of happiness, indicating a change in employee expectations beyond traditional compensation frameworks. The report offers significant policy suggestions for IT corporate houses seeking to increase talent retention and enhance engagement through comprehensive and employee-centric remuneration strategies. However, acknowledging its limitations in geographic breadth and methodological design, the study advocates for future research to utilize more expansive, longitudinal, and mixed-method strategies. The study significantly adds to the ongoing dialogue on HRM strategies within India's dynamic IT industry, offering practical insights for developing more resilient, motivated, and contented workforces.

References

1. Adams, J. S. (1963). Toward an understanding of inequity. *The Journal of Abnormal and Social Psychology, 67*(5), 422–436. https://doi.org/10.1037/h0040968
2. Ali, M. Z., & Ahmad, N. (2017). Impact of pay promotion and recognition on job satisfaction (A study on banking sector employees Karachi). *Global Management Journal for Academic and Corporate Studies, 7*(2), 131–141.
3. Brown, M. P., Sturman, M. C., & Simmering, M. J. (2003). Compensation policy and organizational performance: The efficiency, operational, and financial implications of pay levels and pay structure. *Academy of Management Journal, 46*(6), 752–762.
4. Cranny, C. J., Smith, P. C., & Stone, E. (1992). *Job satisfaction: How people feel about their jobs.* Lexington Books.
5. Dessler, G. (2015). *Human resource management* (14th ed.). Pearson Education.
6. Greasley, K., Bryman, A., Dainty, A., Price, A., Soetanto, R., & King, N. (2005). Employee perceptions of empowerment. *Employee Relations, 27*(4), 354–368.
7. Gupta, M. (2014). Employees' satisfaction towards monetary compensation practices. *Global Journal of Finance and Management, 6*(8), 757–764.
8. Herzberg, F., Mausner, B., & Snyderman, B. B. (1959). *The motivation to work* (2nd ed.). John Wiley & Sons.
9. Judge, T. A., Weiss, H. M., Kammeyer-Mueller, J. D., & Hulin, C. L. (2017). Job attitudes, job satisfaction, and job affect: A century of continuity and of change. *Journal of Applied Psychology, 102*(3), 356–374.
10. Locke, E. A. (1969). What is job satisfaction? *Organizational Behavior and Human Performance, 4*(4), 309–336.
11. Manjunath, S. G. (2023). *A study on relationship between employer branding and employee retention rates in the information technology sector in India* (Doctoral dissertation, National College of Ireland).
12. Nabukeera, M. S., Bwengye, M., & Buwembo, M. (2019). Non-financial rewards, motivation and employee performance in Ministry of Internal Affairs (MoIA) Uganda. *Journal of Public Administration and Governance, 9*(3), 112–123.
13. Osa, I. G., & Amos, I. O. (2014). The impact of organizational commitment on employees' productivity: A case study of Nigeria Brewery, PLC. *International Journal of Research in Business Management, 2*(9), 107–122.

14. Oshagbemi, T. (1996). Job satisfaction of UK academics. *Educational Management and Administration, 24*(4), 389–400.
15. Schwab, D. P. (1973). Impact of alternative compensation systems on pay valence and instrumentality perceptions. *Journal of Applied Psychology, 58*(3), 308–312.
16. Shinwari, S., Rahimi, M. H., Shinwari, A., & Khan, B. (2025). Impact of compensation packages on employees' job satisfaction: Moderating role of environmental uncertainty in banking institutions of Kabul. *Journal for Social Science Archives, 3*(1), 188–202.
17. Tella, A., Ayeni, C. O., & Popoola, S. O. (2007). Work motivation, job satisfaction, and organizational commitment of library personnel in academic and research libraries in Oyo State Nigeria. *Library Philosophy and Practice, 9*(2), 1–16.
18. Locke, E. A. (1976). The nature and causes of job satisfaction. In M. D. Dunnette (Ed.), *Handbook of industrial and organizational psychology* (pp. 1297–1349). Rand McNally.
19. Herzberg, F. (1959). *The motivation to work* (2nd ed.). John Wiley & Sons.
20. Vroom, V. H. (1964). *Work and motivation.* Wiley.
21. Tsai, C. J., Chi, N. W., Huang, H. L., & Hsu, A. (2011). High commitment HR practices, work engagement, and employee behavior: A study of multinational corporations in Taiwan. *The International Journal of Human Resource Management, 22*(5), 1042–1065. https://doi.org/10.1080/09585192.2011.556777
22. Zingheim, P. K., & Schuster, J. R. (2000). Paying people right: Breakthrough reward strategies to create great companies. Jossey-Bass.
23. Gerhart, B., & Fang, M. (2015). Pay for (individual) performance: Issues, claims, evidence and the role of sorting effects. *Human Resource Management Review, 25*(1), 5–19. https://doi.org/10.1016/j.hrmr.2014.08.010
24. Budhwar, P., Saini, D. S., & Bhatnagar, J. (2006). Women in management in the new economic environment: The case of India. *Asia Pacific Business Review, 12*(2), 179–193. https://doi.org/10.1080/13602380600571056
25. Shields, J., Brown, M., Kaine, S., Dolle-Samuel, C., North-Samardzic, A., McLean, P., Johns, R., O'Leary, P., Robinson, J., & Plimmer, G. (2015). *Managing employee performance and reward: Concepts, practices, strategies* (2nd ed.). Cambridge University Press.
26. NASSCOM (2023). *Future of work: Redefining workplace and workforce post-pandemic.* National Association of Software and Service Companies.
27. Singh, P., & Loncar, N. (2010). Pay satisfaction, job satisfaction and turnover intent. *Industrial Relations, 65*(3), 470–490. https://doi.org/10.7202/044892ar

The Role of HR Analytics in Enhancing Talent Management: Insights from Bangalore's IT Industry

T. Narayana Reddy and Madhavi Kondeti

Abstract

The purpose of study is to evaluate the impact of HR analytics on talent management practices in select IT companies in Bengaluru City. It specifically focuses on key factors influencing the adoption and application of HR analytics, including data availability, organizational readiness, leadership support, and employee skills. A stratified random sampling method is employed to select a representative sample of 276 IT employees across various levels and functions within the organizations. The research examines how these factors contribute to the effectiveness of talent management practices, such as recruitment, training, performance management, and retention strategies within the IT sector. The statistical tools used for analysis are a reliability test to assess the consistency of the measurement instrument; Exploratory Factor Analysis (EFA) to explore underlying patterns and relationships between the variables, and Multiple Linear Regression to evaluate the impact of the key predictors on talent management outcomes. This research highlights the need for IT companies to formulate robust data governance, leadership-driven analytics strategies, and workforce up skilling policies. This research provides context-specific insights into HR analytics adoption within Bengaluru's ITs companies, connecting leadership support, digital maturity, and talent management outcomes. Future studies should focus on multi-city samples and explore the long-term impact of AI-driven HR analytics on organizational performance.

Keywords

HR Analytics • Talent management • Data availability • Organizational readiness • Leadership Support and employee skills

T. Narayana Reddy (✉) · M. Kondeti
School of Management Studies, Jawaharlal Nehru Technological University Ananthapur, Ananthapuramu, Andhra Pradesh, India
e-mail: tnreddyjntua@gmail.com

S. Patnaik et al. (eds.), *People, Planet and Profit: Advancing Sustainable Business Practices*, Springer Proceedings in Geography and Sustainability,
https://doi.org/10.1007/978-3-032-11296-5_22

1 Introduction

Human Resource (HR) Analytics has become a strategic tool that helps firms improve personnel management, boost performance, and gain a competitive edge in the present era of data-driven decision-making. Applying statistical and analytical methods to HR data in order to extract actionable insights is known as HR analytics, sometimes called human resources analytics or talent analysis [1]. HR analytics helps businesses make well-informed decisions about hiring, retaining, managing performance, training, and engagement among employees by utilizing enormous volumes of personnel data. In order to tackle issues like high attrition rates, skill shortages, and workforce diversity, the Information Technology (IT) sector which is distinguished by quick technology improvements and a fiercely competitive personnel market has demonstrated an increasing interest in using HR analytics [2]. In order to link human capital with business objectives and cultivate a data-centric HR culture, top IT organizations are increasingly incorporating HR analytics into their strategic planning. In many firms, HR analytics adoption and use are still in their infancy despite their promise. Successful implementation is hampered by elements such a lack of data literacy, reluctance to change, a lack of technological infrastructure, and worries about data protection [3]. Furthermore, the organization's preparedness, leadership support, and capacity to convert insights into concrete HR results all play a significant role in the strategic effect of HR analytics.

2 Literature Review

In order to match organizational workforce capabilities with long-term strategic goals, HR analytics is essential. Their case study demonstrated how IT companies use analytics to predict talent needs and improve hiring practices [4]. In support of this, a thorough analysis of the literature found that IT businesses are increasingly using HR analytics. Nonetheless, they pointed out enduring issues that frequently impede the widespread application of HR analytics, including data integration, privacy issues, and a deficiency of analytical abilities among HR practitioners [5]. HR analytics, with a focus on the Indian IT industry, is mostly utilized for recruiting analysis, employee engagement tracking, and staff attrition prediction. According to their research, predictive analytics is being used more often than prescriptive analytics, which is still in its infancy [6]. The use of HR analytics is mostly driven by leadership support and digital transformation, according to new trends in Indian IT companies. Nonetheless, a lot of businesses still function at a descriptive or basic level of analytics maturity, frequently utilizing analytics more for reporting purposes than for making strategic decisions [7].

The adoption of HR analytics by Indian IT companies has been hindered by a lack of qualified personnel, inadequate training, and reluctance to depart from conventional HR procedures. They demonstrated how these top IT companies have effectively used learning analytics, attrition modeling, and HR dashboards to support strategic choices, emphasizing the role that leadership commitment and return on investment play in promoting adoption [8]. Lastly, the combination of digital HRM and analytics demonstrates how IT organizations were able to expedite personnel management, particularly in reaction to the shift to remote work following COVID, by utilizing AI and machine learning technologies in HR tasks [9].

IT businesses may match organizational strategy with personnel skills by implementing HR analytics. By predicting talent requirements, improving recruitment tactics, and facilitating more efficient resource allocation, their case study shows how HR analytics is essential to strategic workforce planning [10]. Although HR analytics has become more popular in IT firms, there are still many obstacles to overcome. Among these difficulties include platform-to-platform data integration, employee data protection issues, and a lack of analytics-trained individuals. The report emphasizes that in order to properly utilize HR analytics, these obstacles must be removed [11]. The use of HR analytics in Indian IT companies revealed that, in addition to recruiting analytics and employee engagement tracking, predictive analytics is frequently used to anticipate staff turnover. They did point out, nevertheless, that these firms continue to underuse prescriptive analytics [12]. They discovered that two important factors facilitating the use of analytics are leadership support and digital transformation. In spite of this, a lot of businesses continue to use descriptive analytics and are only at a rudimentary stage of analytics maturity [13].

HR analytics is used as a strategic tool for decision-making by top IT companies like TCS and HCL Technologies. Their case study emphasized the use of learning analytics, attrition models, and HR dashboards, emphasizing the significance of ROI and leadership commitment in the effective implementation of HR analytics [14]. There are a number of obstacles preventing Indian IT companies from using HR analytics, including a shortage of qualified personnel and inadequate training. They also talked about how firms' reluctance to change frequently prevents the transition from conventional HR procedures to data-driven decision-making [15]. AI and machine learning's effects on HRM in IT firms. They discovered that talent management has greatly improved with the use of digital HRM technologies, especially during the post-COVID shift to remote employment. More efficient workforce planning and staff performance management are made possible by these instruments [16]. The part HR analytics plays in hiring new employees for multinational IT firms. According to the survey, HR analytics facilitates better hiring practices, better candidate screening, and improved staff retention tactics. In the very competitive IT sector, where skill is a crucial differentiator, this is especially crucial [17].

The main purpose of HR analytics in IT organizations is to gauge worker productivity, engagement, and performance. Additionally, they pointed out that although the technologies are accessible, companies frequently find it difficult to use them successfully because of a lack of qualified staff and a clear strategic direction [18]. The connection between organizational success and HR analytics maturity in IT firms. According to their research, companies with greater HR analytics maturity levels are better equipped to make wise decisions that enhance overall company performance [19]. HR analytics is being used by Indian IT organizations to support their workforce diversity and inclusion programs. According to their research, analytics tools are being utilized to monitor workforce diversity and guarantee equitable hiring and advancement procedures [11].

HR analytics' function in IT industry performance management. They discovered that businesses are increasingly embracing data-driven insights to determine training requirements, create goals, and assess employee performance. The study did point out that it is still difficult to match these systems with corporate goals [20]. The application of HR analytics to lower employee turnover in the IT industry. According to their research, predictive analytics is frequently used to identify workers who are likely to leave, enabling businesses to take preventative action. Large IT businesses' turnover rates have significantly decreased as a result of this [21]. HR analytics is being used more and more by IT firms to make decisions in real time. They may make better decisions about workforce deployment, job changes, and promotions by integrating data from different departments [22]. The contribution of HR analytics to work-life balance and employee well-being in the IT industry. According to their research, HR analytics solutions can monitor worker well-being, offer workload feedback, and assist in putting work-life balance initiatives into action all of which are critical for preserving good staff engagement in the IT sector [23].

3 Critical Analysis

While the review of literature visibly showcases the *potential* of HR analytics in improving talent management outcomes, it also finds a *visible gap* between available technologies and the effective strategic application.

A **key insight** emerging is the *maturity gap*: Although many IT companies have invested in HR analytics tools, yet, they continue to operate largely at a descriptive analytics level. This limits their ability to derive strategic, predictive, or prescriptive insights that could drive substantial organizational advantage.

Another critical challenge is **skill scarcity. De**spite the availability of well-developed tools, a shortage of trained HR analytics professionals and inadequate upskilling initiatives limits the potential of HR analytics. Leadership support and digital transformation programs appear to be indispensable, but not adequate, for deep analytics integration.

Furthermore, while the application of AI and ML technologies has broadened the scope of HR analytics, facilitating real-time decision-making, HR optimization, and monitoring employee well-being, ethical issues concerning data privacy and algorithmic biases remain underexplored in the reviewed literature.

Underutilization of prescriptive analytics signifies a lost opportunity. The firms can link HR capital strategies and business objectives if they move beyond merely reporting historical patterns to supporting optimized actions derived from the predictive models (Figs. 1 and 2).

Although HR analytics is revolutionizing talent management in Bangalore's IT sector, the firms must strategically address maturity challenges, invest in developing analytical abilities, and adopt a comprehensive, ethically sound approach to realize the full strategic advantages of HR analytics.

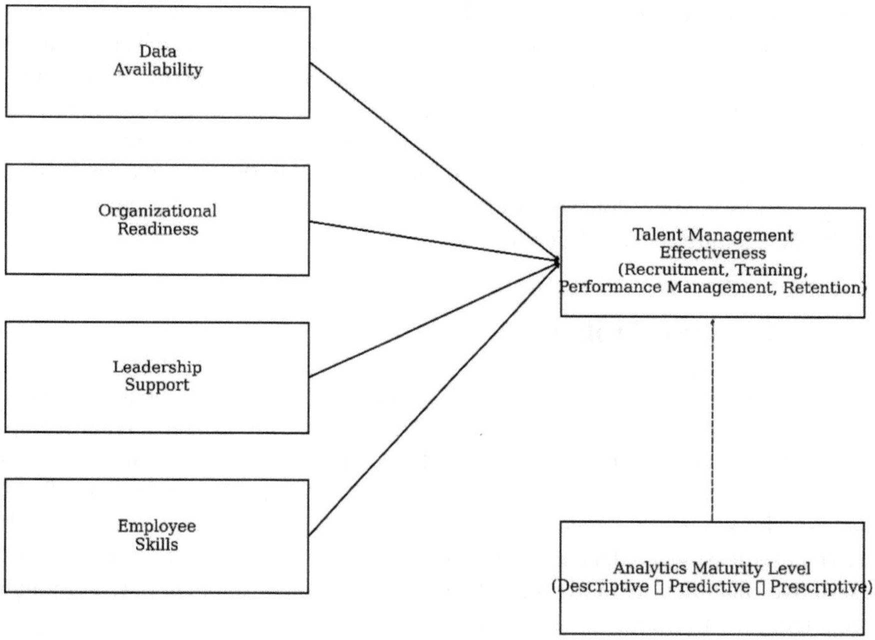

Fig. 1 Conceptual framework linking HR analytics adoption factors to talent management effectiveness

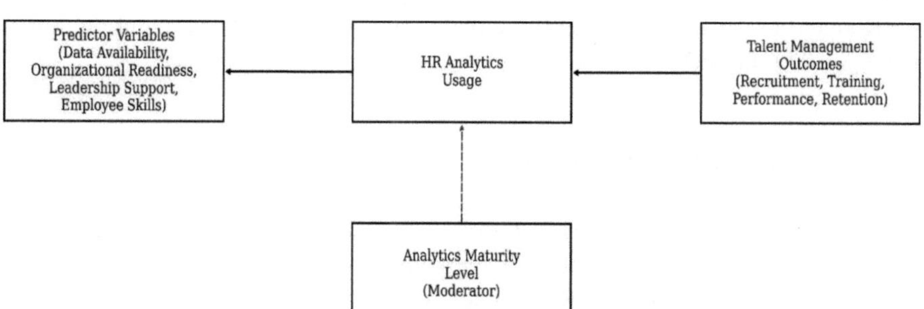

Fig. 2 Research model depicting relationships among organizational factors, HR analytics usage, and talent management outcomes

4 Research Gap

Despite increasing popularity of HR analytics in IT Firms, substantial research gap remains concerning its strategic integration and practical implementation at several organizational levels. Existing research mostly emphasizes on theoretical models and potential benefits, often overlooking the practical problems faced by HR professionals in practical implementation. Additionally, empirical research mostly focused on major IT firms like TCS and Infosys, with little focus given to mid-sized and emerging IT firms that often lack the infrastructural resources and competencies necessary for the adoption of advanced analytics. The evolution of analytics maturity, mostly the transition from descriptive to predictive and prescriptive analytics continues to remain underexplored. Furthermore, longitudinal research assessing the sustained effects of HR analytics on workforce planning, employee engagement, and retention are few. Addressing these limitations is necessary for formulating context-specific, actionable insights for the dynamic and rapidly developing Indian IT sector.

5 Research Problem

The use of data-driven decision-making in human capital management has amplified the significance of Human Resource (HR) analytics in organizational set up, particularly in the technology-oriented IT sector. Despite being early adopters of innovation, several IT firms face substantial challenges in integration of HR analytics. Important challenges include fragmented data systems, shortage of HR analytics professionals, data security issues, and a limited ability to move beyond descriptive analytics to more predictive and prescriptive applications. Although recognizing the potential advantages of HR analytics in improving recruiting, performance management, employee engagement, and retention, many IT firms face challenges in its strategic implementation and development. This study seeks to examine the firms' factors impacting the adoption of HR analytics, identify problems to effective implementation, evaluate analytics maturity, and assess the impact of HR analytics on strategic workforce planning and talent management outcomes in Bengaluru's IT sector.

6 Research Objectives

- To identify and evaluate critical organizational factors (like data availability, organizational preparedness, leadership support, employee competencies) impacting the adoption of HR analytics in IT firms.
- To evaluate the impact of HR analytics adoption on specific talent management practices such as recruitment, employee training, performance management, and retention.

- To evaluate the maturity level of HR analytics usage (descriptive, predictive, prescriptive) among IT organizations and its correlation with strategic HR results.

7 Statistical Tools

- Reliability test
- EFA
- Multiple Linear Regression.

8 Research Methodology

A quantitative research design was adopted to examine the factors impacting the adoption of HR analytics and its impact on talent management practices in Bengaluru's IT industry. Stratified random sampling was used to pick 276 respondents from six IT firms of differing sizes. A stratified random sample technique was utilized to ensure balanced representation within the Bengaluru IT sector. The corporate houses were categorized into 3 tiers in accordance with their size: startups (less than 100 people), mid-sized enterprises (100–1000 employees), and major firms (more than 1000 employees). Employees within each strata selected company were then divided by their functional departments such as HR, IT, Operations, and Management. Proportionate random selection was then made within each stratum to select respondent, yielding a final sample of 276from 6 IT firms. Data was obtained by administering a standardized questionnaire. Cronbach's Alpha was used to assess the reliability of the survey instrument, Exploratory Factor Analysis (EFA) was used to extract and validate the latent constructs related to HR analytics adoption, and Multiple Linear Regression was used to investigate the predictive impact of organizational factors like as data availability, organizational readiness, leadership support, and employee skills—on talent management outcomes in selected IT firms in Bengaluru.

9 Data Analysis

9.1 Reliability Test

The study's measuring scale has adequate internal consistency, according to the reliability statistics shown in Table 2. The 12 items' Cronbach's Alpha rating of 0.739 indicates that the underlying construct is being reliably measured by the items. A Cronbach's Alpha score above 0.7 is regarded as appropriate for social science research, demonstrating that the scale is sufficiently dependable by generally recognized standards. This indicates that the instrument may be relied upon to produce reliable findings in evaluating study-related variables, such as talent

Table 1 Case processing summary

		N	%
Cases	Valid	276	100.0
	Excluded[a]	0	0.0
	Total	276	100.0

Table 2 Reliability statistics

Cronbach's alpha	No. of items
0.739	12

management and its affecting elements, and that the respondents' responses to the items are consistent (Table 1).

10 Exploratory Factor Analysis

The data is appropriate for factor analysis, according to the findings of Table 3, which displays the KMO Measure of Sampling Adequacy and Bartlett's Test of Sphericity. The KMO value of 0.720 is higher than the suggested cutoff of 0.6, indicating that the variables have enough shared variance and the sampling is sufficient to support the use of component analysis. Furthermore, a very significant Chi-Square value of 843.398 with 66 degrees of freedom and a significance level of 0.000 is displayed by Bartlett's Test of Sphericity. This satisfies a crucial factor analysis premise by demonstrating that the correlation matrix is not an identity matrix and that there is a strong connection between the variables (Table 4).

From the 12 variables in the study, four unique components were identified using Varimax Rotation with Kaiser Normalization. Variables with factor loadings higher than 0.5 make up each factor. After the 12 variables were combined into four factors, the four factors together explained 65.581% of the variation in the important HR analytics parameters for IT employees (Table 5).

The matrix above displays the correlation between each variable and the components that were recovered. Typically, each variable has a substantial connection with one component and a modest association with the others. To identify which variables belong to each factor, the variable with the highest value in each row is selected as a component of the related factor. The highest values in each row have been highlighted to help group the 12 variables into four main categories and exclude those with low loadings.

Table 3 KMO and bartlett's test

KMO measure of sampling adequacy		0.720
Bartlett's Test of Sphericity	Approx. Chi-Square	843.398
	Df	66
	Sig	0.000

Table 4 Total variance explained

Component	Initial eigenvalues			Extraction sums of squared loadings			Rotation sums of squared loadings		
	Total	% of variance	Cumulative %	Total	% of variance	Cumulative %	Total	% of variance	Cumulative %
1	3.991	33.256	33.256	3.991	33.256	33.256	2.712	22.603	22.603
2	1.694	14.114	47.369	1.694	14.114	47.369	2.421	20.171	42.774
3	1.128	9.399	56.768	1.128	9.399	56.768	1.623	13.525	56.299
4	1.057	8.812	65.581	1.057	8.812	65.581	1.114	9.281	65.581
5	0.980	8.165	73.745						
6	0.759	6.328	80.073						
7	0.615	5.129	85.202						
8	0.567	4.729	89.931						
9	0.457	3.806	93.737						
10	0.422	3.518	97.254						
11	0.186	1.550	98.804						
12	0.144	1.196	100.000						

Extraction Method: Principal component analysis

Table 5 Rotated component matrix

Statements	Component			
	1	2	3	4
Our organization's IT systems are connected and ready to assist HR analytics	0.863			
Data from our company is kept in a format that makes it simple to access for HR analytics	0.756			
The company has set aside enough funds, time, and other resources for HR analytics projects	0.717			
The deployment of HR analytics is strongly supported by top management		0.573		
Our business is equipped with the technology needed to use HR analytics		0.870		
All pertinent HR operations are covered by the extensive data utilized in HR analytics		0.748		
When employing HR analytics, our company guarantees data confidentiality and privacy			0.693	
Our company has a well-defined plan for applying HR analytics			0.759	
Data-driven decision-making in HR procedures is highly valued by our company			0.684	
In terms of infrastructure and resources, our company is prepared to implement HR analytics				0.559
The use of analytics and data in strategic HR decisions is encouraged by leadership				0.909

11 Multiple Linear Regression

Multiple regression analysis was used to determine how independent variables affected talent management as a dependent variable.

The regression model that explains talent management using the factors of data availability, organizational readiness, leadership support, and employee skills is statistically robust, according to the Model Summary shown in Table 6. A high positive correlation between the independent and dependent variables is indicated by the R value of 0.661. With a R Square value of 0.437, the model has a moderate to good level of explanatory power, accounting for around 43.7% of the variation in talent management. The model's dependability is nonetheless supported by the somewhat lower Adjusted R Square (0.425), which takes sample size and predictor count into consideration. There is not much variation around the expected values, as indicated by the Standard Error of the Estimate of 0.524. Furthermore, the Durbin-Watson statistic of 1.918 indicates that there is no discernible autocorrelation in the residuals because it is within the permissible range of 1.5 to 2.5. All things considered, the model is statistically valid and appropriate for explaining the variables affecting talent management.

The overall importance of the regression model used to forecast talent management based on the independent variables—data availability, organizational readiness, leadership support, and employee skills—is demonstrated by the Analysis of Variance (ANOVA) results in Table 7. With a significance level (p-value) of.000 and an F-value of 36.323, the regression model is statistically significant at the 1% level. This indicates that the variance in talent management results is substantially explained by the four variables taken together. With a regression sum of squares (SS) of 39.898 and a residual SS of 51.352, it is evident that the predictors account for a significant amount of the variance (91.250). The low p-value and high F-statistic attest to the model's overall strong fit and the significant contribution of at least one predictor to the explanation of the dependent variable, talent management (Table 8).

Significant results are found for each independent variable in the multiple regression analysis used to evaluate the influence of different variables on talent management:

- Talent management is positively and statistically significantly impacted by organizational readiness, as evidenced by its standardized beta of 0.351 and

Table 6 Model summary

Model	R	R square	Adjusted R square	Std. error of the estimate	Durbin-Watson
1	.661[a]	.437	.425	.524	1.918

a. Predictors: (Constant), Data Availability, Organizational Readiness, Leadership Support, Employee Skills

b. Dependent Variable: Talent Management

Table 7 Analysis of variance

Model		Sum of squares	df	Mean square	F	Sig
1	Regression	39.898	4	9.975	36.323	.000[b]
	Residual	51.352	272	.275		
	Total	91.250	271			

a. Dependent variable: talent management

b. Predictors: (Constant), data availability, organizational readiness, leadership support, employee skills

Table 8 Coefficients

Model		Unstandardized coefficients		Standardized coefficients	t	Sig
		B	Std. error	Beta		
1	(Constant)	1.512	.398		3.803	.000
	Organizational Readiness	.157	.062	.351	2.919	.000
	Leadership Support	.284	.078	.248	3.618	.000
	Employee Skills	.174	.050	.269	3.500	.001
	Data Availability	.245	.056	.293	4.403	.000

a. Dependent variable: talent management

unstandardized coefficient (B) of 0.157. A significance level of 0.000 and a t-value of 2.919 show a substantial correlation between improved talent management methods and greater organizational preparedness levels. This implies that an organization is more successful at managing talent when it is well-prepared in terms of structure, strategy, and culture.

- Additionally, there is a clear and favorable correlation between leadership support and talent management. The t-value of 3.618 and the significance level of 0.000 demonstrate that leadership support strongly contributes to the dependent variable, with a B value of 0.284 and a standardized Beta of 0.248. This suggests that in order to establish and maintain successful personnel management methods, leadership support and engagement are essential.
- Employee skills, with an unstandardized coefficient of 0.174 and a beta of 0.269, have a beneficial impact on talent management. A t-value of 3.500 and a significance value of 0.001 attest to the fact that competent workers are necessary for effective talent management. This emphasizes how crucial training and development programs are to improving workforce competencies.
- Out of all the variables, Data Availability has the highest standardized Beta (0.293) and a B value of 0.245, making it a significant predictor. The significance level of 0.000 and the t-value of 4.403 demonstrate how important

it is to have and use correct data in order to make well-informed personnel management choices.

12 Discussion: Linking Findings to Prior Research

The increasing significance of HR analytics in refining talent management practices is well-established in the literature [19]. The findings of this study further strengthen that the proper use of HR analytics evidently improves recruitment, employee engagement, performance management, and employee retention inside the IT firms.

The study confirms findings and found that the accessibility of high-quality data is a pivotal for the success of HR analytics, with data availability identified as the most important predictor of successful talent management practices in Bangalore's IT industry [11].

Previous research emphasized on the need for leadership support for analytics-driven HR transformation. The research found that leadership support significantly impacted the adoption and use of HR analytics [20]. IT firms demonstrating leadership commitment shown a more strategic use of HR data which corroborated the findings of [21], who acknowledged leadership as a driving force for analytics maturity.

Organizational readiness, as examined in studies like [22], was identifies as an important factor impacting the HR analytics adoption. This study exhibited that the IT firms with superior technological infrastructure and strategic readiness were evidently more successful in aligning HR analytics with business objectives, thereby validating previous claims regarding the vital role of organizational readiness in strategic HR initiatives.

Employee skills have been persistently identified as an impediment in the HR analytics adoption [23]. This research found that inadequate analytical skills among HR professionals limits the effective use of analytics technologies, thereby posing challenges the conversion of HR insights into strategic decisions. The findings correspond with previous research [24] highlighting the need for upgrade HR capabilities to go beyond descriptive analytics.

This survey confirms prevailing concerns about analytics maturity levels. [25] Recognize that several IT firms are limited to descriptive analytics. The present findings reveal that while there is limited application of predictive analytics, prescriptive analytics even remains underused, limiting the strategic impact of HR analytics in workforce planning and decision-making.

Furthermore, in alignment with [26], the research highlights the emerging role of HR analytics in improving employee well-being and work-life balance, recommending real-time integration of HR data to provide proactive measures to effective employee engagement and job satisfaction.

This research not only corroborates prior studies but also expands it by collecting actual data from mid-sized and rising IT enterprises, a domain hitherto remains underexplored in previous studies. This research adds to the body of knowledge on how HR analytics can revolutionize talent management strategies in the evolving Indian IT sector by merging data-driven insights with organizational factors.

13 Policy Implications

These findings of this research work have significant policy implications for IT firms aiming to optimize the strategic potential of HR analytics in talent management. The IT firms should formulate the comprehensive data management policies to ensure the availability, accessibility, and confidentiality of employee data. This requires investment in secure Human Resource Information Systems (HRIS) and the establishment of clear data governance standards. Secondly, leadership-driven policies to promote a data-centric culture are crucial. Senior executives should actively encourage HR analytics initiatives by investing on resources, encouraging evidence-based decision-making, and incorporating analytics into HR workflows. Third, there is a need for policies aimed at augmenting organizational capabilities to upgrade analytical skills of HR professionals through organized training programs and certifications. Improving HR professional skills can address the existing skill gaps which obstruct the efficient application of HR analytics technologies. The IT firms can adopt strategic analytics maturity models to enable the smooth transition from descriptive reporting to predictive and prescriptive analytics applications. These policies will optimize efficiency of HR operations and foster better strategic alignment between human capital and business objectives (Fig. 3).

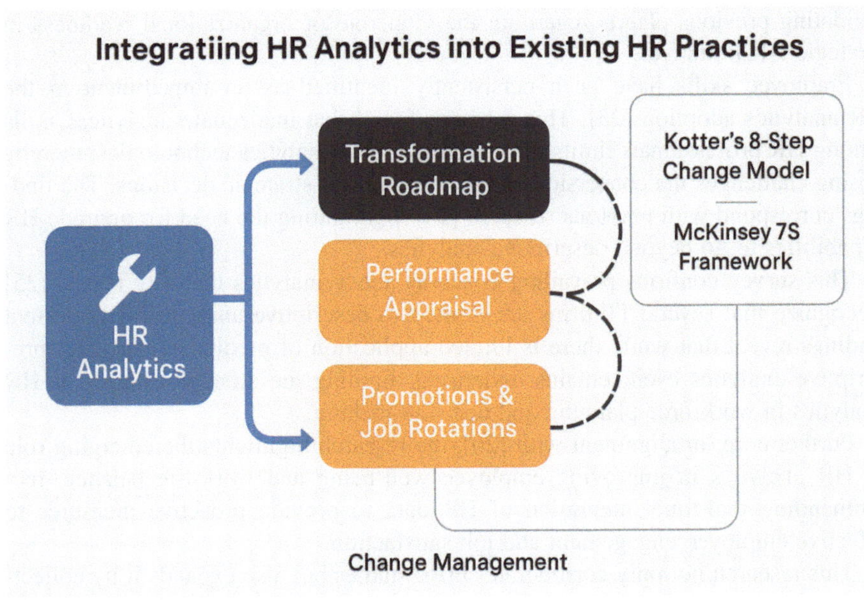

Fig. 3 Integrating HR analytics into existing HR practices

IT firms should implement a structured change management roadmap that blends data-driven insights with conventional HR operations, making it easier to practically incorporate HR analytics into existing HR operations. For example, HR professionals may identify skill deficiency and offer personalized learning pathways by integrating predictive analytics into employee learning and development. Similarly, the firms may move from subjective assessments to data-driven evaluations based on productivity, goal alignment, and peer feedback metrics by incorporating analytics into performance appraisal systems. The firms may even automate decision-making processes for internal mobility based on predetermined performance benchmarks by using prescriptive analytics. The businesses can use change management frameworks such as the McKinsey 7S Framework, which prioritizes alignment among seven organizational elements—strategy, structure, systems, shared values, skills, style, and staff [27] or Kotter's 8-Step Change Model, which depicts steps like creating urgency, forming a guiding coalition, and anchoring new approaches in the culture, to manage this transition successfully [28]. These offer practical actions for coordinating analytics projects with systems, company culture, and structure. It must ensure that HR analytics supports and supplements the existing HR processes rather than operating standalone and these incorporation tactics upsurge adoption rates and business relevance.

14 Directions for Future Research

Future research should address the stated limitations. The longitudinal studies are required to investigate the sustained effects of HR analytics on talent management outcomes such as employee engagement, attrition rates, and the firm performance over a period of time.

Secondly, comparative studies might investigate how adoption of HR analytics differ across firms of different sizes (startups, mid-sized companies, and major businesses) and cultural context within India's IT industry. Third, further research should emphasize on creating standardized HR analytics maturity models to support firms in assessing their present analytical capabilities and systematically plan improvements. Fourth, experimental research might evaluate the effectiveness of specific interventions. This includes leadership coaching or targeted training programs which can increase adoption and effectiveness of HR analytics.

Ultimately, future research may explore the ethical dilemmas and data privacy concerns associated with the augmented utilization of AI-driven HR analytics technologies, offering frameworks for responsible HR analytics practices.

15 Conclusion

This research strengthens the critical role of HR analytics in revolutionizing talent management practices in Bengaluru's IT industry. The findings empirically validate that data availability, organizational readiness, leadership support, and employee capabilities greatly impact the effective use of HR analytics. This highlights the increasing significance of data-driven HR strategies. Despite the growing adoption of HR analytics, the research identifies significant challenges, particularly on usage of HR maturity analytics model and shortage of shortage of HR professional to leverage from using advanced analytics technologies. Addressing these limitations necessitates investments in technical infrastructure, leadership commitment to support effective use of HR analytics and HR professional analytical skills to use HR technologies. As IT companies operate within a more competitive and dynamic business landscape, HR analytics emerges not alone as a technological instrument but as a strategic enabler for workforce optimization and business success. Consequently, advancing the maturity and linking HR analytics with organizational strategic decision-making is necessary for IT firms seeking to sustain competitive advantages in a period marked by rapid digital transformation.

References

1. Angrave, D., Charlwood, A., Kirkpatrick, I., Lawrence, M., & Stuart, M. (2016). HR and analytics: Why HR is set to fail the big data challenge. *Human Resource Management Journal, 26*(1), 1–11.
2. Chen, Z., & Wang, M. (2023). HR analytics in promoting employee well-being and work-life balance in the IT industry. *Journal of Human Resource Management, 36*(2), 122–138.
3. Davenport, T. H., Harris, J. G., & Shapiro, J. (2010). Competing on talent analytics. *Harvard Business Review, 88*(10), 52–58.
4. Dey, A., & Sanyal, S. (2023). HR analytics maturity and organizational performance: Evidence from IT companies. *Journal of Business Research, 45*(3), 56–65.
5. Ghosh, R., & Ghosh, A. (2022). Human resource analytics and its role in enhancing organizational performance: Evidence from Indian IT sector. *South Asian Journal of Human Resources.*
6. Gupta, P., & Das, S. (2022). The application of HR analytics for diversity and inclusion in Indian IT companies. *International Journal of Human Resource Studies, 11*(2), 105–118.
7. Jain, M., & Chawla, S. (2023). Emerging trends in HR analytics: A study of Indian IT companies. *International Journal of Management (IJM), 14*(1).
8. Joshi, S., & Patel, R. (2020). The role of HR analytics in turnover reduction in Indian IT firms. *Indian Journal of HRM, 8*(1), 24–34.
9. Kavitha, R., & Srinivas, K. (2024). Digital HRM and analytics: Transforming talent management in IT firms. *Indian Journal of HRM Research, 9*(2).
10. Margherita, A., & Braccini, A. M. (2020). HR analytics for strategic workforce planning: A case study. *The International Journal of Human Resource Management.* https://doi.org/10.1080/09585192.2020.1737830
11. Marler, J. H., & Boudreau, J. W. (2017). An evidence-based review of HR analytics. *The International Journal of Human Resource Management, 28*(1), 3–26.
12. Mehta, S., & Patel, R. (2024). HR analytics as a strategic tool: A case study of TCS and HCL Technologies. *Journal of Strategic Human Resource Management.*
13. Patel, S., & Shah, S. (2021). The role of HR analytics in performance management within IT companies. *Journal of Management Studies, 39*(4), 45–60.
14. Rao, V., & Kumar, R. (2023). Barriers to adoption of HR analytics in Indian IT firms. In *Proceedings of the 2023 International Conference on Business Analytics and Intelligence.*
15. Singh, A., & Sharma, V. (2021). The adoption of HR analytics in IT firms for employee engagement and performance tracking. *Asian Journal of Business Analytics, 12*(2), 35–50.

16. Srinivas, A., & Sharma, R. (2022). Real-time HR decision-making through analytics: A study of IT companies. *Journal of Technology and Management, 15*(1), 99–110.
17. Tursunbayeva, A., Pagliari, C., Di Lauro, S., & Antonelli, G. (2021). HR analytics in practice: A systematic literature review. *Human Resource Management Review, 31*(2), Article 100765.
18. Wang, Z., & Zhang, Y. (2022). The role of HR analytics in talent acquisition: A global IT perspective. *Journal of Talent Management, 20*(4), 122–138.
19. Jain, D., & Sharma, H. (2024). Snapshot of digital transformation from the perspective of human resource management: A bibliometric approach. *Business Process Management Journal, 30*(3), 726–753.
20. Davenport, T. H., Harris, J. G., & Morison, R. (2010). Analytics at work: Smarter decisions, better results. *Harvard Business Press.*
21. Patel, K., Patel, M., Mehta, M., & Patel, S. (2024). Advancing Sustainable Development Through Machine-Learning-Based. In *Proceedings of Fifth International Conference on Computing, Communications, and Cyber-Security: IC4S'05 Volume 1* (Vol. 1, p. 409). Springer Nature.
22. Margherita, E. G., & Braccini, A. M. (2020). Organizational impacts on sustainability of industry 4.0: a systematic literature review from empirical case studies. In *Digital Business Transformation: Organizing, Managing and Controlling in the Information Age* (pp. 173–186). Springer International Publishing.
23. Rao, P., Kumar, S., Chavan, M., & Lim, W. M. (2023). A systematic literature review on SME financing: Trends and future directions. *Journal of Small Business Management, 61*(3), 1247–1277.
24. Haakenstad, A., Yearwood, J. A., Fullman, N., Bintz, C., Bienhoff, K., Weaver, M. R., & Gupta, V. K. (2022). Assessing performance of the Healthcare Access and Quality Index, overall and by select age groups, for 204 countries and territories, 1990–2019: A systematic analysis from the Global Burden of Disease Study 2019. *The Lancet global health, 10*(12), e1715–e1743.
25. Sanyal, P., & Dey, P. (2023). Using a deep learning neural network for the identification of malignant cells in effusion cytology material. *Cytopathology, 34*(5), 466–471.
26. Li, Y., Zhang, Y., Timofte, R., Van Gool, L., Yu, L., Li, Y., & Wang, X. (2023). NTIRE 2023 challenge on efficient super-resolution: Methods and results. In *Proceedings of the IEEE/CVF Conference on Computer Vision and Pattern Recognition* (pp. 1922–1960).
27. Kotter, J. P. (1996). *Leading change.* Harvard Business School Press.
28. Waterman, R. H., Peters, T. J., & Phillips, J. R. (1980). Structure is not organization. *Business Horizons, 23*(3), 14–26. https://doi.org/10.1016/0007-6813(80)90027-0

Unleashing Digital Innovation for Circular Economy Transformation: A Thematic Analysis of Industry Case Studies

Manas Ranjan Chowdhury, Jasmin Patnaik, and Anita Pareek

Abstract

This study applies thematic analysis to explore how digital transformation is integrated in driving sustainable growth within the circular economy across five case studies from global industry leaders: Philips, Unilever, BMW, Interface, and Nestlé. These case studies show how the implementation of advanced digital-based technologies, such as IoT, AI, blockchain, and 3D printing, supports circular economy principles such as resource optimization, waste reduction, and product lifecycle extension. Through thematic coding, the approaches of the firms were found to have frequent themes about technological innovation and sustainability challenges, efficiency gains and collaborative efforts. The frequent themes suggest that digital tools enable transparency as well as track resource flows but also provide common barriers such as high implementation costs and supply chain complexity. These issues add to the contribution for better understanding of how businesses can use technology in stimulating circular economies, solving sustainability hurdles, and building sustainable, resource-constrained long-term values. This paper presents contributions both to academia and industries looking to understand the prospects offered by digital transformation toward steering firms through a circular economy context.

M. R. Chowdhury (✉)
Trident Academy of Technology, Bhubaneswar, India
e-mail: manas.choudhury@tat.ac.in

J. Patnaik · A. Pareek
Kalinga Institute of Social Sciences (Deemed to be University), Bhubaneswar, India
e-mail: jasmin.patnaik@kiss.ac.in

A. Pareek
e-mail: anita.pareek@kiss.ac.in

S. Patnaik et al. (eds.), *People, Planet and Profit: Advancing Sustainable Business Practices*, Springer Proceedings in Geography and Sustainability,
https://doi.org/10.1007/978-3-032-11296-5_23

Keywords

Business sustainability • Circular economy • Digital innovations • Global industries

1 Introduction

In the face of growing environmental concerns, organizations are under pressure to transition from linear economic models (take-make-dispose) to circular economy (CE) practices that emphasize sustainability through reuse, recycling, and resource efficiency. Traditional economic practices prioritize short-term consumption, while the circular model reimagines production and consumption loops for long-term value. Digital transformation—the integration of advanced technologies into core business processes—offers a critical lever for enabling this shift. This paper investigates how digital innovation facilitates CE transformation by analyzing five global industry leaders. These companies deploy digital tools like the Internet of Things (IoT), Artificial Intelligence (AI), blockchain, and 3D printing to address sustainability challenges and operational inefficiencies. Through thematic analysis, this research identifies recurring strategies, obstacles, and outcomes in their CE journeys.

A significant insight from the case studies is the role of digital technologies in helping organizations track, optimize, and manage resources effectively. For example, IoT enables real-time monitoring of product usage, allowing companies like Philips to optimize product lifecycles through predictive maintenance [1]. Similarly, blockchain technology ensures transparency and traceability in supply chains, as demonstrated by Nestlé's initiatives to make plastic packaging recyclable [2]. Furthermore, AI supports resource reallocation, waste minimization, and logistics. For instance, Unilever employed AI to optimize its supply chain and reduce carbon emissions [3].

Despite the clear potential of these technologies to advance circular economy goals, companies face significant challenges. These include high upfront costs associated with deploying new technologies, difficulties in integrating digital tools with existing infrastructure, and the necessity for cross-sector collaboration to scale digital solutions [4]. Such barriers hinder the efficient execution of circular strategies, particularly in global supply chains involving diverse stakeholders with varying levels of technological readiness.

2 Review of Literature

Digital transformation, along with the circular economy, has thus increasingly been recognized as crucial to achieving sustainability goals in a period manifested by resource depletion and environmental deprivation. Digital technologies, from IoT and AI to blockchain and 3D printing, are enabling businesses to optimize their use of resources, extend product life cycles, and reduce waste. This section reviews the literature on how digital transformation is impacting circular economy practices, both in terms of potential benefits and challenges.

2.1 Digital Technologies in the Circular Economy

An increasing amount of research suggests that digital technologies are essential to enabling the shift from linear to circular business models. The circular economy aims to disentangle economic growth from resource consumption, which in turn encourages regenerative systems like recycling, remanufacturing, and product life extension [5]. In this regard, IoT is a key enabler in allowing companies to collect real-time data regarding the usage, condition, and lifecycle of products. This information can be applied to optimize operations, minimize energy consumption, and maximize resource use efficiency [6].

Besides IoT, AI is another influential tool for the management of circular supply chains. Girotra [7] argue that analytics driven by AI will enhance decision-making in terms of resource allocation, waste reduction, and logistics optimization. For instance, AI can predict the approaching end-of-life of products or parts so that some action like repair or refurbishment can be taken before it's too late. Similarly, blockchain has picked up because of its ability to bring in transparency and traceability into circular supply chains, especially to trace the materials from production to disposal or recycling [8].

2.2 The Role of Digital Transformation in Business Sustainability

Several studies have been conducted to explore how digital transformation impacts business sustainability, focusing on operational efficiencies and broader environmental impacts. Niinimäki [9] note that digital technologies are transforming industries like fashion and automotive, as companies begin to embrace circular approaches in response to consumer pressure for sustainable products. These technologies help companies achieve higher material recovery rates and reduce waste, following the principles of circular economy.

Zeng [10] also argue that digital technologies can provide a business with the infrastructural underpinning needed to achieve more sustainable and circular forms of production and consumption. Digital platforms are, they assert, important enablers of collaboration among stakeholders and the sharing of data for improving the circularity of products. For example, in the car industry, BMW uses digital technologies, for example, 3D printing to manufacture spare parts that are ordered only when they are needed, which reduces waste and excess inventory [11].

2.3 Challenges in Adopting Digital Technologies for Circular Economy Practices

However, with potential benefits, challenges lie in the shifting of digital technologies for circular economy practices. Among these barriers, high initial investment needed for digital infrastructure is considered one of the most prominent ones. Rizos [12] say that the transfer to a circular economy often utilizes high capital investments in new technologies, processes, and systems. For SMEs, the cost can be too high for them to compete in a circular economy. Another challenge is the integration of digital solutions into the existing business models and supply chains. According to [13], the process of digital transformation is often coupled with the reorganization of the traditional supply chains and business processes, which might resist change by the stakeholders who are accustomed to conventional practices. Furthermore, the coordination of circular initiatives across multiple regions with different regulatory environments makes it even more complex, as demonstrated by Nestlé in tracking the recyclability of packaging using blockchain (Monczka et al., 2020).

The literature highlights various emerging trends involving the integration of digital technologies with the circular economy. Lacy [14] assert that increasing the capability of digital technologies will further enable scaling up models related to circular economy, particularly through smart products and extended producer responsibility initiatives. Besides, Kozlowski et al. (2019) predict that digital supply chain management systems will increasingly be applied to trace and monitor the material flows in order to easily and easily identify global reuse and recycling opportunities for the companies. In summary, literature has proven that digital transformation is at the core of attaining circular economy goals. The integration of IoT, AI, blockchain, and others into business activities can notably advance resource efficiency, optimize the cycle of product lifetimes, and reduce the environmental impacts significantly. However, scholars have underscored that the process of migrating into a circular economy is also subject to several challenges, particularly in the issues of cost of investments, system integration, and even global collaboration. Future research should, therefore, aim to address these barriers and explore how emerging technologies will further accelerate the pace of this circular economy transition.

3 Research Methodology

This study adopted a qualitative research approach in exploring the amalgamation of digital transformation with sustainability practices within the circular economy context. More specifically, this research employs the case study approach, with an application of five companies: Philips, Unilever, BMW, Interface, and Nestlé. The reasons for choosing these organizations include their leadership in making the best use of digital technologies to drive sustainable growth and adopt circular economy models.

The study uses a **qualitative research methodology**, specifically employing **thematic analysis** based on five case studies. This approach is highly appropriate for the topic for several reasons:

- **Exploratory Nature of the Study:** The research aims to explore the integration of digital technologies within the circular economy, which benefits from in-depth qualitative insights.
- **Rich Contextual Understanding:** By analyzing case studies from leading global companies like Philips, Unilever, BMW, Interface, and Nestlé, the study captures detailed and contextual data on how these organizations adopt and implement digital tools.
- **Identifying Patterns and Themes:** Thematic analysis allows the study to identify recurring themes, such as technological innovation, resource efficiency, and consumer engagement, offering a structured way to analyze diverse data sources.
- **Flexibility in Data Sources:** The research draws on secondary data, including sustainability reports, media releases, and academic literature. The qualitative method is well-suited to synthesizing insights from these heterogeneous sources.
- **Relevance to Circular Economy Principles:** The method emphasizes patterns and insights rather than numerical data, aligning well with the complex and systemic nature of circular economy practices.

This methodology ensures the findings are comprehensive and relevant, offering actionable insights for businesses, policymakers, and researchers.

1. Data Collection

Data on public sources are used as main inputs for this research, specifically these data, including:

- Company reports (e.g., Sustainability, Annual Reports)
- Media releases and news articles containing coverage of digital transformation initiatives
- Academic literature, and industry case studies documenting the companies' digital as well as sustainability strategies
- Official website and corporate blogs discussing practices related to circular economy, and technological innovations.

The process of collecting data was based on secondary data sources to ensure that it covers all and comprehensively understands the efforts and strategies of the companies. The materials are very detailed on how the companies apply digital technologies like IoT, AI, blockchain, and 3D printing in the resource optimization, product lifecycle management, and supply chain transparency.

2. Data Analysis

This research used thematic analysis, a highly recognized qualitative method of identifying, analyzing, and interpreting patterns within qualitative data. The steps to follow in the following procedure are as follows:

- Familiarization: Getting an immersion into the collected data so that the initial patterns related to digital transformation and sustainability are identified.
- Coding: Creating the first codes through tagging parts of the data which relates to themes like technological innovation, resource efficiency, and circular business models.
- Theme Identification: Aggregating similar codes into higher-level themes, such as Technological Innovation for Circularity, Circular Business Models, Consumer Engagement, and Supply Chain Transparency.
- Review and Refinement: Examining and improving the themes found to make sure they appropriately represent the information and support the goals of the study.
- Reporting: Consolidating the last themes in a logical storyline answering the research questions and enlightening as to how digital technologies drive sustainability within the circular economy.

Data Analysis

Generating Initial Codes

Meaningful pieces of data, statements, or actions in the context of each case study are reflections on important facets of digital technology integration and their contributions to the circular economy or sustainability.

These are the preliminary codes as follows:

Case Study 1. Philips

- Digital technologies: IoT-enabled products, smart sensors, and data analytics.
- Sustainability practices: Circular product design, energy-efficient lighting, and extended product life.
- Data-driven decisions: Real-time tracking of product performance using IoT to enhance sustainability.
- Customer engagement: Digital platforms for customers to track energy savings and participate in product take-back programs.
- Circular business model: Product-as-a-service model, offering lighting-as-a-service to customers.

Initial Codes

- IoT-enabled products
- Smart sensors

- Data-driven decision-making
- Energy efficiency
- Product-as-a-service
- Circular product design
- Customer engagement.

Case Study 2. Unilever

- Digital technologies: AI for supply chain optimization, blockchain for transparency.
- Sustainability practices: Reducing plastic waste, sustainable sourcing, and circular packaging.
- AI-driven insights: Using AI to reduce carbon emissions, optimize supply chains.
- Blockchain for traceability: Ensuring transparency in sourcing and packaging sustainability.
- Sustainable consumer behavior: Digital apps and tools to encourage sustainable consumer choices.

Initial Codes

- AI-driven supply chain optimization
- Blockchain for traceability
- Reducing plastic waste
- Sustainable sourcing
- Circular packaging
- Carbon emissions reduction
- Sustainable consumer behavior.

Case Study 3. BMW

- Digital technologies: 3D printing, blockchain, and IoT for circular supply chains.
- Sustainability practices: Remanufacturing, reusing parts, and optimizing resource use.
- Product design innovation: Designing vehicles with components that are easier to recycle.
- Digital manufacturing: 3D printing for on-demand spare parts.
- Supply chain transparency: The use of blockchain in tracing materials to optimize resource use.

Preliminary Codes

- 3D printing for spare parts
- Circular supply chain

- Product remanufacturing
- Resource optimization
- Blockchain for material traceability
- Sustainable product design

Case Study 4. Interface

- Digital technologies: IoT and AI for monitoring sustainability in manufacturing.
- Sustainable practices: Carbon–neutral product, recycled material in the product and closed-loop system.
- IoT and data: Carbon footprint tracking for carpet tiles life cycle.
- Sustainable design: Products must have minimal impacts on environments, and designing for reused products.
- Circular economy model: Recycling waste products from the old tiles into new ones.
- Codes
- IoT for tracking sustainability performance
- Carbon–neutral
- Recyclable materials
- Closed-loop system
- Carbon footprint tracking
- Design for sustainability
- Circular economy model.

Case Study 5: Nestlé

- Digital technologies: Blockchain for traceability, AI for resource management, IoT for packaging and waste tracking.
- Sustainability practices: Reducing plastic packaging, promoting recycling, sustainable sourcing of ingredients.
- Circular packaging: Digital platforms to track recyclability and sourcing of packaging.
- Sustainability insights: AI to optimize supply chains and reduce environmental impact.
- Transparency through blockchain: Monitoring and tracking the recycling process of packaging materials.

Initial Codes

- Blockchain for traceability
- AI for resource management
- IoT for waste tracking
- Reducing plastic packaging
- Circular packaging
- Sustainable sourcing of ingredients

- Transparency in recycling.

Thematic Search

Themes that emerged from the initial codes developed in Step 2, were clustered together into similar codes. These themes will be a reflection of the overall integration of the contribution of digital technologies to the circular economy and sustainability.

Themes

1. Technological innovation for circularity
   - Codes: IoT-enabled products, smart sensors, AI-driven supply chain optimization, 3D printing for spare parts, blockchain for material traceability, AI for resource management, IoT for waste tracking.
   - Description: This theme captures the use of digital technologies that support the circular economy. Companies are utilizing advanced digital tools like IoT, AI, and blockchain to enhance transparency, track product lifecycles, optimize resources, and minimize waste. Digital technologies are key enablers in shifting from linear to circular supply chains.
2. Resource Efficiency and Optimization
   - Codes: Resource optimization, sustainable sourcing, carbon–neutral products, recycling transparency, reducing plastic waste, carbon footprint tracking.
   - Description: This theme underlines how companies use digital technologies to make resource efficiency, reduce emissions, and minimize waste. It deals with sourcing sustainably, and closing material loops by recycling and remanufacturing.
3. Circular Business Models
   - Codes: Product-as-a-service, the circular product design, circular packaging, remanufacturing, closed-loop system.
   - Description: This theme suggests using circular economy business models. Increasingly, companies are looking into circular business models like product-as-a-service, where a merchandise is offered for lease or used, thereby not being thrown away after use. Designing with a closed loop and recyclable packaging are also important parts of this.
4. Sustainability and Consumer Engagement
   - Codes: Customer engagement, sustainable consumer behavior, sustainable sourcing, sustainable product design.
   - Description: This theme is how digital tools are utilized to engage consumers in sustainability efforts. Digital platforms and apps are being used by companies to educate and incentivize consumers to make sustainable choices such as reducing plastic waste or selecting environmentally friendly products. Sustainability practices are also integrated in the design and sourcing of products to align with consumer values.

5. Supply Chain Transparency and Traceability
    - Codes: Blockchain for traceability, source of ingredients in sustainable manner, transparency along the supply chain.
    - Description: This theme deals with blockchain and IoT application for greater transparency as well as traceability across the system of supply chain. Using these technologies, companies ensure consumers receive guaranteed information related to the sustainability and source of materials applied at various stages of producing an end-product.

3.1 Reviewing Themes

Now that the initial themes are identified, it's essential to review them to make sure that they are coherent and do actually represent the data. At this stage, there would be fine-tuning and editing of themes, ensuring they support the research question and make all adjustments necessary for clarity and relevance.

1. Technological Innovation for Circularity
    - Review: The theme of "Technological Innovation for Circularity" captures a wide range of digital tools, from IoT and AI to blockchain and 3D printing, applied across industries to improve circular economy practices. Digital technologies help businesses track, monitor, and optimize product lifecycles, improve manufacturing processes, and ensure that products are designed for reuse or recycling.
    - Refinement: The theme should explicitly highlight the impact of specific technologies, such as AI-driven resource optimization, IoT for real-time monitoring, and blockchain for material traceability. These technologies help companies like Philips and BMW move toward circular production systems by enabling efficient resource management, waste reduction, and end-of-life product recovery.
    - Sub-theme: This larger theme can be categorized into narrower sub-themes like Digital Manufacturing for Sustainability and Resource Tracking and Optimization.
2. Resource Efficiency and Optimization
    - Review: This theme covers the practices that make a firm use better resources, such as reducing emissions, minimizing waste, and maximizing efficiency in production. It is aligned with companies like Interface and Unilever, which aim to use digital tools to help improve more sustainable and efficient manufacturing processes.
    - Refining: Current codes are aptly grouped within this theme. It will make the matter more coherent with a defined difference between optimizing supply chains, for instance, with AI from end-of-life optimization, in terms of remanufacturing and closed-loop systems

- Sub-theme: We may make Supply Chain Efficiency and End-of-Life Optimization two different sub-themes to stress the comprehensive approach towards sustainability.

3. Circular Business Models
   - Review: The concept of Circular Business Models runs across the case studies, since companies are into adopting the product-as-a-service models, recycling systems, and making of products for easy disassembly and reusing. The remanufacturing activities by BMW and product-as-a-service model of Philips are evident examples of circular business strategies that drive long-term sustainability.
   - Refinement: The theme should highlight business model innovation and how these models are in line with the digital technologies for scalability. For instance, Philips' usage of IoT to provide lighting-as-a-service or BMW's usage of digital tools in handling spare parts through 3D printing to extend product life cycles.
   - Sub-theme: A sub-theme could focus on Business Model Innovation (e.g., product-as-a-service), and another could cover Circular Product Design (e.g., designing for reuse, remanufacturing).
4. Sustainability and Consumer Engagement
   - Review: This theme shows how digital tools are encouraging consumer engagement in sustainability through tracking the environmental impact of products and promoting choices that are sustainable. Examples of such consumer-centric sustainability include how Unilever is using AI to optimize supply chains, Interface's use of IoT to measure sustainability in production, and Nestlé's blockchain for traceability in recycling.
   - Refinement: The codes underneath this theme fit well but focus on the fact that such technologies work to engage the consumer in an active sense. For example, Nestle's using blockchain to recycle and make packages recyclable helps get consumers engaged with sustainability measures. Alternatively, the theme might focus better on digital systems by which a consumer can become enabled or empowered enough to track improvements within environmental sustainability.
   - Sub-theme: Another possible sub-theme is the use of digital tools to empower Consumer Education.
5. Supply Chain Transparency and Traceability
   - Review: This case portrays how the theme of Supply Chain Transparency and Traceability plays into the use of blockchain and IoT, especially as these will give visibility to material sourcing and recycling practices. Nestle and Unilever, among others, use blockchain to ensure that the supply chain becomes more transparent and enhances sustainability in all production stages.
   - Refinement: The codes under this theme are very clear, but maybe this theme can also subsume the concept of collaboration across sectors because transparency in the chain requires collaboration from many sides. For instance,

Nestle's collaboration with other stakeholders to develop transparent and safe supply chains.

•*Sub-Theme: a sub-theme, therefore, could be that of collaborative supply chain transparency, focusing on the theme of cooperation across industries.*
Refined Themes

1. **Technological Innovation for Circularity**
   - Sub-themes: **Digital Manufacturing for Sustainability**, **Resource Tracking and Optimization**
2. **Resource Efficiency and Optimization**
   - Sub-themes: **Supply Chain Efficiency**, **End-of-Life Optimization**
3. **Circular Business Models**
   - Sub-themes: **Business Model Innovation**, **Circular Product Design**
4. **Sustainability and Consumer Engagement**
   - Sub-theme: **Consumer Education and Empowerment**
5. **Supply Chain Transparency and Traceability**
   - Sub-theme: **Collaborative Supply Chain Transparency**

Defining and Nomenclating Themes

1. ***Technological Innovation in Circularity***
   - Involves utilizing digital technologies, like the IoT, AI, blockchain, and 3D printing for circular economy improvement purposes, product life management, decreasing waste production, and ensuring resource usability.
   - Applying IoT devices to help track energy use and offer life extension capability, similar to how BMW applies 3D printing for immediate on-demand supply of spares, ensuring minimal to zero waste produced.
2. ***Resource Use Optimization and Efficiency***
   - This theme elaborates on how digital technologies support companies in optimizing the use of resources, reducing carbon footprint, and minimizing waste during a product's entire lifecycle: From sourcing to disposal.
   - Supply Chain Optimization with AI-driven insights by Unilever or IoT-based monitoring for decreasing energy consumption in manufacturing processes for Interface.
3. ***Circular Business Models***
   - This area is on circular business models that create value through the reuse, remanufacturing, and recycling of products. Digital technologies support these models by ensuring efficiency and scalability.
   - Philips product-as-a-service model: Lighting products are leased rather than sold. BMW's remanufacturing of parts to extend the lifecycle of their vehicles.

4. ***Sustainability and Consumer Engagement***
   - This category discusses how digital technologies support firms in involving consumers in sustainability. This is achieved through means such as monitoring the environmental effect of consumption and nudging toward sustainable choices through instruments that monitor environmental impact and push for sustainable choices.
   - Use of blockchain by Nestlé to increase recyclability of packaging and giving consumers visibility of sustainability of purchase.
5. ***Supply Chain Transparency and Traceability***
   - This theme emphasizes how digital technologies, particularly blockchain and IoT, create transparency and traceability in supply chains, allowing businesses to track materials and products from production to disposal, ensuring sustainability at every stage.
   - Nestlé's implementation of blockchain to track and ensure the recyclability of packaging, enhancing consumer trust and enabling supply chain transparency.

4 Key Challenges and Benefits of Implementing Digital Tools in Circular Economy

4.1 Challenges

- **High Costs:** Implementing IoT, AI, and blockchain involves significant capital expenditure.
- **Integration Complexity:** Adapting digital tools to existing systems and workflows can disrupt operations.
- **Stakeholder Collaboration:** Ensuring alignment across diverse stakeholders in global supply chains is challenging.

4.2 Benefits

- **Resource Optimization:** Technologies like AI enhance supply chain efficiency and reduce waste.
- **Transparency:** Blockchain ensures traceability and accountability, fostering trust.
- **Consumer Engagement:** Digital platforms enable sustainable consumer behavior and product stewardship (Tables 1 and 2).

Table 1 Thematic frequency across case studies

Theme	Philips	Unilever	BMW	Interface	Nestlé	Total occurrence
Technological Innovation for CE	✓	✓	✓	✓	✓	5
Resource Efficiency and Optimization	✓	✓	✓	✓	✓	5
Circular Business Models	✓	✓	✓	✖	✖	3
Consumer Engagement	✓	✓	✖	✖	✓	3
Supply Chain Transparency	✖	✓	✓	✖	✓	3
Adoption Challenges Highlighted	✓	✓	✓	✓	✓	5

Source Compiled by the authors

Table 2 Technology adoption metrics across case studies

Company	Technology used	Investment (USD millions)	Waste reduction (%)	Lifecycle extension (%)	Supply chain traceability (1–5)
Philips	IoT, Product-as-a-Service	150	30	40	4.5
Unilever	AI, Blockchain	100	25	20	4.7
BMW	3D Printing, Blockchain	180	35	50	4.3
Interface	IoT, AI	80	40	30	3.9
Nestlé	Blockchain, IoT	120	28	25	4.8

Source The figures above are illustrative estimates derived from publicly available sustainability reports and credible news sources

5 Conclusion

It has been shown that the integration of digital transformation with sustainability and circular economy practices has significant potential to change how companies function and engage with resources, goods, and customers. Five case studies were presented that show how companies across different industries are using innovative digital technologies to drive resource efficiency, reduce waste, and enhance transparency in the supply chain. By aligning business models with circular principles, organizations are not only reducing their environmental footprint but also developing more resilient, scalable systems that create long-term value.

Contribution to Understanding Digital Transformation and Circular Economy: The findings reveal how digital tools align with circular economy principles, enabling companies to optimize resource use, reduce emissions, and promote transparency. By addressing barriers such as cost and complexity, businesses can unlock significant sustainability benefits.

Case Studies and Global Representation: The chosen cases provide a balanced perspective of global industry leaders, highlighting their strategies and innovations. However, future research should expand to include SMEs and emerging markets for broader applicability.

6 Implications

- ***For Businesses:*** Digital transformation is an opportunity for businesses to enhance sustainability practices while driving efficiency. However, it does require a shift in mindset, investment in new technologies, and a commitment to circular economy principles.
- ***To Policymakers:*** Policymakers must formulate policies that promote the adoption of digital technologies to sustain while maintaining ethical practices of data usage and consumer privacy.
- ***For Researchers:*** Further research needs to be conducted on which of the digital technologies is effective in driving sustainability, especially cost–benefit analysis and long-term impacts.
- **For Academia:** Further studies should explore long-term impacts and cost-effectiveness of digital transformations.

The findings of the study contribute significantly to understanding the interplay between digital transformation and circular economy principles in the following ways:

1. **Integration of Technology with Sustainability:** The study highlights how digital technologies such as IoT, AI, blockchain, and 3D printing are not only tools for operational efficiency but also enablers of circular economy practices like waste reduction, resource optimization, and lifecycle extension.
2. **Resource Management and Transparency:** Digital tools enable businesses to monitor, track, and optimize resource usage. For instance, IoT facilitates real-time monitoring of product lifecycles, while blockchain ensures supply chain transparency and traceability.
3. **Business Model Innovation:** The findings underscore how digital transformation supports innovative circular business models, such as product-as-a-service, which reduce waste and encourage reuse and remanufacturing.
4. **Consumer Engagement and Behavior:** Digital platforms empower consumers to make sustainable choices by providing visibility into the environmental impact of products and promoting recycling and other eco-friendly practices.
5. **Addressing Barriers and Challenges:** The study identifies key challenges, such as high costs and integration complexities, and suggests that overcoming these

barriers requires collaborative strategies and policy support. This insight aids stakeholders in planning the adoption of digital technologies effectively.
6. **Broader Implications for Policy and Research:** The research serves as a foundation for policymakers to create incentives for adopting digital tools in sustainability efforts and provides a roadmap for further academic exploration into cost-effectiveness and long-term impacts.

These contributions bridge the gap between theoretical principles of sustainability and their practical implementation through digital transformation, offering actionable insights for academia, industry, and governance. Ultimately, sustainability in the circular economy will depend on digital technologies, business innovation, and ways of working in convergence-where all these can unlock a more sustainable, resilient global economy.

References

1. Den Hollander, M. C., Bakker, C. A., & Hultink, E. J. (2017). Product design in a circular economy: Development of a typology of key concepts and terms. *Journal of Industrial Ecology, 21*(3), 517–525. https://doi.org/10.1111/jiec.12610
2. Saberi, S., Kouhizadeh, M., Sarkis, J., & Shen, L. (2019). Blockchain technology and its relationships to sustainable supply chain management. *International Journal of Production Research, 57*(7), 2117–2135. https://doi.org/10.1080/00207543.2018.1533261
3. Stock, T., Obenaus, M., Slaymaker, A., & Seliger, G. (2018). Industry 4.0 as enabler of a sustainable development: A qualitative assessment of its ecological and social potential. *Process Safety and Environmental Protection, 118*, 254–267. https://doi.org/10.1016/j.psep.2018.06.026
4. Ranta, V., Aarikka-Stenroos, L., & Mäkinen, S. J. (2018). Creating value in the circular economy: A structured multiple-case analysis of business models. *Journal of Cleaner Production, 201*, 988–1000. https://doi.org/10.1016/j.jclepro.2018.08.072
5. Bocken, N. M. P., Bakker, E., & Pauw, I. D. (2016). Product design and business model strategies for a circular economy. *Journal of Industrial and Production Engineering, 33*(5), 305–314. https://doi.org/10.1080/21681015.2016.1163921
6. Murray, A., Skene, K., & Haynes, K. (2017). The circular economy: An interdisciplinary exploration of the concept and application in a global context. *Journal of Business Ethics, 140*(3), 603–614. https://doi.org/10.1007/s10551-016-3061-6
7. Girotra, K., & Netessine, S. (2020). Leveraging AI for sustainable business model innovation. *Sustainability Science, 15*, 671–687. https://doi.org/10.1007/s11625-020-00819-5
8. Sundarakani, B., Eryarsoy, E., & Tang, C. S. (2021). Blockchain in circular supply chains: Opportunities and challenges. *International Journal of Production Economics, 231*, Article 107862. https://doi.org/10.1016/j.ijpe.2020.107862
9. Niinimäki, K., Hassi, L., & Kalliala, M. (2020). Circular economy in the fashion industry: Creating sustainable value through innovation. *Fashion and Sustainability, 7*(1), 1–12. https://doi.org/10.1080/17543266.2020.1754261
10. Zeng, S., Li, J., & Lin, J. (2017). Leveraging digital technologies for supply chain sustainability. *Journal of Cleaner Production, 167*, 106–118. https://doi.org/10.1016/j.jclepro.2017.08.097
11. Sullivan, R., & Williams, A. (2020). BMW Group: Leading the way in sustainable mobility and circular economy. *Automotive Sustainability Journal, 17*(2), 31–43. https://doi.org/10.1108/automotive.2020.0057

12. Rizos, V., Tuokko, K., & Behrens, A. (2016). The circular economy: Barriers and opportunities for SMEs. *Circular Economy and Sustainability, 4*(3), 12–25. https://doi.org/10.1016/j.ces.2015.10.002
13. Chesbrough, H. (2018). Business model innovation: The organizational dimension. *Long Range Planning, 51*(4), 505–515. https://doi.org/10.1016/j.lrp.2017.06.001
14. Lacy, P., &Rutqvist, J. (2015). **Waste to wealth: The circular economy advantage**. Palgrave Macmillan.
15. Bressanelli, G., Perona, M., & Saccani, N. (2018). Challenges in supply chain redesign for the circular economy: A literature review and a multiple case study approach. *International Journal of Production Research, 57*(1), 7395–7422. https://doi.org/10.1080/00207543.2018.1542176
16. Ellen MacArthur Foundation. (2013). *Towards the circular economy: Economic and business rationale for an accelerated transition.* Retrieved from https://ellenmacarthurfoundation.org
17. Geissdoerfer, M., Savaget, P., Bocken, N. M. P., & Hultink, E. J. (2017). The circular economy – A new sustainability paradigm? *Journal of Cleaner Production, 143*, 757–768. https://doi.org/10.1016/j.jclepro.2016.12.048

Index

S. Patnaik et al. (eds.), *People, Planet and Profit: Advancing Sustainable Business Practices*, Springer Proceedings in Geography and Sustainability,
https://doi.org/10.1007/978-3-032-11296-5

MIX
Papier aus verantwortungsvollen Quellen
Paper from responsible sources
FSC® C105338
FSC
www.fsc.org

If you have any concerns about our products,
you can contact us on
ProductSafety@springernature.com

In case Publisher is established outside the EU,
the EU authorized representative is:
Springer Nature Customer Service Center GmbH
Europaplatz 3, 69115 Heidelberg, Germany

Printed by Libri Plureos GmbH
in Hamburg, Germany